Chemical Bonds Outside Metal Surfaces

PHYSICS OF SOLIDS AND LIQUIDS

AMORPHOUS SOLIDS AND THE LIQUID STATE
Edited by Norman H. March, Robert A. Street, and Mario P. Tosi

CHEMICAL BONDS OUTSIDE METAL SURFACES
Norman H. March

CRYSTALLINE SEMICONDUCTING MATERIALS AND DEVICES
Edited by Paul N. Butcher, Norman H. March, and Mario P. Tosi

ELECTRON SPECTROSCOPY OF CRYSTALS
V. V. Nemoshkalenko and V. G. Aleshin

HIGHLY CONDUCTING ONE-DIMENSIONAL SOLIDS
Edited by Jozef T. Devreese, Roger P. Evrard, and Victor E. van Doren

MANY-PARTICLE PHYSICS
Gerald D. Mahan

THE PHYSICS OF ACTINIDE COMPOUNDS
Paul Erdös and John M. Robinson

POLYMERS, LIQUID CRYSTALS, AND LOW-DIMENSIONAL SOLIDS
Edited by Norman H. March and Mario P. Tosi

SUPERIONIC CONDUCTORS
Edited by Gerald D. Mahan and Walter L. Roth

THEORY OF THE INHOMOGENEOUS ELECTRON GAS
Edited by Stig Lundqvist and Norman H. March

Chemical Bonds Outside Metal Surfaces

Norman H. March
Coulson Professor of Theoretical Chemistry
University of Oxford
Oxford, England

Plenum Press • New York and London

Library of Congress Cataloging in Publication Data

March, Norman H. (Norman Henry), 1927–
Chemical bonds outside metal surfaces.

(Physics of solids and liquids)
Includes bibliographical references and index.
1. Surfaces (Physics) 2. Surface chemistry. 3. Chemical bonds. 4. Metals—Surfaces. I. Title. II. Series.
QC173.4.S94M37 1986 530.4′1 86-5022
ISBN 0-306-42059-7

A Division of Plenum Publishing Corporation
233 Spring Street, New York, N.Y. 10013

Printed in the United States of America

Preface

The problem of molecules interacting with metal surfaces has for a very long time been recognized to be of considerable technological as well as fundamental importance. Thus in the former category, a substantial number of important synthetic reactions for industrial purposes make use of metal surfaces as catalysts. Or again, problems of corrosion of metals are of great practical importance, such as in nuclear-reactor technology [see, for instance, my earlier articles, in: *Physics Bulletin*, Volume 25, p. 582, Institute of Physics, UK (1974); and in: *Physics and Contemporary Needs* (Riazuddin, ed.), Vol. 1, p. 53, Plenum Press, New York (1977)]. It is therefore of significance to strive to gain a more fundamental understanding of the atomic, and ultimately the electronic, processes that occur when a molecule is brought into the proximity of a metal surface.

The present volume focuses mainly on the theory and concepts involved; however, it is intended for readers in chemistry, physics, and materials science who are not specialists in theory but nevertheless wish to learn more about this truly interdisciplinary area of theoretical science. The aim of the book is to present the way in which valence theory can be synthesized with the understanding of metals that has been gained over the last half century or so. While advanced theory has at times been necessary, the detail is largely presented in an extensive set of Appendixes. Very occasionally, when such theory appears in the main text, the reader has usually been advised what to omit if he is merely interested in the main predictions of the theoretical arguments.

My indebtedness to many other workers will rapidly become evident to the readers of this book. Special thanks are due to my former research students: Drs. Richard C. Brown [see, for instance, *Phys. Rep.* **24C**, 77 (1976)], the late Ian D. Moore, Ilana Gabbay, John S. McCaskill, and Kathie R. Painter, as well as to Messrs. Peter J. Hiett and Kevin Joyce, who were Part II students in Theoretical Chemistry at Oxford. Invaluable help and advice in the general area covered by this book were provided by my friends and colleagues: Drs. Peter Grout, Don Richardson, and Gaetano Senatore, and also Professors Fernando Flores, "Jay" Mahanty, and Mario Tosi. Regular visits to the Theoretical Physics Division at AERE Harwell over many years have kept me in touch with the materials-science problems related to nuclear technology and thanks are due to Dr. A. B. Lidiard and his colleagues for the stimulation and motivation provided by these visits.

While I must, of course, accept sole responsibility for the contents of this book, and indeed no other person has read more than an individual chapter, it is a pleasure to thank Mr. Richard G. Chapman in Oxford, Dr. C. J. Wright (Harwell, UK), and Professor Z. W. Gortel at the University of Alberta, Edmonton, for helpful comments on parts of the manuscript. Should any readers find my book sufficiently interesting and be willing to trouble themselves, I would be most grateful to be told where I might improve the presentation, or where, I trust only very occasionally, I have misunderstood or erred in emphasis or judgment.

It is also a pleasure to thank all the authors who have generously allowed the reproduction of figures from their papers (the references are cited in the caption in each such case), and especially Dr. R. M. Lambert, Cambridge, UK, for supplying the original of Figure 6.3. Last, but by no means least, Professor H. Suhl very kindly acted as host for my visit to the University of California, San Diego, La Jolla, during a sabbatical leave granted by the University of Oxford. To him, his group of research students, and his colleague Dr. R. Kariotis, go my warm thanks for the opportunity to complete this book in an ideally quiet and scholarly atmosphere.

N. H. March

Contents

Appendixes

Chapter 1

Background, Phenomenology, and Motivation

The description of a chemical bond in interaction with a metal surface has long been a matter of concern for both fundamental theory and technology. Ideally, of course, such a description should take account of both the detailed electronic structure of the metal, including its surface, and the valence theory of the molecule involved. There has been a good deal of progress in implementing this program, but so far it has proved possible to carry out this task only by very lengthy numerical procedures.

Therefore, it is plainly of interest to seek simpler methods, starting from the most basic approximations of valence theory, and then incorporating the proximity of the metal surface by means of approximate concepts, such as image theory applied to a perfect electric conductor. Such a procedure, as discussed in Chapter 2 for diatoms, makes sense when the molecule is well outside the electron cloud spilling out from the metal surface, i.e., in the physisorbed regime, but needs refinement and extension in the face of chemical interaction between molecule and metal (e.g., the case of H_2O on ruthenium metal, treated at some length in Chapter 3).

However, even if such a program, combining the electronic band theory (allowing for Coulomb interactions between electrons) of the metal surface with the finest available wave functions for the molecular electrons, were tractable eventually by largely analytical theory, it would still not make very close contact with many of the important questions for chemical physics and materials science. This is because dynamics, rather than static conformational theory, is a major ingredient in any attempt to answer many salient questions in this area, such as desorption rates from metal surfaces and the way the rates of chemical reactions can be dramatically changed in the presence of catalytic metal surfaces.

To emphasize this perspective, we shall first discuss the background to the kinetics of adsorption and desorption after an elementary introduction to the problem of chemisorption in two limiting cases. In addition to their intrinsic interest, such investigations of kinetics frequently lead to valuable information concerning the nature of bonding in the adsorbed phase. A short introduction to reaction mechanisms outside surfaces follows.

One of the long-term aims of any theory of chemical bonds outside metal surfaces must be a comprehensive first-principles theory of catalytic reactions, as already implied above. Though no such complete theory exists at the time of writing, it is important for motivating and furthering work in the area covered by the present book to recognize both the types of problem that will be encountered, as well as the salient features that any useful theory must eventually embrace. Therefore, we felt it important in this introductory chapter, and following the lead given by Spencer and Somorjai in their important review on catalysis, to present the "case history" of one important catalytic process: the hydrogenation of carbon monoxide. This is then complemented in the final chapter by, first, an introduction to definitions and concepts of importance in catalysis, and second, a discussion of some selected topics in the modern theory of catalysis, chosen to throw light on basic principles (such as orbital symmetry conservation), which may perhaps, eventually, be taken over successfully from free-space chemical-reaction theory (e.g., Woodward–Hoffmann rules) into reaction theory outside metal surfaces. We have in this way endeavored to include in this book enough exemplary material for scientists in the boundary regions between physics, chemistry, and materials science to recognize just what factors one may eventually be forced to incorporate into the theory in order to understand fully the rich variety of phenomena encountered in the processes of catalysis.

With this general introduction, we shall next briefly review a little of the relevant historical background to the quantitative study of molecules outside metal surfaces, to be treated in detail in Chapters 2 and 3.

1.1. CHEMISORPTION IN IONIC AND COVALENT LIMITS

As a prelude to the more quantitative studies of diatoms and polyatomic molecules adsorbed on metal surfaces, to be discussed in Chapters 2 and 3, it is natural that the early work focused on the limiting cases of purely ionic and purely covalent bonding.[1]

1.1.1. Ionic Bond Formation

The formation of an ionic bond was thought to comprise two stages:

1. An electron transfer from the gas molecule to the surface.
2. The gas ion approaches the surface until the equilibrium separation is attained.

The transfer of an electron from the gas molecule to the surface involves an energy gain of $W - I$, where W is the work function and I the ionization potential. For a pure ionic bond, the heat of adsorption, Q_0 say, is given approximately by

$$Q_0 = N_A(W - I + e^2/4r_0) \tag{1.1.1}$$

where N_A denotes Avogadro's number. The final term in equation (1.1.1) already invokes image theory (also utilized in some detail in the next two chapters) to write the energy of attraction needed to bring the ion within the ionic radius r_0 of the surface. In Table 1.1, the values calculated in this way are compared with the experimental values for alkali-metal atoms on tungsten. There is fair agreement, and the same is found for molybdenum, etc., such systems having been treated, for example, by Moesta.[2]

1.1.2. Covalently Bonded Cases

When one turns to the adsorption of gases such as hydrogen or oxygen on metallic surfaces, an equation such as (1.1.1) is no longer of any use. Here one must consider the other limiting case, namely that of purely covalent bonding.

Early work, summarized for instance by Wedler,[3] expressed the heat

TABLE 1.1. Calculated and Experimental Heats of Chemisorption for Alkali Metal Atoms on Tungsten[a]

System	$N_A W$	$N_A I$	$N_A e^2/4r_0$	Q_0^{calc}	Q_0^{exp}
Na on W	104	118	45	31	32
K on W	104	100	36	40	—
Cs on W	104	89	31	46	64

[a] The model for calculation is the purely ionic limit, based on equation (1.1.1), and units are kcal mol^{-1}.

of adsorption for, say, the case of hydrogen as

$$Q_0 = 2E_{\text{Me-H}} - E_{\text{H-H}} \tag{1.1.2}$$

where $E_{\text{Me-H}}$ and $E_{\text{H-H}}$ are the bond energies for the adsorbed phase and for hydrogen, respectively. Then $E_{\text{Me-H}}$ can be evaluated following Pauling[4] in the form

$$E_{\text{Me-H}} = \tfrac{1}{2}(E_{\text{Me-Me}} + E_{\text{H-H}}) + 23.06(x_{\text{Me}} - x_{\text{H}})^2 \tag{1.1.3}$$

where the last term involving the electronegativity difference allows for the ionic contribution of the bond. The quantity $E_{\text{Me-Me}}$ is obtained from the heat of sublimation H_s, say, by setting

$$E_{\text{Me-Me}} = \tfrac{2}{12}H_s \tag{1.1.4}$$

in the case of a face-centered-cubic metal with twelve nearest neighbors. The heat of adsorption is then derived by combining equations (1.1.2) and (1.1.3) to yield

$$Q_0 = E_{\text{Me-Me}} + 46.1(x_{\text{Me}} - x_{\text{H}})^2 \tag{1.1.5}$$

The difference of electronegativities will be discussed further in Chapter 2. For the moment it will be noted that earlier work by Eley[5] took the difference of electronegativity from the dipole moment M of the Me–H bond. In turn, M was estimated from measurements of the surface potential. As an alternative approach, Stevenson[6] estimated the electronegativity of the metal from the work function, using Pauling's value of 2.1 for the electronegativity x_{H}. The results of this approach along the two routes of Eley and of Stevenson are compared with experiment in Table 1.2. This work will be developed in Chapter 2 by appealing to results from band-structure calculations of the various transition-metal series.

TABLE 1.2. Measured and Calculated Heats of Chemisorption (units are kcal mol^{-1})

System	Q_0, exp	Q_0, Eley	Q_0, Stevenson
H_2 on Ta	45	34	50
H_2 on W	45	37	46
H_2 on Cr	45	16	24
H_2 on Ni	18.4	19	29
H_2 on Fe	32	19	32

1.2. KINETICS OF ADSORPTION AND DESORPTION

It is frequently true that investigation of the kinetics of an adsorption or desorption process leads to useful information on the state of binding in the adsorbed phase.[7]

From kinetic theory of gases, the number of molecules impinging on unit area of a surface per unit time at gas pressure p is given by

$$\frac{dN}{dt} = \frac{p}{(2\pi m k_B T)^{1/2}} \qquad (1.2.1)$$

If s denotes the probability that an impinging molecule is adsorbed, then this so-called sticking probability will clearly influence the rate of adsorption r_a according to

$$r_a = \frac{sp}{(2\pi m k_B T)^{1/2}} \qquad (1.2.2)$$

A variety of factors influences the sticking probability:

1. If the adsorption is an activated process, only those molecules possessing the necessary activation energy can be adsorbed.
2. Steric considerations of the special configuration of an activated complex can still hinder adsorption even when the activation energy is available.
3. The impinging molecule possesses kinetic energy and, additionally, the heat of adsorption is liberated when adsorption occurs. For adsorption to be effected, the energy must be dissipated sufficiently rapidly; otherwise the molecule will be immediately desorbed again.
4. For adsorption on a heterogeneous surface, s will assume different values on the various sites for adsorption.
5. The impinging molecule must find a suitable adsorption site. In chemisorption, for example, the molecule can only be taken up at sites not already occupied by adsorbed molecules. However, in physisorption, a multimolecular layer is possible.

It is therefore in general a difficult matter to calculate the sticking probability directly and thence determine the rate of adsorption.* However, for an activated adsorption process one can write

$$s = \kappa(\theta) f(\theta) \exp[-\Delta E(\theta)/RT] \qquad (1.2.3)$$

* Some recent progress is reported in 'Notes added in proof' on p. 259.

where $\Delta E(\theta)$ denotes the activation energy and $f(\theta)$ takes into account the necessary available surface. These quantities as well as $\kappa(\theta)$, which can be regarded as a condensation coefficient, depend on the coverage θ.

The rate of adsorption from equation (1.2.2) is then given by

$$r_a = \frac{p}{(2\pi m k_B T)^{1/2}} \kappa(\theta) f(\theta) \exp[-\Delta E(\theta)/RT] \tag{1.2.4}$$

To deal with $f(\theta)$ one can proceed as follows. If the adsorbate molecule is only adsorbed at one point, the unoccupied part of the surface is characterized by a factor $(1-\theta)$ and the probability of a molecule striking an unoccupied site is

$$f(\theta) = 1 - \theta \tag{1.2.5}$$

If dissociative adsorption takes place on two sites, however, one must take into account whether or not the adsorbate can meander across the surface.

For mobile adsorption

$$f(\theta) = (1-\theta)^2 \tag{1.2.6}$$

On the other hand, for immobile adsorption the number of nearest neighbors z enters the treatment, as discussed in detail, for instance, by Miller,[8] because unoccupied sites must have an unoccupied neighbor for dissociative adsorption to occur. In this case one may write

$$f(\theta) = \frac{z}{z-\theta}(1-\theta)^2 \tag{1.2.7}$$

There is a good deal of variation possible in the manner in which the activation energy depends on the coverage. It is frequently useful on homogeneous surfaces to approximate the variation by a linear dependence, namely

$$\Delta E = \Delta E_0 + a\theta \tag{1.2.8}$$

When the surfaces under consideration are energetically heterogeneous, one must integrate over the distribution of activation energies.

In order to attempt an interpretation of the condensation coefficient κ, transition-state theory may be invoked. According to this theory, it would be necessary for the free adsorbate molecule to pass through the activated complex, which we denote by (AS)*, before becoming the bound adsorbate. This process involves passing over the potential-energy barrier of height E, which is depicted in Figure 1.1 as a function of an appropriate reaction coordinate.

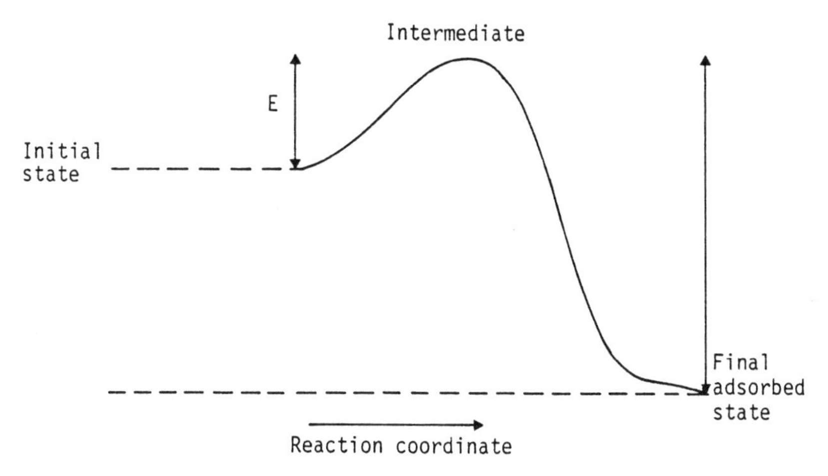

FIGURE 1.1. Potential-energy barrier as a function of an appropriate reaction coordinate.

The molecules of the activated complex are in statistical equilibrium with both the reactants and the products. The reactants are the molecules S in the gas phase and the unoccupied adsorption sites A, the product being the bound adsorbate. For the first equilibrium

$$A + S \rightleftharpoons (AS)^* \tag{1.2.9}$$

the rate of adsorption being given by

$$r_a = k^* c_A c_S \tag{1.2.10}$$

that is, the rate of formation of the activated complex assuming that an immobile adsorbed layer is formed. The velocity constant k^* of the reaction that forms the activated complex is, according to transition-state theory, Z denoting partition function,

$$k^* = \frac{k_B T}{h} \frac{Z_{(AS)^*}}{Z_A Z_S} \exp(-\Delta E/RT) \tag{1.2.11}$$

The rate of adsorption follows from equations (1.2.10) and (1.2.11) as

$$r_a = c_A c_S \frac{k_B T}{h} \frac{Z_{(AS)^*}}{Z_A Z_S} \exp(-\Delta E/RT) \tag{1.2.12}$$

An expression for the condensation coefficient can be obtained by utilizing equation (1.2.12) in conjunction with equation (1.2.4). In doing so, it should be noted that the partition function for the transition complex is composed of the specific properties of the complex, namely vibration and rotation. The adsorbed layer was assumed to be immobile in the above treatment, so translation is not included.

Though in the above treatment of adsorption kinetics an activation energy has been assumed, in fact nonactivated adsorption is possible. With desorption, however, an activation energy is always involved.

A treatment of desorption kinetics corresponding to that which led to equation (1.2.4) for adsorption kinetics is, in principle, possible. The free adsorbate molecule is produced by desorption, so that the desorption rate will be proportional to the concentration of bound adsorbate; this latter quantity is in turn proportional to the coverage θ or, in more general terms to a function denoted by $f'(\theta)$. Of course, only those adsorbed complexes that possess the required activation energy can dissociate. Thus the rate of desorption r_d can be expressed in the form

$$r_d = \delta(\theta) f'(\theta) \exp[-\Delta E_d(\theta)/RT] \tag{1.2.13}$$

Specific properties of the system under study are thereby subsumed into the desorption coefficient $\delta(\theta)$. As for the condensation coefficient discussed above, an interpretation of $\delta(\theta)$ is afforded by, for example, transition-state theory.

The quantity $f'(\theta)$ is basically determined by whether or not desorption and recombination occur together; this happens, for instance, when hydrogen atoms desorb as molecules.

1.3. THERMODYNAMICS OF ADSORPTION

1.3.1. Langmuir Isotherm

Langmuir[9] derived his celebrated adsorption isotherm* by considering the kinetics of adsorption. This isotherm is valid for monomolecular adsorption that is ideally localized. According to the theory of the rate of adsorption given in Section 1.2, when every gas molecule is adsorbed on a fixed site one has

$$r_a = \kappa \frac{p}{(2\pi m k_B T)^{1/2}} (1-\theta) \exp(-E_a/RT) \tag{1.3.1}$$

* A number of highbrow statistical-mechanical derivations of Langmuir's isotherm are now available. We have preferred here the intuitive approach, not the formal ones.

where κ is the condensation coefficient and E_a the activation energy for adsorption. Since neither κ nor E_a depends on the coverage, one may assume that the adsorption of a molecule is in no way influenced by the presence of another adsorbed molecule (an obvious oversimplification at high coverage). The corresponding equation for desorption is, from equation (1.2.13),

$$r_d = \delta\theta \exp(-E_d/RT) \tag{1.3.2}$$

At equilibrium

$$r_a = r_d \tag{1.3.3}$$

Setting

$$b = \frac{\kappa}{8(2\pi m k_B T)^{1/2}} \exp[-(E_a - E_d)/RT] \tag{1.3.4}$$

one finds

$$\theta = bp/(1 + bp), \qquad b = b(T) \tag{1.3.5}$$

where θ is the coverage n/n_m; here n molecules adsorbed corresponds to coverage θ and n_m to the monolayer coverage. At very low pressures $1/b \gg p$, so that equation (1.3.5) simplifies to $\theta = bp$. At high pressures, on the other hand, $1/b \ll p$ and it follows from equation (1.3.5) that $\theta \to 1$, i.e., according to the Langmuir adsorption isotherm the coverage approaches the monolayer asymptotically at high pressures and independently of the temperature.

The statistical derivation given by Fowler, which evaluates the properties of the equilibrium state, gives for the constant b:

$$b = \frac{h^3}{(2\pi m)^{3/2}(k_B T)^{5/2}} \frac{Z_s(T)}{Z_g(T)} \exp(q/k_B T) \tag{1.3.6}$$

where $Z_s(T)$ and $Z_g(T)$ are the partition functions of a molecule in the adsorbed state and in the free gaseous state,* respectively, the energy difference between these two states being the heat of adsorption, written above as q.

Frequently, dissociative adsorption occurs: this means that a molecule is not chemisorbed in its molecular form but only after dissociation. In these cases the Langmuir adsorption isotherm becomes (see Hill[11])

$$n = \frac{n_m (bp)^{1/2}}{1 + (bp)^{1/2}} \tag{1.3.7}$$

* Anomalies in adsorption equilibrium can occur near phase transitions in the substrate, as discussed in Appendix 5.2.

because one has to take into account that two sites are involved both for the adsorption and for the desorption of one gas molecule. The terms $(1-\theta)$ and θ, therefore become $(1-\theta)^2$ and θ^2, respectively, when the isotherm is set up by equating the right-hand sides of equations (1.3.1) and (1.3.2) for r_a and r_d.

To check whether adsorption is described by a molecular or a dissociative form for the Langmuir adsorption isotherm, one rewrites equation (1.3.5), for the case of molecular adsorption on single adsorption sites as

$$\frac{p}{n} = \frac{1}{n_m b} + \frac{p}{n_m} \tag{1.3.8}$$

and a plot of p/n vs. p should give a straight line. For the dissociative two-point adsorption, equation (1.3.7) is equivalent to

$$\frac{p^{1/2}}{n} = \frac{1}{n_m b^{1/2}} + \frac{p^{1/2}}{n_m} \tag{1.3.9}$$

More often than not, adsorbates can be adsorbed in a number of different states. If the adsorption of each of these states can be described individually by the Langmuir adsorption isotherm, then for the general case of several molecular and dissociative states one may write

$$n = \sum_i \frac{n_{mi} b_i p}{1 + b_i p} + \sum_j \frac{n_{mj} (b_j p)^{1/2}}{1 + (b_j p)^{1/2}} \tag{1.3.10}$$

Here n_{mi} and n_{mj} are the quantities of the adsorbates in states i and j, respectively, which altogether constitute the monolayer; b_i and b_j take into account that the different states display distinguishable heats of adsorption.

Below, we shall discuss the derivation by Brunauer *et al.*[10] of an adsorption isotherm for multilayer adsorption that can be applied practically. This isotherm represents a generalization of the ideal localized monolayer adsorption. The end result, whether derived by kinetic arguments such as given in de Boer's book,[7] or by statistical methods, is

$$\frac{p}{n(p_0 - p)} = \frac{1}{n_m C} + \frac{C-1}{n_m C} \frac{p}{p_0} \tag{1.3.11}$$

where n is the number of moles adsorbed at an equilibrium pressure p, n_m is the number of moles adsorbed in the monolayer, and p_0 is the saturated vapor pressure of the adsorbing gas at the temperature of the measurement. The meaning of the constant C will emerge from the derivation given in the following section.

1.3.2. Multilayer Adsorption Theory

Brunauer *et al.*[10] developed a multilayer adsorption theory. A basic assumption of their work is that the Langmuir isotherm (1.3.5) applies to every adsorption layer: a molecule encountering an occupied site on the surface does not leave that site immediately but forms a short-lived adsorption complex. With increasing vapor pressure, when p approximates the saturated vapor pressure p_0 the number of free sites on the surface decreases as well, since double, triple, etc., adsorption complexes are formed.

Hill[11] has given a statistical-mechanical derivation of the result of Brunauer *et al.*; the original kinetic-theory arguments will be followed here. Disregarding interactions between adsorbate molecules along the metal surface, one can consider multilayer adsorption of vapor on a homogeneous surface as a series of equilibrium steps:

1. Vapor + free surface $\rightleftarrows$ single complex
2. Vapor + single complex $\rightleftarrows$ double complex
3. Vapor + double complex $\rightleftarrows$ triple complex, etc.

Let $\theta', \theta'', \theta''', \ldots$ be the fractions of the metal-surface area covered with single, double, triple, etc., adsorption complexes. The total amount of adsorbed vapor is then given by

$$n = n_{\mathrm{m}}(\theta' + 2\theta'' + 3\theta''' + \cdots) \tag{1.3.12}$$

where n_{m} is the monolayer capacity.

The various equilibrium constants are

$$k' = \frac{\theta'}{p\theta_0}, \qquad k'' = \frac{\theta''}{p\theta'}, \qquad k''' = \frac{\theta'''}{p\theta''}, \cdots \tag{1.3.13}$$

where θ_0 is the fraction of the adsorbent area not occupied by adsorbate molecules. The value of k' is usually much greater than that of k'', since the surface–adsorbate interaction rapidly decreases with increasing distance from the surface. While the constants $k'', k''', \ldots$ are not precisely equal in practice, their differences are usually much smaller than that between k' and k''. Thus Brunauer *et al.* make the approximation $k'' \simeq k''' \simeq \cdots \simeq k_{\mathrm{L}}$, where k_{L} is the equilibrium constant for the saturated vapor $\rightleftarrows$ liquid system and equals $1/p_0$. (If, for the vaporization equilibrium, liquid $\rightleftarrows$

saturated vapor, $k_p = p_0$, then for the condensation equilibrium $k_L = 1/k_p$.) Hence equation (1.3.13) yields

$$\theta' = k'p\theta_0$$

$$\theta'' = k''p\theta' = k_L p\theta' = \frac{p}{p_0}\theta'$$

$$\theta''' = k'''p\theta'' = (k_L p)^2\theta' = \left(\frac{p}{p_0}\right)^2 \theta', \quad \text{etc.}$$

Substitution of these results into equation (1.3.12) gives

$$n = n_m k'p\theta_0 \left[1 + 2\frac{p}{p_0} + 3\left(\frac{p}{p_0}\right)^2 + \cdots\right] \tag{1.3.14}$$

However, also

$$\theta_0 + \theta' + \theta'' + \theta''' + \cdots = \theta_0 \left\{1 + k'p\left[1 + \frac{p}{p_0} + \left(\frac{p}{p_0}\right)^2 + \cdots\right]\right\} = 1 \tag{1.3.15}$$

Since $p/p_0 \leq 1$, the sum of the series in square brackets in equation (1.3.14) is $(1 - p/p_0)^{-2}$ while that in equation (1.3.15) is $(1 - p/p_0)^{-1}$. After setting $C = k'/k_L$, one obtains the isotherm of Brunauer *et al.*,

$$\frac{n}{n_m} = \theta = \frac{Cp/p_0}{(1 - p/p_0)[1 + (C - 1)p/p_0]} \tag{1.3.16}$$

for multilayer vapor adsorption.

It is noteworthy that both Langmuir and Brunauer *et al.* isotherms neglect horizontal interactions between adsorbed molecules in the surface layer. The Brunauer *et al.* treatment allows for interactions between adsorbate molecules solely in the direction vertical to the surface.

The effect of such horizontal interactions on the adsorption process can often be important, e.g., for large molecules or for molecules forming associated compounds via hydrogen bonds on the surface. Adsorbate–adsorbate horizontal interactions can result in bending of the adsorption isotherm toward the pressure axis, and affect the heat of adsorption on a homogeneous surface with coverage (see Chapter 5).

1.4. REACTION MECHANISMS OUTSIDE SURFACES

Before embarking on the final objective of this introductory chapter, namely to motivate what follows by referring to a specific example of catalysis in some detail, it will be useful to briefly record here two proposed specific reaction mechanisms outside surfaces. The first of these is associated with the names of Langmuir and Hinshelwood, and the second with Rideal and Eley. To conclude this section, it will be useful to give a discussion of the Redhead equation for desorption rate, so that the way is cleared for the detailed treatment of desorption in Chapter 5.

1.4.1. Langmuir–Hinshelwood and Rideal–Eley Mechanisms

We shall very briefly introduce the basic ideas, referred to as Langmuir–Hinshelwood and Rideal–Eley reactions.

Let us consider a simple reaction, A + B = C, which is known to proceed via the following steps: (1) adsorption of the components, (2) reaction on the catalyst surface, and (3) product desorption.

The transport of mass and heat is assumed to be so rapid that it does not interfere with the kinetics. By using this simple approach and obvious notation, all reactions may, in principle, be classified into two basic types:

$$\text{Type I:}\quad A_g + B_g = A_{ads} + B_{ads} = C_{ads} = C_g \tag{1.4.1}$$

$$\text{Type II:}\quad \text{(a)}\ A_g = A_{ads} \tag{1.4.2}$$

$$A_{ads} + B_g = C_{ads} = C_g \tag{1.4.3}$$

$$\text{(b)}\ A_{\text{chem ads}} + B_{\text{molecular phys ads}} = C_{ads} = C_g \tag{1.4.4}$$

Type I reactions, in which all components involved in the reaction pass through the chemisorption stage, are often called Langmuir–Hinshelwood reactions.* Type II reactions are usually associated with the names of Rideal and Eley.

1.4.2. Redhead Equation for Desorption Rate

Following Redhead,[12] the rate of desorption from unit surface area may be written

$$N(t) = -\frac{d\theta}{dt} = \nu_n \theta^n \exp(-E/RT) \tag{1.4.5}$$

where n is the order of the reaction, θ the surface coverage (molecules per cm^2), ν_n the rate constant, and E the activation energy of desorption (cal/mol).

* A specific example is given in the 'Notes added in proof' on p. 260 in which Type I and Type II are contrasted.

Redhead considered a linear change of sample temperature with time given by

$$T = T_0 + \beta t \tag{1.4.6}$$

and assumed E to be independent of θ. He then solved equation (1.4.5) to find the temperature (T_p say) at which the desorption rate is a maximum. For a first-order reaction he thus obtained

$$E/RT_p^2 = (\nu_1/\beta)\exp(-E/RT_p) \tag{1.4.7}$$

while for $n = 2$, a second-order reaction, his results are

$$\begin{aligned} E/RT_p^2 &= (2\theta_p\nu_2/\beta)\exp(-E/RT_p) \\ &\div \frac{\theta_0\nu_2}{\beta}\exp(-E/RT_p) \end{aligned} \tag{1.4.8}$$

where θ_0 is the initial surface coverage and θ_p the coverage at $T = T_p$.

Equation (1.4.7) shows that T_p is independent of coverage for a first-order reaction with constant E and thus E can be found directly from a measurement of T_p provided a value of ν_1 is assumed. Redhead observed that the relation between E and T_p is very nearly linear and for $10^{13} > \nu_1/\beta > 10^8$ (K^{-1}) is given to within $\pm 1.5\%$ by

$$E/RT_p = \ln\frac{\nu_1 T_p}{\beta} - 3.64 \tag{1.4.9}$$

In fact, the activation energy can be determined without assuming a value of the rate constant by varying β and plotting $\ln T_p$ vs. $\ln \beta$. The value of E can then be found from the relation

$$E/RT_p + 2 = d\ln\beta/d\ln T_p \tag{1.4.10}$$

The rate constant can then be determined by substituting E into equation (1.4.7). For reasonable accuracy, β must be varied by at least two orders of magnitude.

Returning to the second-order case, it can be seen from equation (1.4.8) that T_p now depends on the surface coverage. The quantity θ_0 may be found from the area under the curve of desorption rate as a function of time and $\ln(\theta_0 T_p^2)$ plotted against $1/T_p$. A straight line of slope E/R is then obtained, while ν_2 is then calculated by substitution in equation (1.4.8). Thus the order of the desorption reaction can be determined from the behavior of the maximum in the desorption-rate curves with coverage. A first-order reaction with a fixed activation energy of desorption gives rise to a peak in the desorption-rate curve that does not change in temperature with coverage (see also note on p. 259).

If the temperature of the peak decreases with increasing coverage, the reaction may be either second order with fixed activation energy, or first order with an activation energy dependent on coverage. These two cases can be distinguished by plotting $\ln(\theta_0 T_p^2)$ against $1/T_p$; as noted above, a second-order reaction with fixed activation energy yields a straight line.

In concluding this chapter, which presents relevant background information and outlines the directions to be followed in the subsequent treatment, it hardly need be stressed that a major reason for interest in chemical bonds outside metal surfaces is the industrial importance of numerous catalytic processes. Consequently, a very practical introduction to the theoretically oriented material in the final chapter will be a detailed study of the catalytic hydrogenation of carbon monoxide, one of the six "case histories" of important industrial catalytic processes reviewed by Spencer and Somorjai.[13]

1.5. CASE HISTORY OF CATALYTIC HYDROGENATION OF CARBON MONOXIDE

This was first reported in 1902, when Sabatier and Senderens[14] produced methane from CO and H_2 using a nickel catalyst. However, it was two decades later when Fischer and Tropsch[15] succeeded in producing measurable quantities of higher hydrocarbons at atmospheric pressure and temperatures in the range 250–350 °C using a catalyst consisting of iron and cobalt with K_2CO_3 and copper as promoters, the so-called Fischer–Tropsch synthesis. As this process became economically viable in the 1970s, following the lapse of its usefulness in the late 1950s, synthetic fuel research began again and, at the time of writing, continues to be actively pursued on both a technological and fundamental level in many places.

The case history below, closely following the review of Spencer and Somorjai,[13] indicates how the advent of surface-science techniques are leading to increasing understanding of the basic atomic mechanisms for the catalytic hydrogenation of CO.

1.5.1. Thermodynamics

The most convenient source of both CO and H_2 is the gasification of coal by steam. For thermodynamic purposes, coal, though of varying composition, will be considered as graphitic carbon. The reaction of graphite with steam is

$$C_{graphite} + H_2O \rightarrow CO + H_2, \qquad \Delta H = 32 \text{ kcal mol}^{-1} \tag{1.5.1}$$

This is an endothermic reaction and must therefore be carried out at high temperatures (about 1100 K). The product mixture can subsequently be enriched with hydrogen via

$$CO + H_2O \rightarrow CO_2 + H_2, \qquad \Delta H = -9.5 \text{ kcal mol}^{-1} \tag{1.5.2}$$

This is useful, since varying the $CO : H_2$ ratio of the reactant gases can alter the product distribution of the catalytic hydrogenation reaction.

Alkane-forming reactions are of two overall types:

(i)
$$(n+1)H_2 + 2nCO \rightarrow C_nH_{2n+2} + nCO_2 \tag{1.5.3}$$

(ii)
$$(2n+1)H_2 + nCO \rightarrow C_nH_{2n+2} + nH_2O \tag{1.5.4}$$

both reactions being thermodynamically feasible. It is generally found that if catalysts which are effective for reaction (1.5.2) are used for CO hydrogenation, both CO_2 and H_2O are produced. Catalysts that are poor for reaction (1.5.2) tend to produce either CO_2 or H_2O as a byproduct.

Since the CO hydrogenation reactions are all exothermic, they are thermodynamically favored by lower temperatures. However, the reaction on the catalysts used at present is typically rather slow, and hence temperatures in the range 500–700 K are required to optimize the rate of product formation.

By Le Chatelier's principle, high pressures favor more associative reactions, so if the reactant pressure is increased the production of higher-molecular-weight products is enhanced. In the case of methanation, for example,

$$3H_2 + CO \rightarrow CH_4 + H_2O \tag{1.5.5}$$

The value of ΔG_f at 730 K and 1 atmosphere is -11.4 kcal mol^{-1}, yielding an equilibrium constant of 4×10^3. However, at 10^{-4} torr pressure, the equilibrium constant is approximately 10^{-10}, leading to a negligible production of methane. As the molecular weight of the desired product increases, the reactant pressure must also be increased. For example, in the case of benzene synthesis (cf. Figure 1.2) pressures in excess of 20 atm must be employed to synthesize significant amounts of product.

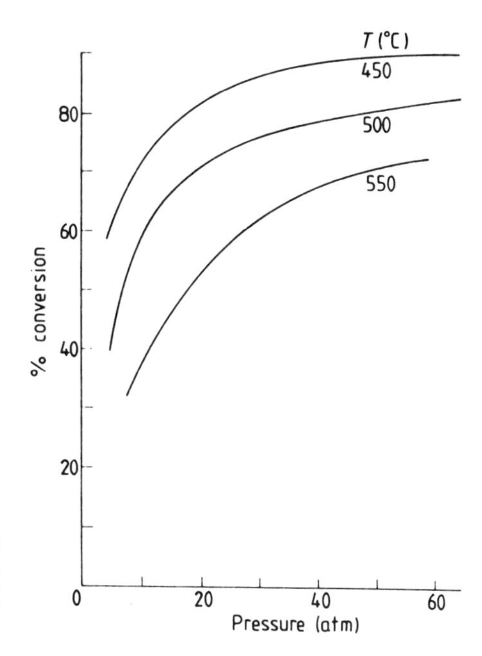

FIGURE 1.2. Effect of increasing reactant pressure on the hydrogenation of carbon monoxide to benzene (after Spencer and Somorjai[13]).

1.5.2. Chemisorption Studies of CO and H_2

Many surface studies using ultraviolet, photoelectron, and infrared spectroscopies have shown that CO adsorbs molecularly on many transition metals, its bond being perpendicular to the surface with carbon directly bonded to the metal.

This bonding with the surface is widely thought to be synergic in nature, with mutual enhancement occurring between donation of electrons from the $CO_{5\sigma}$ orbital into unoccupied metal orbitals and back-donation from other metal orbitals into the $CO_{2\pi^*}$ (antibonding) system (cf. Figure 1.3). Both linear bonding (one metal site) and multiple bonding (several metal sites) of CO to the metal surface have been observed (see Figure 1.4).

On many transition metals, the molecularly adsorbed CO species serves merely as a precursor to the dissociative chemisorption of the molecule. This is an activated process, the transition state presumably involving the simultaneous interaction of both C and O with the metal surface. This interaction is achieved during vibrational deformation and, in particular, is likely to occur in the case of CO bonded to several metal atoms, as discussed by Richardson and Bradshaw.[16] The preferential dissociation of multiply-bonded CO was demonstrated by Araki and Ponec,[17] who adsorbed CO on nickel–copper alloys. The main effect of

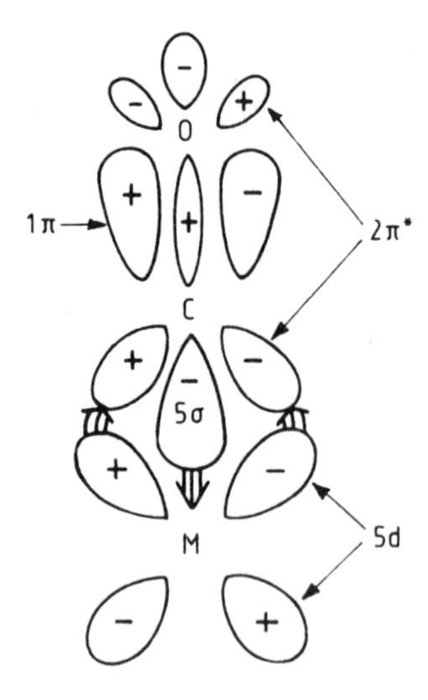

FIGURE 1.3. Bonding of carbon monoxide to platinum metal, showing donation of electrons from the 5σ orbital of CO into metal, and back-donation of electrons from the $5d$ metal orbitals into the CO $2\pi^*$ antibonding orbitals (after Spencer and Somorjai[13]).

alloying the nickel with copper is to merely diminish the surface concentration of large clusters of nickel atoms. It was found that both the concentration of bridge-bonded CO molecules and the extent of CO disproportionation were reduced on addition of copper, while the rate coefficient of disproportionation remained constant.

H_2 has been found to dissociate on many transition metal surfaces, such as Pd,[18] Fe,[19] and Ni.[20] A mutual enhancement of adsorption has been observed when CO and H_2 are simultaneously present in the gas phase. This suggests some type of interaction between the adsorbed H and CO species, but presently there is controversy surrounding the existence and importance of such an interaction. Many infrared studies have prompted the suggestion of the formation of an H_2CO species[21] on iron after simultaneous adsorption of CO and H_2. The precise bonding in this species has not been established at the time of writing, however, and it is no longer thought to be central to the mechanism of the Fischer–Tropsch reaction.

1.5.3. Selectivity

Although cobalt and iron-based catalysts can be used under conditions that lead principally to the formation of products of a particular molecular-weight range, they are not selective catalysts *per se*. Selective catalysts do exist for the methane and methanol syntheses, however, and surprisingly these were among the earliest CO hydrogenation catalysts to be formu-

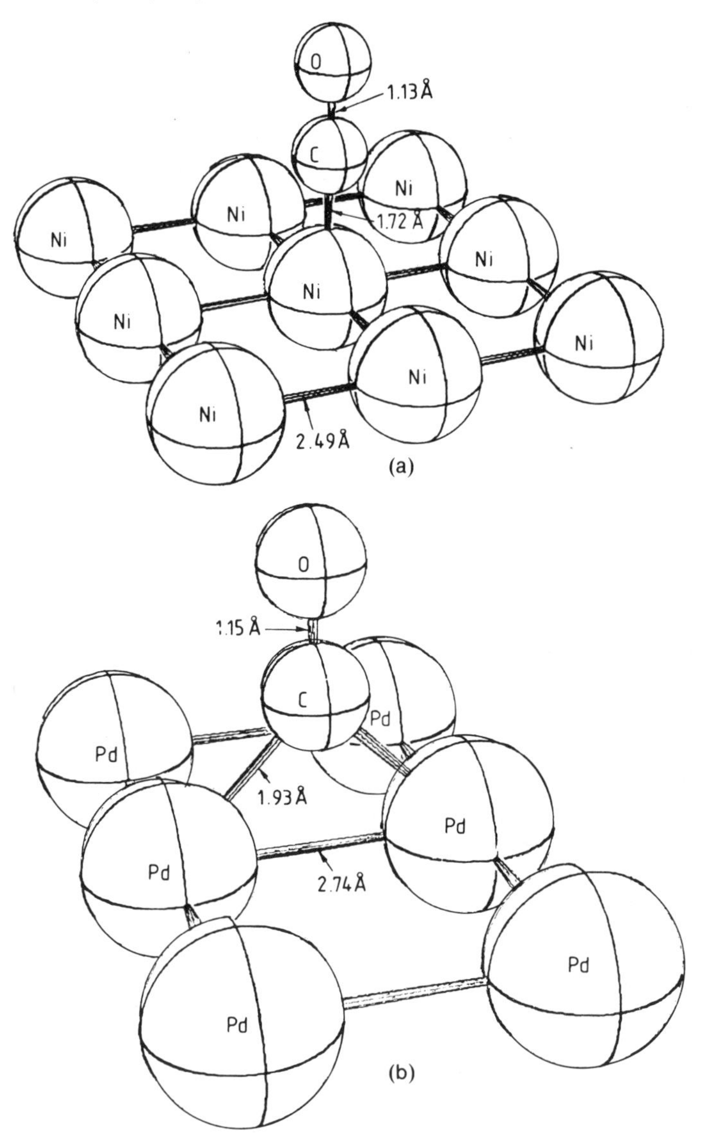

FIGURE 1.4. (a) Linear bonding of carbon monoxide on a nickel surface. (b) Multiple (bridge) bonding of carbon monoxide on a palladium surface (after Spencer and Somorjai[13]).

lated. Nickel[22,23] is an excellent methanation catalyst,

$$2CO + 2H_2 \xrightarrow[500\text{–}600\,K]{Ni} CH_4 + CO_2 \qquad (1.5.6)$$

Palladium, for example, has been shown by Poutsma *et al.*[24] to produce methanol selectively at about 12 atm pressure.

The oxidation state of the active metal seems to affect the product distribution of a catalyst very markedly. This is manifested in the effects of promoters and catalyst supports, which may stabilize particular oxidation states of the catalyst metal. For example, Anderson[25] has demonstrated that the product distribution on CO hydrogenation over iron nitrides is very different from the distribution obtained over both clean iron and iron oxides, suggesting that the oxidation state of the active metal has a significant effect on the selectivity of the catalyst. Sexton and Somorjai[26] have shown that while clean unsupported rhodium does not lead to the formation of oxygenated products, preoxidation of the surface leads to the generation of alcohols, aldehydes, and acids. Moreover, the use of basic oxide supports with the rhodium catalyst results in the formation of alcohols, while more acidic supports favor the production of olefins and methane. The effect of manganese and alkali metals on rhodium catalysts is to enhance the production of oxygenates.

The ability of alkali metals to stabilize an oxidation state is well illustrated by the observation of Somorjai,[27] that the presence of potassium can prevent the reduction of iron oxides in the reducing conditions encountered during CO hydrogenation. In this system, potassium also plays the useful role of removing inactive graphitic carbon from the iron-oxide surface, probably via the formation of K_2CO_3[28] and subsequent desorption of CO_2. If this carbon were allowed to build up, the Fischer–Tropsch activity of the oxide would be greatly diminished.

Another approach to selectivity in the Fischer–Tropsch synthesis has been developed by Meisel *et al.*[29] and involves the incorporation of zeolite shape-selective catalysts into the same reaction chamber as the metal catalyst. This has been found to produce aromatic molecules (and therefore high-octane petrol) with a high degree of selectivity.

1.5.4. Mechanism

It is now widely believed that the initial step in many CO-hydrogenation reactions is the disproportionation of CO, otherwise known as the Boudouard reaction:

$$2CO \rightarrow C + CO_2 \tag{1.5.7}$$

Experiments on nickel,[17] rhodium,[26] and iron[30] have shown the formation of an active carbon overlayer during the synthesis reaction. This carbon has been found to react with hydrogen to produce methane at the same rate as does carbon monoxide. The role of surface carbon has also been emphasized by Biloen *et al.*,[31] who dissociated ^{13}CO on cobalt, nickel, and ruthenium, and found that subsequent ^{12}CO hydrogenation over these surfaces resulted in the formation of some $^{13}CH_4$. The CO-derived active carbon is prone to graphitization, however,[32] when heated over 700 K or in an inert environment. Graphitic carbon appears to be unreactive with hydrogen, making 700 K an upper temperature limit for CO hydrogenation.

Methane has been formed on surfaces that do not readily dissociate CO, such as Pd,[24] Ni at 300 K,[33] Pt and Ir,[23] and it is thought that another mechanism operating via an enol species may be effective in these cases.[34–36] This process is believed to be less common, under industrial conditions, than that involving prior CO dissociation.

The above examples should make quite clear the complexity of the mechanisms involved. It will clearly be some time before first-principles theory will be able to make direct contact with such complicated detail. However, along with some other examples,[13] a suggestion for the reaction sequence for the Fischer–Tropsch synthesis by Bell[37] is reported in Appendix 6.4. Unfortunately, the precise mechanistic roles of the various site reactions are not well understood at present, but it would appear that their control, via suitable catalyst formulation, can lead to the selective synthesis of almost any hydrocarbon,[13] starting from CO and H_2.

We may summarize by saying that catalysis is at the heart of most chemical and energy-production technologies. Application of the methods of surface science continually permits scrutiny of a variety of catalyst surfaces on the atomic scale. The structure, composition, and oxidation state of surface atoms may now be determined* and correlated with the reaction parameters, such as rate, activation energy, and product distribution. The molecular understanding gained permits the design of improved catalyst formulations and holds promise that one may devise new catalysts for carrying out specific chemical functions. Spencer and Somorjai[13] conclude their review with the following words: "The art of catalysis is rapidly being converted into catalysis science."

* Most of the techniques available are summarized in Appendix 1.1.

Chapter 2

Diatomic Molecules

Following the above discussion of some important realistic examples of catalysis, we turn to the theory of chemical bonds outside metal surfaces. In this chapter, we shall be concerned primarily with the effect of the proximity of a planar metal surface on the binding energy of a diatomic molecule. Of course, it is clear that when the charge clouds of metal and molecule overlap strongly, then significant perturbations of the molecular valence electrons can and do occur. However, even before the molecule enters the "spill-out" region of itinerant electrons from the metal, one may expect some influence of the metal on the bond strength and length. This is because there will be an additional interaction between the electrons in the molecule through the response of the metallic electrons to the charge fluctuations in the molecule. Though a full calculation of this effect is not an easy matter, it will be demonstrated below, for instance in a diatomic molecule parallel to the metal surface, that this additional interaction will tend to weaken the bond.

2.1. PHYSISORBED DIATOMS WITH WELL-DEFINED CORES

Mahanty and March[38] studied the case of a diatomic molecule with well-defined core electrons at the equilibrium separation, when the molecule is well outside metal surfaces.* This is the physisorption regime, characterized by negligible overlap between the electron clouds of

* **Note added in proof.** See also the more formal discussion of A. D. McLachlan, *Proc. Roy. Soc.* **A271**, 387; **A274**, 80 (1963). More recent work is by E. A. Power and T. Thirunamachandran, *Phys. Rev.* **A25**, 2473 (1982).

molecule and metal. In a model of diatomic molecules in which the long-range part of the interatomic potential-energy curve is due to the London nonretarded dispersion interaction between the constituent atoms, these workers demonstrated the result referred to above, namely that the proximity of a metal surface leads to a weakening of the chemical bond when the molecule is in the parallel configuration relative to the metal. Their treatment, based on image theory in the approximation in which the metal is taken as a perfect conductor, will be developed below. For a Lennard-Jones type of interatomic potential, it will be shown that the maximum reduction of bond strength for the molecule lying parallel to the planar metal surface is $\frac{4}{9}$.

With the cores of the two constituent atoms identifiable at their equilibrium distance, to a useful approximation the interatomic potential-energy curve can be expressed as the sum of a short-range repulsive part and a long-range attractive tail that is basically the London nonretarded dispersion interaction of the form $-A/r^6$, where

$$A \equiv A_{\text{London}} = \tfrac{3}{4}\hbar\alpha_1(0)\alpha_2(0) \tag{2.1.1}$$

Here $\alpha_j(\omega)$ is the polarizability of the jth constituent atom at frequency ω. This approximation will be poorest for molecules like H_2 in which the electron density is large in the interatomic region; therefore H_2 will be discussed separately below.

If the repulsive part of the potential-energy curve is assumed to have the Lennard-Jones B/r^{12} form, one readily finds that the minimum in the potential occurs at

$$r_{\min} = (2B/A)^{1/6} \tag{2.1.2}$$

the value of the potential energy at this point being

$$V(r_{\min}) = -A^2/4B \tag{2.1.3}$$

The short-range repulsive part, which arises basically from the exchange repulsions of the electron clouds of the two atoms, can be taken to be relatively insensitive to the proximity of the metal. The constant A determining the dispersion interaction part, on the other hand, depends on the proximity of the metal surface. By anticipating the result to be derived below, for a diatomic molecule parallel to the metal surface, with interatomic separation ρ and distance from the surface z, the effective dispersion-force constant becomes

$$A'(s) = A_{\text{London}}F(s) \tag{2.1.4}$$

where $s = 2z/\rho$ and, when the metal is treated as an ideal conductor, $F(s)$

is given by

$$F(s) = 1 + \frac{1}{(1+s^2)^3} - \frac{4}{3}\frac{(1+s^2/4)}{(1+s^2)^{5/2}} \tag{2.1.5}$$

The reduction factor $F(s)$ varies monotonically from $\frac{2}{3}$ at $z=0$ to 1 at $z=\infty$. It can be noted from equation (2.1.3) that the well depth or binding energy can be reduced by a factor $F(s)^2$, i.e., by a value between $\frac{4}{9}$ and 1 depending on the relative values of z and ρ.

2.1.1. Image Theory of the Dispersion Force*

The desired result is readily obtained in terms of the quantity $\mathbf{G}$ connecting the electric field $\mathbf{E}(\mathbf{r})$ at $\mathbf{r}$ to a dipole source $\boldsymbol{\mu}(\omega)$ at $\mathbf{r}'$ oscillating with frequency ω, namely

$$\mathbf{E}(\mathbf{r}) = \mathbf{G}(\mathbf{r}\mathbf{r}'; \omega)\boldsymbol{\mu}(\omega) \tag{2.1.6}$$

In technical terms, $\mathbf{G}$ is a "dyadic Green function." What is important in the present context is that in the presence of a dielectric surface, taken to be the (x, y) plane, when both $\mathbf{r}$ and $\mathbf{r}'$ are outside the dielectric (i.e., $z, z' > 0$), $\mathbf{G}$ can be expressed as the sum of two parts:

$$\mathbf{G}(\mathbf{r}\mathbf{r}'; \omega) = \mathbf{G}_{\mathrm{D}}(\mathbf{r}-\mathbf{r}') - \Delta(\omega)\mathbf{G}_{\mathrm{I}}(\mathbf{r}\mathbf{r}') \tag{2.1.7}$$

Here the direct part $\mathbf{G}_{\mathrm{D}}$ follows from elementary electrostatics and is given by

$$\mathbf{G}_{\mathrm{D}}(\mathbf{r}-\mathbf{r}') = -(\nabla\nabla')\frac{1}{|\mathbf{r}-\mathbf{r}'|} \tag{2.1.8}$$

while $\mathbf{G}_{\mathrm{I}}$ is the indirect part, arising through the mediation of the semi-infinite dielectric medium with dielectric constant $\varepsilon(\omega)$. Using image theory one finds, with $\Delta(\omega) = [\varepsilon(\omega)-1]/[\varepsilon(\omega)+1]$,

$$\mathbf{G}_{\mathrm{I}}(\mathbf{r}\mathbf{r}') = -(\nabla\nabla')\frac{1}{|\mathbf{r}-\mathbf{r}'_{\mathrm{im}}|} \tag{2.1.9}$$

where $\mathbf{r}'_{\mathrm{im}}$ is the position of the image of the point $\mathbf{r}'$. In Ref. 42, it is shown that the dispersion interaction energy of the two atoms, to leading order in the polarizabilities $\alpha_1(\omega)$ and $\alpha_2(\omega)$, can be written as

$$E_{\mathrm{D}}(12) = -\frac{\hbar}{4\pi}\int_{-\infty}^{\infty} d\xi\,\alpha_1(i\xi)\alpha_2(i\xi)\,\mathrm{Tr}[\mathbf{G}(21)\mathbf{G}(12)] \tag{2.1.10}$$

* The reader interested in the results, but uninitiated in Green function methods, should proceed to Section 2.1.1.1.

where, in the trace of the Green function product, the shorthand notation $1 \equiv \mathbf{r}_1$, $2 \equiv \mathbf{r}_2$ has been employed.

Following Mahanty and March,[38] the cases of the diatomic molecule parallel and perpendicular to the metal surface are considered below. Furthermore, the metal is taken to be an ideal conductor, so $\Delta = 1$, corresponding to $\varepsilon \to \infty$. This approximation is a little drastic for a real metal, since it ignores the details of the contributions to the dielectric response of the metal from electronic excitations, and in particular from surface plasmons. However (cf. Mahanty and March[38] for comment), a more detailed analysis that takes such contributions into account, while altering some of the coefficients in $E_D(12)$, will not affect the general trends discussed here. With $\Delta = 1$, equation (2.1.10) becomes

$$E_D(12) = \tfrac{1}{6}\mathrm{Tr}[\mathbf{G}(21)\mathbf{G}(12)]\left[\frac{3\hbar}{2\pi}\int_{-\infty}^{\infty} \alpha_1(i\xi)\alpha_2(i\xi)d\xi\right] \qquad (2.1.11)$$

The term in the large brackets on the right-hand side of this equation is the constant A of equation (2.1.1); the equivalence of the two is known from the work of Margenau and Kestner.[39]

The traces required to evaluate E_D are given explicitly in equations (11) and (12) of Ref. 38. One can thus immediately determine E_D for parallel and perpendicular configurations, the general-orientation case being recorded also in Ref. 38.

2.1.1.1. Parallel Configuration

For a molecule parallel to the metal surface, the traces entering equation (2.1.11) simplify. Then equation (2.1.11) leads to the results given in equations (2.1.4) and (2.1.5). It is interesting to note that the effect of the ideal conducting surface is that, for any finite $\bar{z} = z_1 + z_2$, at sufficiently large $\rho = [(x_1 - x)^2(y_1 - y_2)^2]^{1/2}$ but in the nonretarded region, the long-range interaction becomes two-thirds the London value. If one takes the "spill-out" region of nonvanishing electron density to be about 1 Å, then $\bar{z}$ must be greater than 2 Å in order that the above analysis remains valid. If $\bar{z}$ is approximately 5 Å and the atoms are 10 Å apart, then $F(s) \sim 0.7$ for $s = \frac{1}{2}$.

If ρ is very large one enters the retarded regime. The method given by Mahanty and Ninham[40–42] enables it to be shown that, in this case,

$$E_D(12)\big|_{z_1 \to 0} \simeq -\frac{26\hbar c}{4\pi\rho^7}\,\alpha_1(0)\alpha_2(0) \qquad (2.1.12)$$

which is a slight enhancement over the Casimir–Polder free-space result of

$-(23\hbar c/4\pi r^7)\alpha_1(0)\alpha_2(0)$. In practice, the separation distance at which retardation effects appear would far exceed the molecular bond length.

2.1.1.2. Perpendicular Configuration

When the molecule is perpendicular to the surface, let the atoms be at z_1 and z_2. Then equations (2.1.10) and (2.1.11) yield

$$E_D(12) = E_{\text{London}}(z_{12})F_\perp(z_1/z_{12}) \tag{2.1.13}$$

where

$$F_\perp(t) = 1 + \frac{1}{(1+2t)^6} + \frac{2}{3(1+2t)^3} \tag{2.1.14}$$

Here $F_\perp(z_1/z_{12})$ is an enhancement factor, decreasing monotonically from 8/3 at $(z_1/z_{12})=0$ to 1 at $(z_1/z_{12})=\infty$. Of course, in the perpendicular configuration this is not simply the energy difference between one atom fixed at z_1, the other first at z_{12} and then $z_{12}\rightarrow\infty$, because, unlike the parallel configuration, the interaction of the individual atoms with the surface must also be included. For fixed z_1, the variable part of the interaction energy would be

$$E_{12}(z_1, z_2) = -\frac{K}{2z_2^3} + V(z_{12}) \tag{2.1.15}$$

It is easy to show that, for any finite z_1, the bond length is reduced from its value in free space, in contrast to the situation for the parallel configuration. In the latter case, it is seen from equation (2.1.2) that the bond length would be rather insensitive to the proximity of the metal. The bond strength will depend more on the details of the individual molecules in the perpendicular configuration than in the parallel configuration.

The conclusions to be drawn from the above analysis may be summarized as follows. Within the framework of the model of the diatomic molecule near the metal surface described by equations (2.1.1)–(2.1.3), in the parallel configuration there can be considerable weakening of the chemical bond and in the perpendicular configuration the opposite effect occurs. Changes in the vibration frequencies could also be estimated from the equation presented above. The errors in the approximate potential functions set up above will, of course, be magnified in computing the curvature at the minimum, which determines the vibrational frequency. Finally, the weakening of the chemical bond in the parallel configuration when the molecule is still outside the spill-out region can be regarded as a prelude to the more dramatic changes as the molecule enters the density profile of the spilled-out electrons.

Even outside the density profile, asymmetry in the atomic charge clouds will obviously occur at close proximity to the surface, leading to small permanent dipoles on the atoms. In fact, this effect will lead to repulsion between the atoms, adding to the lengthening and weakening of the chemical bond for the parallel configuration.

2.2. INTERACTION BETWEEN TWO HYDROGEN ATOMS

Two regimes of H–H interaction have been investigated in the work of Flores *et al.*[43] These are:

1. Beyond the spill-out region of the metal electrons (physisorption).
2. Embedded protons in the electron spill-out.

These workers used an extension of the Heitler–London method to treat (1). In line with the work of Mahanty and March[38] discussed in Section 2.1 for molecules with well-defined cores, they found that the H_2 binding energy for the molecule parallel to the surface is reduced, while for the perpendicular direction stronger binding obtains. In region (2), the asymptotic form of the screening charge around a single proton is considered in a linear-response framework and for an infinite-barrier model of the surface. If this screened potential is taken to be analytic in **k**-space (no singularities) then the interaction energy, $\Delta E(X)$ say, falls off as X^{-5} times an oscillatory function, for H_2 parallel to the planar metal surface. The effect of self-consistency on this result is also examined. The conclusion is that the asymptotic interaction energy is unchanged in form, though the amplitude is altered.

2.2.1. Physisorbed H_2

In the case of H_2 outside the planar metal surface and relatively far from the spill-out region, Flores *et al.*[43] make the following assumptions:

1. A Heitler–London theory is a useful starting point.
2. The Coulomb interaction $1/|r_a - r_b|$ between two charges a and b is replaced by $1/|r_a - r_b| - 1/|r_a - r_b^{\mathrm{I}}|$, or equivalently by $1/|r_a - r_b| - 1/|r_a^{\mathrm{I}} - r_b|$, where the notation is explained in Figure 2.1, I referring to an image distance.

With these two assumptions one can follow the traditional Heitler–London approach in order to find the energy of the complete assembly of H_2 plus metal surface. It is straightforward to show that this

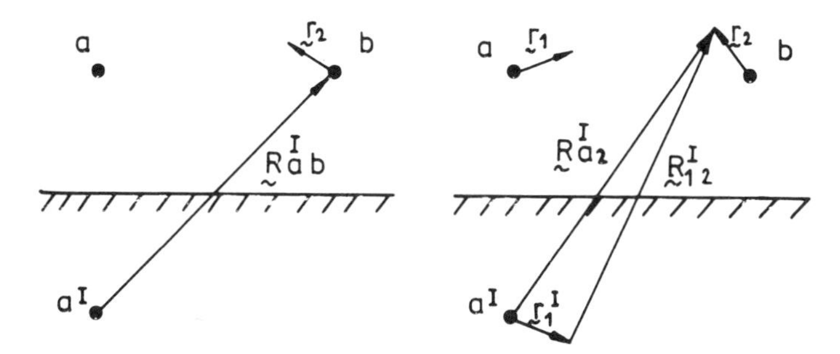

FIGURE 2.1. Left-hand part of diagram shows two protons, a and b, in a configuration parallel to the planar metal surface; a^{I} is the image of proton a. Right-hand part of diagram illustrates the notation, showing vector r_1, its image r_1^{I}, and r_2, defining R_{a2}^{I} and R_{12}^{I}.

energy can be expressed, as in the Heitler–London treatment, in the form

$$E = 2E_{1s} + Q + \alpha \tag{2.2.1}$$

where

$$Q = J/(1+S^2), \qquad \alpha = K/(1+S^2) \tag{2.2.2}$$

the notation being that of Eyring *et al.*[44]

However now, in the presence of the surface, the behavior of J and K must be reexamined. In principle, J is the total Coulomb energy of the system. This can be obtained as the sum of two terms: the Coulomb energy of the atoms without the surface present, and the correction due to the image potential. For instance, there is the interaction between one atom and the image of the other; however, due to spherical symmetry this interaction is identically zero.

While the conclusion therefore is that the Coulomb energy of the new system with the surface coincides with that of the free H_2 molecule, a more interesting situation arises when one considers K. One can then set

$$K = K_0 + \delta K \tag{2.2.3}$$

where K_0 is the usual Heitler–London value and δK is given by

$$\delta K = -\frac{S^2}{R_{ab}^{\mathrm{I}}} + 2S\int \frac{\psi_a(2)\psi_b(2)d2}{R_{a2}^{\mathrm{I}}} - \int \frac{\psi_a(1)\psi_b(2)\psi_b(1)\psi_a(2)d1\,d2}{R_{12}^{\mathrm{I}}} \tag{2.2.4}$$

Here S is the overlap integral between the two $1s$ hydrogen-like wave functions $\psi_a(1)$ and $\psi_b(1)$. The distances R_{a2}^{I} and R_{12}^{I} are shown in Figure 2.1.

Therefore the correction to the Heitler–London approximation introduced by the presence of the surface is given by

$$\delta E = \delta K/(1 + S^2) \tag{2.2.5}$$

The calculation of δK can be carried out be developing $(R^{\mathrm{I}}_{a2})^{-1}$ and $(R^{\mathrm{I}}_{12})^{-1}$ in series expansions. Hence

$$\frac{1}{R^{\mathrm{I}}_{a2}} = \frac{1}{R^{\mathrm{I}}_{ab}} - \frac{r_2 \cdot R^{\mathrm{I}}_{ab}}{(R^{\mathrm{I}}_{ab})^3} + \frac{1}{2}\left[\frac{3(R^{\mathrm{I}}_{ab} \cdot r_2)^2}{(R^{\mathrm{I}}_{ab})^5} - \frac{r_2^2}{(R^{\mathrm{I}}_{ab})^3}\right] + \cdots \tag{2.2.6}$$

$$\frac{1}{R^{\mathrm{I}}_{12}} = \frac{1}{R^{\mathrm{I}}_{ab}} - (r_2 - r^{\mathrm{I}}_1) \cdot \frac{R^{\mathrm{I}}_{ab}}{(R^{\mathrm{I}}_{ab})^3} + \frac{1}{2}\left\{3\left[R^{\mathrm{I}}_{ab} \cdot (r_2 - r^{\mathrm{I}}_1)^2\right] - \frac{(r_2 - r^{\mathrm{I}}_1)^2}{(R^{\mathrm{I}}_{ab})^3}\right\} + \cdots \tag{2.2.7}$$

the notation being as in Figure 2.1.

2.2.1.1. Parallel and Perpendicular Configurations

These series developments can be used to easily show that, in lowest order, for the H_2 molecule parallel to the planar surface,

$$\delta K = \tfrac{3}{4} d^2 S^2/(R^{\mathrm{I}}_{ab})^3 \tag{2.2.8}$$

where d is the separation of protons a and b. This yields the change in energy as

$$\delta E = \tfrac{3}{4}[d^2/(R^{\mathrm{I}}_{ab})^3] S^2/(1 + S^2) \tag{2.2.9}$$

The total energy as a function of d is evidently given by adding the above energy δE to the usual Heitler–London energy E_0; the form of the correction is shown in Figure 2.2.

When estimating the importance of δK, it is useful to evaluate it at the minimum, 1.64 atomic units (au), of the Heitler–London curve. This approach yields

$$\delta K = [0.64/(R^{\mathrm{I}}_{ab})^3] \quad \text{au} \tag{2.2.10}$$

The Heitler–London (H–L) minimum energy is now -0.115 au so $\delta E + E_{\min} = 0$ at $R^{\mathrm{I}}_{ab} = 1.77$ au.

The above distance is unphysical, however, since then an atom and its image are strongly overlapping. In general the approximations are only

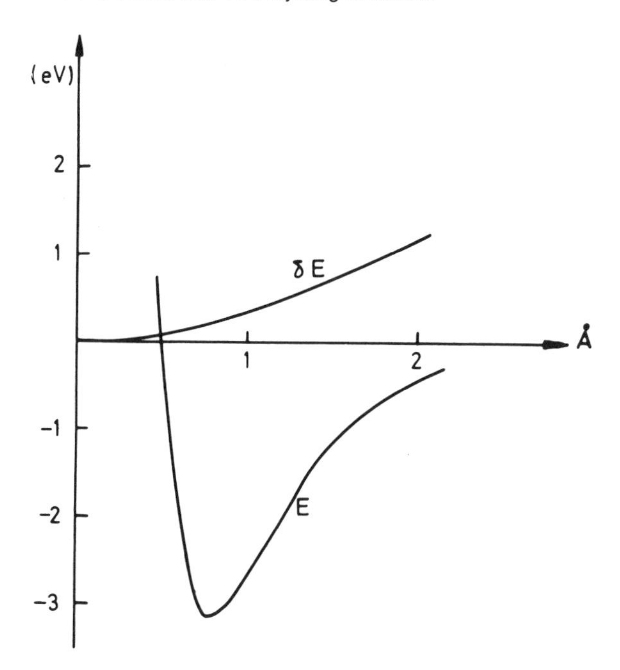

FIGURE 2.2. Heitler–London energy E and correction δE of equation (2.2.9) as a function of H–H separation in Å. The scale of δE is arbitrary, depending on the H_2-surface distance $R^{\mathrm{I}}_{ab}/2$ [see equation (2.2.9)].

appropriate for $R^{\mathrm{I}}_{ab} \gtrsim 11$ au, as can be deduced from other theoretical calculations, such as those of Smith *et al.*[45]

The conclusion is that δE, for the physisorbed region, is much less than $E_{\mathrm{H\text{-}L}}$ and naturally no dramatic effects are expected in the molecular configuration. However, as mentioned above, the general findings are in accord with the work of Mahanty and March[38] for molecules with well-defined cores, discussed in Section 2.1, since the bond in the parallel configuration is weakened in the presence of the surface, though the increase in bond length is not expected to be important.

Flores *et al.*[43] also considered the perpendicular configuration, but we shall not give the details here. The binding is found to be somewhat stronger in the presence of the surface. However, the simplicity of the parallel configuration is, of course, that the interaction of each atom with the surface is the same in that case.

2.2.2. Protons Embedded in the Spill-Out Metal Electron Cloud

While the above approach of "atoms in molecules" seems appropriate in the physisorbed region, we shall next discuss the case when the two protons are embedded in the spill-out of the metal electrons beyond the positive-ion surface.

More specifically, we shall first consider carefully the screening of a single proton embedded in the inhomogeneous electron cloud at the surface. Of course, this starting point implies that the proton sits in an electron gas of sufficiently high density to (1) lose the 1s bound state and (2) leave the screened proton as a relatively weak perturbation.

Flores *et al.*[43] stress that (1) and (2) are quite restrictive assumptions, which will probably limit the applicability of their results to certain high-density transition metals.

The change in electron density due to a weak impurity potential $V(\mathbf{r})$ can be written in the form

$$\Delta\rho(\mathbf{r}) = \int F(\mathbf{r}\mathbf{r}')V(\mathbf{r}')d\mathbf{r}' \tag{2.2.11}$$

where the linear response function F can be calculated exactly for the Bardeen model of a metal surface (see Appendix 2.1) as

$$F = \begin{cases} -\dfrac{k_f^2}{(2\pi)^3}\left\{\dfrac{j_1(2k_f s)}{s^2} + \dfrac{j_1(2k_f s')}{s'^2} - \dfrac{2j_1[2k_f(s+s')]}{ss'}\right\}, & z, z' > 0 \\ 0, & \text{otherwise} \end{cases} \tag{2.2.12}$$

In equation (2.2.12), s is written for $|\mathbf{r}-\mathbf{r}'|$ and s' for $[|\mathbf{X}-\mathbf{X}'|^2 + (z+z')^2]^{1/2}$, where $\mathbf{X}$ and $\mathbf{X}'$ are vectors in the plane parallel to the surface.

Provided $r \gg r' \simeq z$ (see Figure 2.3), one has

$$s' \simeq s + 2zz'/s \simeq s + 2z'\cos\theta \tag{2.2.13}$$

where $\cos\theta = z/r$ and then, assuming that the cylindrically symmetric impurity potential $V(X, z)$ in equation (2.2.11) is localized and correspondingly that it has no singularities in its Fourier transform $(\mathbf{k})$ space, the asymptotic form of the displaced charge $\Delta\rho$ is given by[43]

$$\Delta\rho \sim r^{-3} \times \text{oscillatory function} \tag{2.2.14}$$

for $\theta \simeq 0$. However, for $\theta \simeq \pi/2$

$$\Delta\rho \sim X^{-5} \times \text{oscillatory function} \tag{2.2.15}$$

X denoting the distance between the protons for the parallel configuration.

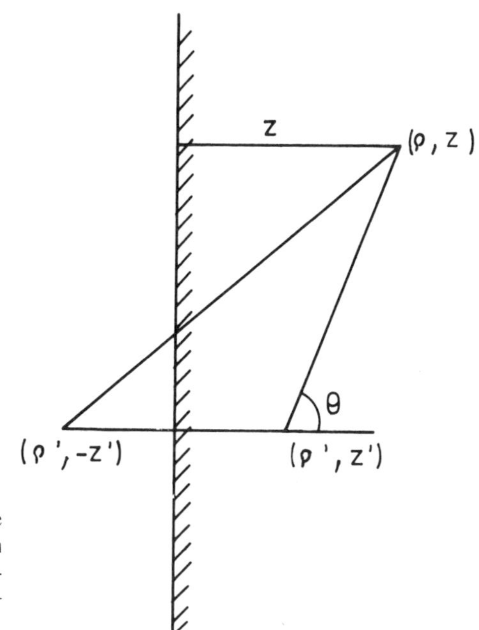

FIGURE 2.3. Point (ρ, z) at which the induced charge is obtained as a function of the potential at (ρ', z'). In the asymptotic limit it is assumed that $[(\rho - \rho')^2 + (z - z')^2]^{1/2} \gg (\rho'^2 + z'^2)^{1/2}$.

This latter result (2.2.15) is in contrast to the displaced electron cloud round a test charge in a bulk electron gas that falls off as the inverse cube of the separation, as is also the case in equation (2.2.14).

Two final steps are needed to establish that the interaction energy ΔE between the protons has precisely the same asymptotic form as $\Delta\rho$. These are:

1. To calculate an improved potential energy $V(\mathbf{r})$ from Poisson's equation using the above forms of $\Delta\rho(\mathbf{r})$. This turns out to have precisely the same asymptotic form as $\Delta\rho(\mathbf{r})$, when one proceeds to complete internal self-consistency between $\Delta\rho$ and V.
2. To use the so-called electrostatic model (a consequence of the Hellmann–Feynman theorem), which yields the interaction energy as the charge of one proton times the screened potential of the other proton in which it sits.

In particular, for the parallel configuration one obtains the asymptotic form of $\Delta E(X)$ as

$$\Delta E(X) \sim X^{-5} \times \text{oscillatory function} \tag{2.2.16}$$

In addition to treating independently the free-electron substrates, with asymptotic results identical to those derived above, Lau and Kohn[46] also applied their techniques to adsorption on the simplest model of transition metals, namely the single-band tight-binding model for a simple cubic crystal studied, following pioneering work by Grimley,[47] by Einstein and Schrieffer.[48] Their approach will be outlined below and considered in relation to the comments of Einstein.[49]

2.3. INTERACTION ENERGY: TIGHT-BINDING MODEL

This section follows the presentation of Lau and Kohn,[46] following early work of Grimley and co-workers[47] and that of Einstein and Schrieffer.[48]

As in the latter work, Lau and Kohn[46] deal with a tight-binding, simple cubic semi-infinite solid with a (100) face. The adatoms A and B, each with a single energy level E_0, sit on top of the substrate atoms at surface sites 0 and n, respectively. The Hamiltonian of the system is taken to be*

$$H = H_0 + H_1 + \lambda H_2 \tag{2.3.1}$$

Here, H_0 is the Hamiltonian for the solid and has nearest-neighbor matrix elements only, denoted by T, so that the bandwidth of the solid is $12T$, centered at zero energy. Quantity H_1 is the Hamiltonian for the adatoms, with eigenvalue E_0, and λH_2 represents the interaction between the adatoms and the solid, its only nonzero matrix elements being $V_{\mathrm{A}0} = V_{\mathrm{B}n} = V^*_{0\mathrm{A}} = V^*_{n\mathrm{B}} = V$; λ is treated as a small parameter, but will eventually be set equal to unity.

To obtain the indirect interaction between adatoms A and B, it is necessary to perform a perturbation calculation to fourth order in λH_2. Lau and Kohn found it more convenient to use the method of canonical transformation as set out, for instance, by Slichter,† i.e., to construct

$$\tilde{H} = e^{-i\lambda S} H e^{i\lambda S} \tag{2.3.2}$$

In this expression S is chosen so as to remove to lowest order in λ the matrix elements of $\tilde{H}_2$ that transfer electrons between the adatom and the substrate atoms. Hence

$$H_2 + i[H_0 + H_1, S] = 0 \tag{2.3.3}$$

* Though we felt it was of interest to give the main steps in the derivation below, the reader who wishes only to understand the nature of the results should immediately consult Figure 2.4.

† See, for example, his book, *Principles of Magnetic Resonance* (Harper and Row, New York, 1963) Appendix D.

or

$$\langle l\alpha \mid S \mid l'\alpha' \rangle = -\frac{i\langle l\alpha \mid H_2 \mid l'\alpha' \rangle}{E'_{l'\alpha'} - E_{l\alpha}} \qquad \text{for } l \neq l'$$

$$\langle l\alpha \mid S \mid l\alpha' \rangle = 0 \tag{2.3.4}$$

where $|l\rangle$ and $|l'\rangle$ are eigenstates of H_0, which are Slater determinants of single-particle eigenstates $|\mathbf{k}\rangle$, $|\alpha\rangle$ and $|\alpha'\rangle$ are eigenstates of H_1 and are of the form $|n_A, n_B\rangle$, where $n_A, n_B = 0$ or 1.

With this choice of S one has

$$\tilde{H} = H_0 + H_1 + \lambda^2 \tilde{H}_2$$

where the relevant matrix element of $\tilde{H}_2$ is now given by

$$\langle l'\alpha' \mid \tilde{H}_2 \mid l\alpha \rangle = \tfrac{1}{2} \sum_{l''\alpha''} \langle l'\alpha \mid H_2 \mid l''\alpha'' \rangle \langle l''\alpha'' \mid H_2 \mid l\alpha \rangle \times \left(\frac{1}{E_{l'\alpha'} - E_{l''\alpha''}} + \frac{1}{E_{l\alpha} - E_{l''\alpha''}} \right) \delta_{\alpha\alpha'} \tag{2.3.5}$$

One has two cases, depending on whether the unperturbed eigenvalue E_0 is greater or less than the Fermi level E_f. For the case $E_0 < E_f$, equation (2.3.5) describes a virtual excitation from the ground state $|l\alpha\rangle$ to the excited state $|l''\alpha''\rangle$ in which the electron on one of the adatoms is transferred to the conduction band above the Fermi level. The state $|l'\alpha\rangle$ is formed when another electron in the band below the Fermi level is then transferred to the adatom, creating an electron–hole pair. The situation is similar for the case $E_0 > E_f$.

The single-particle eigenstates $|\mathbf{k}\rangle$ of H_0 are of the form

$$|\mathbf{k}\rangle = (2/N)^{1/2} \sum_{\mathbf{m}} e^{i\mathbf{p}\cdot\mathbf{m}} \sin[k_z(m_z + a)] \, |\mathbf{m}\rangle \tag{2.3.6}$$

where $|\mathbf{m}\rangle$ is a localized wave function centered at $\mathbf{m}$, m_z is the component of $\mathbf{m}$ perpendicular to the surface and $m_z = 0$ for surface sites, N is the number of lattice atoms, a is the lattice constant, and $\mathbf{k} = (\mathbf{p}, k_z)$.

If second-order perturbation theory is applied regarding the term $\lambda^2 \tilde{H}_2$ as the perturbation, one has

$$\Delta E_{\text{int}} = \frac{16|V|^4 \lambda^4}{N^2} \sum_{\substack{\mathbf{k}',\mathbf{k}'' \\ E_{\mathbf{k}'} \leq E_f}} \sin^2(k'_z a) \sin^2(k''_z a) \exp[i(\mathbf{p}' - \mathbf{p}'') \cdot \mathbf{n}a] \times \left(\frac{1}{E_{\mathbf{k}'} - E_0} + \frac{1}{E_{\mathbf{k}''} - E_0} \right)^2 \frac{1}{E_{\mathbf{k}'} - E_{\mathbf{k}''}} \tag{2.3.7}$$

When $E_0 = E_f$, this equation becomes singular and there is a resonance. This corresponds to virtual transfer of electrons between the adatom and the conduction band, leading to a very strong interaction. If E_0 is not too close to E_f, and E_f is near either extremity of the energy band, so that the Fermi surface is nearly spherical (see Appendix 2.2 for a discussion of the influence on interaction range of the Fermi-surface topology), then equation (2.3.7) can be evaluated to yield, after putting $\lambda = 1$,

$$\Delta E_{\text{int}} = 16|V|^4 \frac{(a^3)^2}{(2\pi)^6} a^4 \times \text{Re} \int d^3\mathbf{k}' d^3\mathbf{k}'' \left(\frac{1}{E_{\mathbf{k}'} - E_0} + \frac{1}{E_{\mathbf{k}''} - E_0}\right)^2$$

$$\times \frac{k_z'^2 k_z''^2 \exp[i(\mathbf{p}' - \mathbf{p}'') \cdot \mathbf{n}a]}{Ta^2[2k_f(k_x' - k_x'') + k_z'^2 - k_z''^2]}$$

$$= \frac{1}{4} \frac{|V|^4}{T^2[E_f - E_0]^2} \left(\frac{k_f a}{\pi}\right)^2 \frac{\cos 2k_f na}{n^5} \times 2T$$

where na is the distance between adatoms A and B.

Plots of ΔE_{int} for $|V| = T$ and different values of E_f and E_0, taken from the work of Lau and Kohn,[46] are shown in Figure 2.4.

Lau and Kohn[46] also discuss the interaction when the substrate has a surface band that is partly filled. Virtual excitations and de-excitations of electrons between empty and occupied states in the band due to the presence of the adatoms then contribute to the interaction energy. The reader is referred to their paper for the details.

2.3.1. Range of Validity: Anisotropic Effects, Adatom–Substrate Coupling, and Energy-Level Matching

The Lau–Kohn approach described above involves the following simplifications: (1) substrate isotropy, (2) weak adatom–substrate coupling, and (3) poor matching of adatom and Fermi energies. In subsequent work, Einstein[49] has sketched a more general approach to the calculation of the asymptotic pair interaction that allows restrictions (1)–(3) to be relaxed. He also noted that for some physically reasonable choices of the model parameters, the general asymptotic expression derived above may only describe the actual pair-interaction energy for the separations at which adatoms interact significantly at rather order-of-magnitude level. Einstein further argued that the asymptotic limit is often somewhat oversimplified in physical content because, as he points out, essentially a single wave function (the highest occupied) dominates the response of the substrate.

Einstein further emphasized that approaches which rely on a relatively weak adatom–substrate coupling exclude the strong-adorption regime in which a "surface molecule" is formed. The Lau–Kohn approach

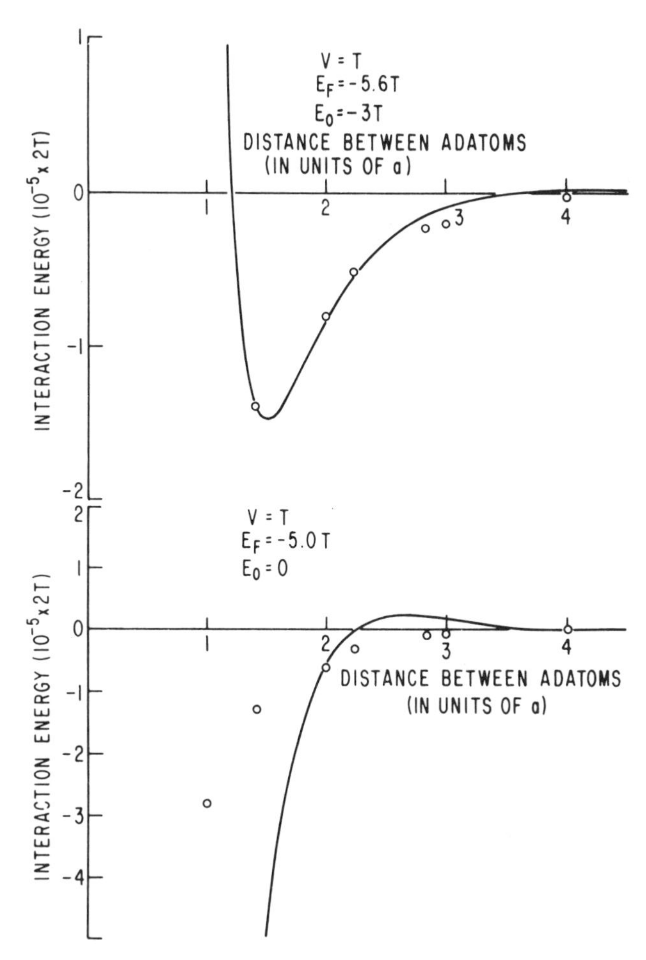

FIGURE 2.4. Plots of interaction energy ΔE_{int} for the case $|V| = T$ and for different values of E_f and E_0 (after Lau and Kohn[46]). Open circles denote results of Ref. 48.

(point 3 above) also requires the substrate Fermi energy to be far from the characteristic adatom energy level E_0. This would need relaxing to embrace some cases of interest for chemisorption.

We consider now in a little more detail the asymptotic expression discussed above. Lau and Kohn generated such a form for arbitrary substrate Fermi energy and adatom–substrate coupling in high-symmetry directions, and compared it with the exact model results of Einstein and Schrieffer.[48] The asymptotic expression adequately reproduces the exact behavior for large separations R between adatoms (several lattice spacings) but, not surprisingly, works less well for small R. As expected, the Lau–Kohn expression, which requires an isotropic Fermi surface (cf. however Appendix 2.2), agrees well with the general asymptotic expression near the bottom of the band. Near the band edge, the asymptotic expression becomes a poorer approximation for the small-R behavior as E_f further approaches the edge. Einstein pointed out that this regime is of some physical interest (the interaction is always attractive).

In summary, however, the R^{-5} times an oscillatory function behavior of the interaction energy $\Delta E_{int}(R)$ at large R is valid for the outer third of the band, according to Einstein. In the central third, evidence suggests a decay proceeding at least this fast, with probably a more complicated phase factor. This phase may be further complicated by broad adsorption resonances. Einstein argued that in practically all physically important situations on transition metals, as represented by a tight-binding model, the asymptotic expression is not fully quantitative in its representation of the indirect interaction.

2.4. MOLECULAR VERSUS DISSOCIATIVE ADSORPTION

The behavior of small molecules like CO, N_2, NH_3, etc., on transition metal surfaces is a subject of considerable interest. A general pattern emerges from analysis of data on various adsorbate-substrate systems, namely, that for a given molecule, metals to the left of a line dividing the Periodic Table dissociate the molecule, while those to the right adsorb it with its chemical bond(s) intact.

In this section, we shall discuss a simple chemical approach, which is based on Pauling's treatment[50] of bond energies in free molecules.

2.4.1. Pauling's Treatment of Bond Energies

This treatment was adapted by Eley[5] and other workers to the chemisorbed state. In particular, Eley obtained estimates of adsorption heats by employing Pauling's equation in calculating the metal atom–adsorbate atom bond energy, E(M–A) say. This quantity is expressed in the form (cf. equation (1.1.3) in appropriate units)

$$E(\text{M–A}) = \tfrac{1}{2}[E(\text{M–M}) + E(\text{A–A})] + (x_{\text{M}} - x_{\text{A}})^2 \qquad (2.4.1)$$

where E(M–M) and E(A–A) are the single bond energies of metal and diatomic molecule, respectively, while x_{M} and x_{A} are the corresponding atomic electronegativities. Equation (2.4.1) is written in electron volts.

In spite of the limitations of equation (2.4.1) as a route to the calculation of accurate bond energies, because of its sensitivity to the electronegativity difference, a point we shall return to below, we shall follow here the work of Flores *et al.*,[51] who demonstrate that equation (2.4.1) is adequate for the purpose of interpreting the above general pattern of dissociative versus molecular adsorption, provided a term due to metal–metal bond breaking is added. It is clear that to give a fully quantitative basis for such a term from wave mechanics is, at the time of writing, just not feasible. Therefore, Flores *et al.* use as input data into this term representing the breaking of metal bonds, a measure of valence as obtained from wave mechanical band-structure calculations. With this addition, Flores *et al.* demonstrate that equation (2.4.1) is adequate to expose a pattern of behavior for the chemisorption of a given adsorbate on a variety of transition metals.

2.4.1.1. Application to Chemisorption

If geometric factors are ignored, as well as metal bond breaking for the time being, the reactivity of a metal toward a molecular adsorbate is determined by:

1. The dissociation energy of the molecule in free space.
2. The strength of the bonds formed between its constituent atoms and the metal.
3. Molecular chemisorption energies.

For most of the systems of interest here, (3) is considerably smaller than (1) or (2), and for the purposes of exposing a pattern rather than making fully quantitative estimates, Flores *et al.* neglect it. Then, for instance in the case of a diatomic molecule, one must obviously compare the energies involved in (1) and (2) above. To study (2), it is important to consider next the way in which bonding in transition metals is to be handled.

2.4.2. Bonding and Valence in Transition Metals

A useful description of this bonding can be given in terms of overlapping hybrids on neighboring atoms. These hybrids are formed from the *s*, *p*-, and *d*-orbitals in such a way that they are directed toward nearest and next-nearest neighbors in the crystal.[52]

If one takes the case of a face-centered-cubic (fcc) transition metal for which the 12 nearest neighbors make the major contribution to the bonding, then in principle the number of hybrids is six or less. This calls for a description of the ground electronic state in terms of a number of

TABLE 2.1. Average Valences Assigned to Transition Metals

Ti	V	Cr	Mn	Fe	Co	Ni	Cu
3.4	3.3	3.4	3.5	3.5	3.2	2.9	2.6
Zr	Nb	Mo	Tc	Ru	Rh	Pd	Ag
4.0	4.0	4.2	4.3	4.1	3.6	3.1	2.7

resonating states* so that, on average, there is a hybrid pointing toward each of the 12 near-neighbor atoms.

Of course, it is a direct consequence of this picture that only a fraction of an electron can be shared in each bond between atoms; this is expressed in Pauling's concept of metallic valence.[50] One then wishes to associate the fractional occupation of the hybrids with some average valence, defined as the total number of electrons used for bonding to the neighboring atoms.

Though the above discussion referred specifically to a fcc crystal, in fact for a hcp metal the local coordination number is also 12 while in a bcc structure there are eight near neighbors and six next-near neighbors. Thus, even for the bcc structure, the assumption of 12 hybrids around each atom should provide a fair average description.

Returning to the matter of the average valence, it is important that this is chosen to be consistent with the available knowledge about the electronic charge distribution in the energy bands of transition metals. For, in spite of the strong electron–electron correlation that Pauling's ideas correctly incorporate, it is known from density functional theory that correlation can be subsumed into a one-body potential in calculating, in principle exactly, the electronic charge density in the ground state from energy band theory.

Therefore Flores *et al.* used the augmented plane wave (APW) calculations of Moruzzi *et al.*[53] for the energy bands of transition metals, to obtain the number of interstitial electrons per atom in the crystal. This number they defined from the APW calculation as the number of electrons outside the muffin-tin sphere. Their proposal was to correlate this quantity with the number of bonding electrons, the reason being that the localized electrons are not shared in bonds and contribute only to the charge in the immediate vicinity of one atom.

Their procedure was then to assume that the average valence is proportional to the number of interstitial electrons. By choosing the proportionality constant to yield a valence of 1 for Na and 3 for Al with bcc and fcc structures, respectively, they obtained the average valences recorded in Table 2.1.

* **Note added in proof.** The work of M. H. McAdon and W. A. Goddard (*Phys. Rev. Lett.* **55**, 2563, 1985) lends quantitative support to these ideas of Pauling (see also N. H. March and M. P. Tosi, *Phys. Chem. Liquids* **10**, 113, 1980).

TABLE 2.2. Bond Energies, in eV[a]

Ti	V	Cr	Mn	Fe	Co	Ni	Cu
2.89	3.00	2.26	1.66	2.57	2.81	3.12	2.70
Zr	Nb	Mo	Tc	Ru	Rh	Pd	Ag
3.20	3.60	3.06	3.07	3.30	3.18	2.55	2.24
Hf	Ta	W	Re	Os	Ir	Pt	Au
(3.14)	(3.86)	(3.88)	(3.77)	(4.01)	(3.80)	(3.82)	(2.85)

[a] Numbers in parentheses were found using valences as for the second-row series; this applies also in Tables 2.3–2.7.

There seem, at the present time, not to be similar quantitative calculations for the number of interstitial electrons in the third transition series. Therefore, in order to extend their analysis to these elements, Flores *et al.* assigned the valences of the second-row elements to them since, in general, properties of the third transition series resemble more closely the second series than the first. However, they point out that this assumption is already such that one ought to consider only the general pattern when including the third transition series.

It is of some interest to comment on the average valences recorded in Table 2.1 in relation to Pauling's maximum valences. Pauling assigned valences of 6 to the elements Cr to Ni and $5\frac{1}{2}$ to Cu: these are 2–3 units larger than the proposals of Flores *et al.* Hume-Rothery *et al.*[54] earlier argued that the metallic valence is less for the elements of the first long periods than it is in the later periods. On the basis of the physical properties of the elements, they suggest that V and Cr do not involve more than four bonding electrons per atom and that Mn has a relatively low valence of around two, followed by a rise in passing to Fe. A decrease in valence through group VIII is more evident in the second and third periods. Flores *et al.* drew attention to the fact that their proposed scheme is supported by the relative importance of the valences exhibited in the chemistry of the transition elements.

Table 2.1 for the average valences can now be used to estimate metal–metal bond energies by dividing twice the cohesive energy[55] by the average valence. Flores *et al.* note that the values thus obtained, which are listed in Table 2.2, differ substantially from those of Eley *et al.*, who obtained metal–metal bond energies by dividing twice the heat of sublimation by the number of nearest neighbors.

2.4.3. Admolecule Bonding to a Metal Surface

With the concept of average valence as above, and the estimated metal–metal bond energies, Flores *et al.* discussed the bonding at the surface along the following lines. They regarded each surface atom as having a number of dangling bonds, determined by the crystal structure and the particular crystal face, while the number of electrons occupying these bonds is governed by the average valence v of Table 2.1. For example, each surface atom on the (111) face of a fcc crystal has three dangling bonds with $v/12$ electrons each. Due to the similarities of local coordination in fcc, hcp, and bcc crystals referred to above, Flores *et al.* assume that this is true for a dangling bond on any face. This implies that in order to form one bond with the adsorbate, other $(1-v/12)$ metal electrons are required for sharing. More generally, a number of metal bonds, n say, depending on the particular crystal face and on the adsorbate, will have to be broken in order to saturate the metal–adatom bond. Thus equation (2.4.1) can be modified to read

$$E(\mathrm{M{-}A}) = \tfrac{1}{2}[E(\mathrm{M{-}M}) + E(\mathrm{A{-}A})] + (x_{\mathrm{M}} - x_{\mathrm{A}})^2 - nE(\mathrm{M,M}) \quad (2.4.2)$$

which is the starting point for the calculations of Flores *et al.*

2.4.4. Chemisorption of N_2, O_2, CO, and NO

The energetics of N_2, O_2, CO, and NO chemisorption on transition metals will now be discussed. Without appeal to specific crystal structures, for the reasons given above, Flores *et al.* assume that for oxygen and nitrogen, every hybridized atomic orbital can form a bond with the metal surface. For carbon, in contrast, it is not clear how many bonds can be formed with the surface, this question involving detailed knowledge of the valence of carbon in interaction with the metal surface.

2.4.4.1. N_2

Figure 2.5, reproduced from Flores *et al.*, shows the specific example of nitrogen in a triangular site on the (111) face of a bcc crystal, where the three nitrogen hybrids overlap with hybrids directed from three metal

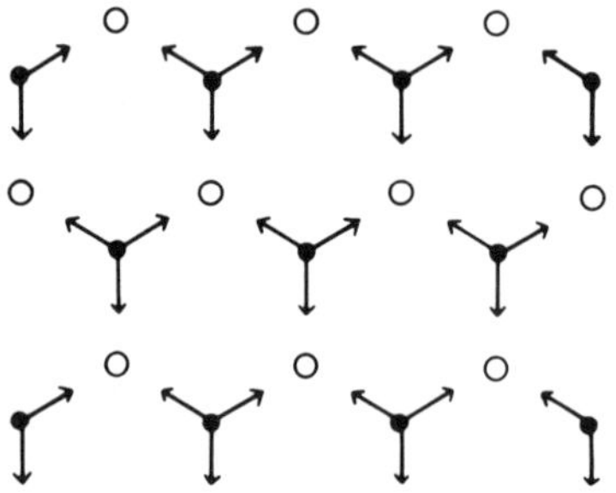

FIGURE 2.5. Adsorption sites for nitrogen atoms represented by open circles, on the (111) face of a body-centered-cubic metal. Solid circles represent metal atoms, arrows denoting appropriate hybrid atomic orbitals pointing out of the surface.

atoms. This was chosen as a favorable case where the directional bonding is evident. However, for a different surface, Flores *et al.* argue that one can expect from the chemistry of nitrogen that the tendency would be for three bonds still to be formed with other hybrids in the surface. Thus, it should not be assumed that the treatment under discussion would lead, as might appear at first sight from Figure 2.5, to strong variation of the metal–adsorbate bond strength by varying the crystallographic orientation of the surface. There is no such strong variation according to available experimental evidence.

The energy of chemisorption of one nitrogen atom can be written as

$$E(\text{M–N}) = 3\{\tfrac{1}{2}[E(\text{M–M}) + E(\text{N–N})] + (x_\text{M} - x_\text{N})^2 - (1 - v/12)E(\text{M–M})\} \tag{2.4.3}$$

the factor 3 accounting for the number of bonds formed between nitrogen and the metal, while $(1 - v/12)$ replaces n in equation (2.4.2). Clearly, equation (2.4.3) is sensitive to the difference $x_\text{M} - x_\text{N}$, as considered in the calculations summarized below.

Equation (2.4.3) must only be applied when $E(\text{M–N})$ is greater than that energy obtained by assuming a fractional bond order given by v and no breaking of the metal bonds. Anticipating the results in Table 2.3 below, it is to be noted that this latter energy is in fact higher for Pt and Ir, and in these two cases it is then the appropriate values that are recorded. In these cases, instead of $E(\text{M–N})$ given by equation (2.4.3), the bond energy $\{\tfrac{1}{2}[E(\text{M–M}) + E(\text{N–N})] + (x_\text{M} - x_\text{N})^2\}$ must be multiplied by $v/12$, the fractional bond order referred to above.

The metal–nitrogen bond energies obtained from equation (2.4.3) together with data from Tables 2.1 and 2.2 are collected in Table 2.3. These should be compared with 4.8 eV, half the dissociation energy of N_2, in

TABLE 2.3. Energy of Metal–Nitrogen Atom Bond, in eV[a]

Cr^D	Mn^D	Fe^D	Co	Ni^M	Cu^M
7.0	8.3	5.8	5.2	3.8	4.2
Mo^D	Tc	Ru	Rh^M	Pd^M	Ag
6.2	—	4.1	3.4	3.1	5.0
W	Re	Os^M	Ir^M	Pt^M	Au^M
(5.2)	(4.3)	(3.5)	(3.5)	(2.6)	(2.5)

[a] Estimated errors are given by Flores *et al.*[51] M denotes molecular adsorption while D refers to dissociation. Absence of labeling means no decisive conclusion. Values in parentheses indicate that the same valences were used in series 3 as in series 2.

order to determine whether dissociation or molecular adsorption is likely to occur.

The solid line in Table 2.3 shows the experimental boundary according to Brodén *et al.*[56] As in Tables 2.4–2.7, Tc has been excluded because of the uncertainties that Flores *et al.* argue are due to the choice of electronegativity. Though Ag has been retained in the tables, they argue that uncertainties in electronegativity remain there also. Agreement between theory and experiment seems to be satisfactory when one bears in mind the large errors originating from the theoretical estimates due to the wide extremes of electronegativity quoted by Gordy and Thomas.[57] For example, binding energies of 5.1 and 6.7 eV have been reported[58] for nitrogen atoms on the (110) and (100) W surfaces while a value of 6.1 eV was assigned to the Fe–N bond.[59] It must be pointed out though that for Pt and Ir in Table 2.3 the values are not in good agreement with experiment. For Pt, though we have no quantitative value of the valence from direct band-structure calculations, comparison of bandwidths of Pt and Pd suggests that the valence of Pt should be somewhat larger than for Pd, while the value for Pt in Table 2.3 was obtained using the same valence (3.1) for both these metals. If, for instance, the valence of Pt were increased to 4, an increase from the value of 2.6 in Table 2.3 to 4.3 ± 0.3 eV would result, indicating sensitivity to the choice of valence in this metal. Thus it is not difficult to see how theory and experiment might be reconciled in the case of Pt. Clearly, the uncertainties in the valences mean that quantitative agreement is not to be expected in the third series, and further band-structure data is awaited here, at the time of writing.

2.4.4.2. O_2

For oxygen, with valence 2, the (110) face is most suitable for chemisorption. Figure 2.6 shows the preferred adsorption site for an oxygen atom on the surface, where favorable interaction with the metal hybrids can be anticipated.

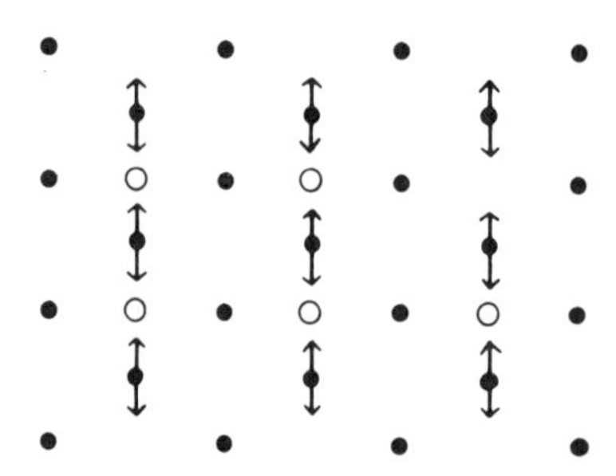

FIGURE 2.6. Adsorption sites for oxygen atoms represented by open circles, on the (110) face of a body-centered-cubic metal. Solid circles represent metal atoms, arrows denoting hybrid atomic orbitals pointing out of the surface.

The metal–oxygen bond energy in this case is given by

$$E(\text{M–O}) = E(\text{M–M}) + E(\text{O–O}) + 2(x_\text{M} - x_\text{O})^2 - 2(1 - v/12)E(\text{M–M}) \quad (2.4.4)$$

Comparison of this quantity, calculated for various metals, with half the dissociation energy of O_2, namely 2.57 eV, leads to the results in Table 2.4.

2.4.4.3. CO

Following the discussion of the homonuclear diatoms N_2 and O_2, two heteronuclear molecules, CO and NO, will be considered briefly in the above context. For CO, with dissociation energy 11.09 eV, there arises the problem of ascribing a valence for C in interaction with the metal surface. The results obtained by Flores *et al.* for a valence of 2 are recorded in Table 2.5.

TABLE 2.4. Energy of Metal–Oxygen Bond in eV[a]

Fe^D 6.7	Co^D 6.3	Ni^D 5.0	Cu^D 5.4
Ru^D 4.9	Rh^D 4.4	Pd^D 4.1	Ag^D 6.0
Os^D (4.4)	Ir^D (4.2)	Pt^D (3.2)	Au (3.6)

[a] Errors are given by Flores *et al.*[51] Solid line indicates experimental borderline between dissociative and molecular adsorption at room temperature. With the exception of Au, which does not adsorb oxygen appreciably,[60] all the transition elements[61] including the noble metals appear to dissociate molecular oxygen at room temperature.

TABLE 2.5. Energy of Dissociated CO Molecule Interacting with Metal Surface, in eV[a]

Cr 11.5	Mn^D 13.7	Fe 10.4	Co^M 9.7	Ni^M 7.7
Mo 11.0	Tc —	Ru^M 8.0	Rh^M 7.2	Pd^M 6.7
W (9.6)	Re^M (8.4)	Os^M (7.2)	Ir^M (6.7)	Pt^M (5.2)

[a] Errors are given by Flores *et al.*[51] Solid line indicates boundary between molecular adsorption and dissociation as found experimentally. The assumption of a valence of 2 for C leads to a boundary line more to the left than the solid line in the table. Agreement with experiment would be restored if C were ascribed a valence close to 2.5.

2.4.4.4. NO

The chemisorption energy for the dissociated nitric-oxide molecule interacting with the metal surface can be obtained by using the previous results for N and O. In order to examine whether the dissociated state is the equilibrium configuration, Flores *et al.* compare this with the dissociation energy of 6.51 eV for NO. In Table 2.6, the results for NO adsorbed on different transition metal surfaces are recorded, while Table 2.7 is for H_2.

Of course, the above discussion of molecular versus dissociative adsorption has inevitably been in gross terms. In individual cases, a good deal is known about the detailed chemical bonding and this we discuss now, in order to deepen understanding of the chemical interaction of molecules with metal surfaces.

Table 2.6. Energy of Dissociated NO Molecule Interacting with Metal Surface, in eV[a]

Cr^D	Mn^D	Fe^D	Co^D	Ni^D	Cu^D
14.7	17.1	12.5	11.5	8.8	9.6
Mo^D	Tc	Ru^D	Rh	Pd	Ag^D
13.2	—	9.0	7.8	7.2	11.0
W^D	Re	Os	Ir	Pt^M	Au
(11.3)	(9.5)	(7.9)	(7.2)	(5.1)	(6.1)

[a] Errors are given by Flores *et al.*[51] M and D correspond to molecular and dissociative adsorption, respectively. Absence of labeling means no decisive conclusion. Solid line indicates experimental boundary according to Brodén *et al.*

TABLE 2.7. Energy of Metal–Hydrogen Atom Bond, in eV[a]

Fe^D	Co^D	Ni	Cu
2.8(110)	—	2.7(111)	2.4(111)
2.9	2.9	2.6	2.4
Ru^D	Rh^D	Pd	Ag
—	—	2.7(111)	—
3.5	3.1	2.6	2.5
Os^D	Ir^D	Pt	Au
—	2.7(111)	2.5(111)	—
(3.7)	(3.2)	(2.8)	(2.5)

[a] First entries are experimental values for the surfaces indicated.[62] Lower entries (errors to these are given by Flores *et al.*[51]) are predictions of the model described in detail in Ref. 51.

2.5. DETAILED BONDING STUDIES

2.5.1. Nitrogen on Metal Surfaces

Hendrickx *et al.*[63] have studied the influence of the surface structure on the adsorption of nitrogen on Rh in comparison with Pt, Ir, Pd, and Ni. It is known that the surface structure of a metal may have a significant influence on the adsorption of simple gases. For example, Albers *et al.*[64] found that, at room temperature, dissociative adsorption of oxygen takes place on a Ag(110) surface but not on a smooth Ag(111) surface.

We shall now discuss the nature of group VIII metal–nitrogen bonding. In this context it is generally accepted that dissociative adsorption of nitrogen does not occur on Pt, Ir, Rh, and Pd surfaces at temperatures below 300 K. Wilf and Dawson[65] reported a slow dissociative adsorption of polycrystalline Pt, with a rate essentially independent of pressure and temperature. In other investigations, dissociative adsorption was not observed on various surfaces of Pt with different surface structure. The presence of atomic nitrogen on Pt and Pd surfaces has been observed following activation of nitrogen by an electron flux, a high-frequency discharge, or by the use of a hot tungsten filament. These N adatoms are known to be strongly bound to the surface.

Hendrickx *et al.* argued that, qualitatively, the bonding of nitrogen to a metal surface can be understood in terms of an interaction of the electron donating $2\sigma_u$ and $3\sigma_g$ orbitals with accepting metal orbitals accompanied by metal back donation into the empty N_2 $1\pi_g$ orbital. Horn *et al.*[66] concluded from their results that the bonding is primarily through a N_2 σ-metal interaction, with the $2\sigma_u$ orbital the largest contribution to the bond and with a small amount of Ni-$1\pi_g$ back bonding.

If this bonding mechanism also holds for nitrogen on Pt, Ir, and Rh surfaces, the variation in the work function may suggest that the relative contribution of back donation increases in the order Pt, Ir, and Rh. A similar trend has been observed by Nieuwenhuys[67] for CO adsorption.

It cannot be excluded that the bonding may be different on other surfaces, such as (111) faces. An indication may be found in the observation by Horn *et al.*,[66] who concluded that nitrogen is only physically adsorbed on Pd(111) as judged by its photoelectron spectrum, which appears to be similar to gas-phase nitrogen, and also by its low desorption temperature. They also reported that the N–N molecular axis is random for this state on Pd(111).

On the other hand, however, Shigeishi and King[68] found an intense infrared absorption band for nitrogen on a Pt(111) foil with a peak located at 2240 cm^{-1}, consistent with the adsorption mode for Ni(110).

It does not seem possible to exclude the existence of another type of bonding, such as that of nitrogen molecules lying parallel to the surface for the (111) surfaces. Hendrickx *et al.*[63] argue that a standing-up position is more likely, certainly at high coverages because of (1) similarities between N_2 and CO adsorption for Rh and Ir surfaces and (2) all well-characterized complexes N_2 have the end-on structure although the side-on structure has been postulated for the transition state (cf. Chapter 5) in reactions of nitrogen.

2.5.2. Cluster Modeling of NO on Ni

Bauschlicher and Bagus[69] have studied NiNO as a model for NO on a metal surface. In spite of the criticisms along the lines of Grimley and Mola, to be discussed below, there is some evidence from earlier work of Bagus *et al.*[69] that many of the qualitative features of the bonding with metal surfaces are present in these single metal atom systems. Since they were primarily interested in the interaction of NO with a metal surface, Bauschlicher and Bagus have restricted themselves to consideration of the $3d^9 4s^1$ state of Ni in order to represent the unfilled d-band of Ni metal.

The resulting conclusions are that two different mechanisms of bonding of NO to Ni are apparent:

1. For a linear geometry, a $3d$–$2\pi^*$ bond is involved and has a short Ni–N bond length of about 1.7 Å.
2. The more weakly bound, second mechanism involves a Ni $4s$–N $2p$ bond in a bent structure.

The relationship of these two states to NO chemisorption on metals is then discussed by these workers. For both mechanisms (1) and (2) above, there are no open-shell electrons on NO, consistent with the fact that multiplet splitting is not definitely observed in the photoemission spectrum of adsorbed NO; this point will soon be discussed.

It should be made clear that the binding energy of NiNO is not that for NO chemisorbed on Ni. However, the bonding mechanisms for the linear π-bonded and the bent state are very different. The above workers claim that it is likely that the major features of these bonds will remain for adsorbed NO.

The linear state with the covalent $3d$–$2\pi^*$ bond is clearly the most stable structure. This type of bonding, according to Bauschlicher and Bagus,[69] is appropriate for NO chemisorbed on first-row open d-shell transition metal surfaces. For Cu and Ag, with closed d-shells, the bonding could arise from the bent $4s$ bonded state.

For the second and third transition rows, these workers argue that the nd and $(n+1)s$ orbitals are diffuse and the nature of the bonding can change from that for the first row. Relativistic effects for the third transition row can also have important consequences for the nature of the bonding.

As already mentioned, for both the linear and the bent models of the bonding, there is no open-shell character associated with NO. Bauschlicher and Bagus predict that there will be no multiplet splitting in the NO-derived photoemission spectrum of chemisorbed NO. This is consistent with the results of Breitschafter *et al.* for Ni(111)/NO.[69]

In concluding this brief discussion, it is worth adding the following, more chemically oriented, observations. The stable linear state of interest has a $3d - 2\pi^*$ bond, which is formed by combining the metal $3d(\pi)^3$ and the NO $2\pi^1$ to form a closed shell π^4 and has some double-bond character. This bond length is shorter than that for the bent $^2A''$ state as a result of the d-contribution to the bond. The open-shell $4s$-orbital is not involved in the bonding but polarizes away from the 5σ to reduce the repulsion.

2.5.3. Limitations and Convergence Properties of Cluster Calculations

Many cluster calculations have been carried out relating to chemisorption. A few of these will be selected below, inevitably somewhat arbitrarily, to give some idea of the type of information that can be obtained.

The examples are taken from the work of Grimley and Mola,[70] who investigated the ability of small substrate clusters to represent adequately the chemisorption of hydrogen by the semi-infinite substrate for three models:

1. The Grimley–Pisani model[71] of chemisorption by a solid treated in the tight-binding approximation. This model uses Dyson's equation[71] to embed an adsorbate in the surface of a semi-infinite solid. The embedding problem is thereby formulated correctly, but to have a practical scheme the model uses a tight-binding electron Green function in the adsorbent beyond the cluster, but the correct self-consistent Green function in the adsorbate/adsorbent cluster.
2. The same model, but with self-consistency extended to the whole solid using the Hubbard Hamiltonian (cf. Appendix 2.3).
3. The CNDO (complete neglect of differential overlap) model for chemisorption by Li metal.

Fully converged results are known, by direct calculation for the semi-infinite substrate, only for model (1), where the work of Grimley and Mola shows in fact the slowest convergence rate with increasing cluster size. Extending the self-consistency, as in (2) and (3), enhances the rate of convergence, but the CNDO model still needs at least eighteen Li atoms to represent adequately the on-site chemisorption of hydrogen by (100) Li.

By way of introducing the conclusions of Grimley and Mola, it should be noted that such cluster models of chemisorption have been widely used, their popularity being partly due to the fact that standard and readily available computational techniques of quantum chemistry can therefore be used in chemisorption theory.

However, Grimley and Mola note that the theoretical model chosen to treat the cluster may sometimes lack the electron-interaction terms necessary to screen the disturbance caused by the admolecule and hence to ensure that chemisorption is a rather localized phenomenon for metal surfaces (cf. Section 1.1 of Chapter 1). These workers claim that such a situation obtains with the extended Hückel theory, for instance. With the Hartree–Fock model (and even with its CNDO/INDO approximate versions) the situation is much improved because there is a self-consistent determination of the electronic charge distribution. However, whatever theoretical model is used to treat the cluster, the ability of the chosen cluster to represent adequately those aspects of the semi-infinite substrate that are important in chemisorption has to be demonstrated.

Grimley and Mola then make some observations on the type of approach in which tight-binding models are employed. Since the second moment of the density of electronic states on an atom in a cluster coincides with its value in the semi-infinite substrate if the atom in the cluster has the correct number of nearest neighbors, it would seem prudent, in making a cluster model of chemisorption, to ensure that the substrate atoms that are directly involved in the surface bond have at least their correct nearest-neighbor environments.

In conclusion their calculations confirm that the ability of a cluster calculation to model chemisorption by a semi-infinite substrate is governed by the electronic bandwidth of the substrate and the extent to which the chemisorption bond is associated with electrons in split-off, and therefore localized, states far removed from the substrate bands. Systems comprising substrates with narrow bands and adsorbates giving split-off states should be modeled very well by small clusters, but the minimum cluster size must depend on the theoretical model used. As already mentioned, they show that more than eighteen Li atoms are needed in the CNDO model of the on-site chemisorption of hydrogen by (100) Li.

2.6. ELECTRONICALLY EXCITED STATES OF CHEMISORBED DIATOMS

So far, the structure and bonding to the substrate of diatomic molecules in their electronic ground state has been the focal point.

Nevertheless, it is quite clear that electronically excited states of adsorbates also play an important role in a number of surface processes. Among these may be listed:

1. Desorption induced by electronic transitions.
2. Photochemistry from valence or core-excited adsorbates.
3. Resonance photoemission.
4. Surface Raman scattering.

A survey of topic (1) can be found in the book edited by Tolk *et al.*,* while with regard to (4) and other optical processes one may consult the article by Otto[73] on light scattering in solids. Suffice it to add at this stage that in understanding such surface processes, it is important to know both the nature of the excited states involved and the mode of their (nonradiative) decay. (For a model calculation of lifetime, see Appendix 2.4.)

2.6.1. EELS Study of Weakly Adsorbed Systems

Using EELS (see Appendix 1.1), a systematic study has been reported of the excitations of weakly adsorbed (e.g., physisorbed) diatomic molecules by Avouris *et al.*[74] on metal surfaces at low temperatures. It was found that the intrinsic excitations of the adsorbates were preserved with only small energy shifts (about 0.1 eV) and broadening (0.1–0.2 eV) relative to the corresponding free (gaseous) adsorbate values. The nonradiative decay of such excitations has been discussed by Avouris and Demuth in terms of dipole coupling to the electron–hole pair excitations of the metallic substrate. In addition to the perturbed intrinsic excitations of the adsorbate, it was shown that the substrate–adsorbate interaction can lead to excitations involving dynamic charge transfer from the substrate to empty adsorbate orbitals.

Avouris *et al.*[74] have presented a general discussion of excited states of adsorbates, the importance of intra-adsorbate many-particle effects in such excitations, and the role of the metallic substrate on the screening of the intra-adsorbate Coulomb and exchange interactions (cf. the image treatment in Section 2.1.1).

Specifically, these workers used EELS to study the electronic excitations of CO adsorbed on Ni(100) and on polycrystalline Cu, and of NO on Ni(100). They also made use of UPS (cf. Appendix 1.1) to obtain information regarding the electronic structure of the ground state of chemisorbed CO and to delineate monolayer and multilayer coverage regimes.

* **Note added in proof.** The reference is to *Desorption Induced by Electronic Transitions*, Springer Series in Chemical Physics, eds. N. H. Tolk, M. M. Traum, J. C. Tully and T. E. Madey (Springer, New York, 1983). For inelastic ion–surface collisions, see Ref. 72.

2.6.2. Summary of Assignments

For NO on Ni(100), the electronic excitations of the adsorbate have been assigned and in particular a broad band at ~5eV is attributed to a $2\pi - 3s$ Rydberg excitation.[74]

Avouris *et al.* have also assigned the low-energy excitations of chemisorbed CO at approximately 6 and 8 eV to triplet and singlet coupled $5\tilde{\sigma} \rightarrow 2\tilde{\pi}^*$ transitions, respectively.

It is quite clear that here is a field in which a great deal of progress is going to be made in the future, and which may well provide a stringent test of many of the concepts of valence theory.

2.7. ORIENTATION OF MOLECULAR ADSORBATES

Umbach and Hussain[75] have presented a source of experimental information, by employing angle-resolved, X-ray-induced Auger spectroscopy for adsorbed molecules. Prior to their work, the observed strong angle-dependent changes of the Auger line shape seem to have not been stressed in adsorbate studies. They are due to the anisotropy of the Auger decay and are not markedly influenced by scattering effects.

The changes can be explained by symmetry arguments and thus provide a direct indication of the symmetry character of the observed transitions. This facilitates the interpretation of molecular Auger spectra and comparison between theory and experiment. It also clarifies the role of the substrate, particularly concerning the screening mechanism, which is rather important for photoemission or photon-stimulated desorption (see Chapter 5) and which has been a topic of some interest.

In addition, angle-resolved Auger electron spectroscopy provides a direct method to determine the geometry (such as the orientation) of molecular adsorbates. Umbach and Hussain emphasize that such drastic angular effects are important for Auger spectroscopy in general, even if it is used only as an analytical tool or as a probe for other techniques (e.g., surface-extended X-ray absorption fine structure, or near-edge X-ray absorption fine structure).

Umbach and Hussain have also noted that different angle dependences have been found in some previous Auger studies. In the gas phase the observed anisotropies are small (about 5%) and are due to angular correlation between the initial excitation and the Auger electron. Some Auger lines of solid samples show azimuthal- and polar-angle dependences, which are dominated by diffraction effects. In the study of Umbach and Hussain, diffraction effects can be excluded because only relative changes in the Auger line shape are considered, and the relative wavelength difference $\Delta\lambda/\lambda$ of the main peaks is only about 0.5%.

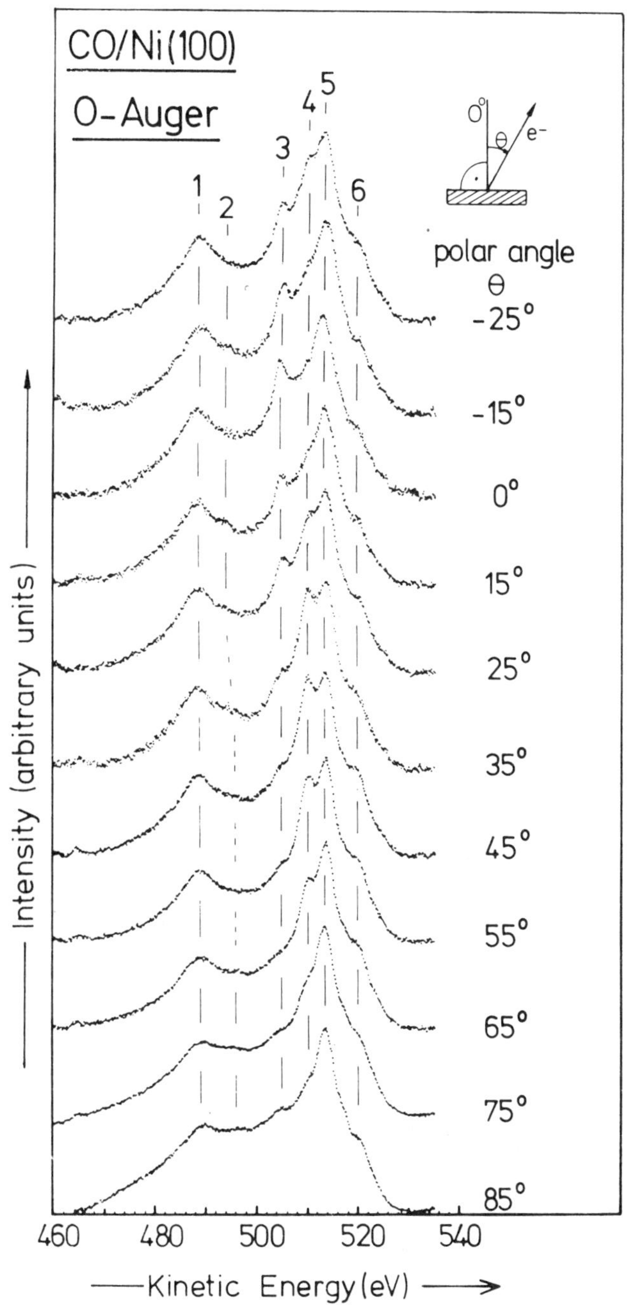

FIGURE 2.7. Oxygen Auger spectra from molecularly adsorbed CO on Ni(100) at 105 K, taken at various polar angles (after Umbach and Hussain[75]).

While most Auger studies are performed with electron excitation, for high-resolution experiments excitation by X-rays is preferable, because these cause less background problems and less beam damage to molecular adsorbates. Umbach and Hussain used radiation in the soft X-ray range (450–1500 eV).

Figure 2.7 shows oxygen Auger spectra from molecularly adsorbed CO on Ni(100) at 105 K, taken at various polar angles. Drastic changes of line shape occur between 500 and 525 eV even for an angular variation of 10° or less. For instance, the peak at 505 eV (labeled 3) has negligible intensity at polar angles above 30°, but increases more than 10 times for normal emission.

The variations of relative peak intensities are shown in Figure 2.8, which presents the quantitative evaluation of the spectra of Figure 2.7. These angular effects can be understood by symmetry considerations, without calculational effort.

Following Siegbahn *et al.*,[76] one can write the intensity I_{jk} of an Auger transition $c \rightarrow jk$ in terms of Coulomb, J_{ij}, and exchange, $K_{jk} = J_{kj}$, matrix elements of the form

$$J_{jk} = \langle \phi_j \phi_k \mid e^2/r \mid \phi_c \phi_e \rangle$$

where c and e refer to the initial core hole and the emitted Auger electron, respectively, and where j and k represent valence orbitals. The proper symmetry group for chemisorbed linear molecules is $C_{\infty v}$ (C_{4v} would yield

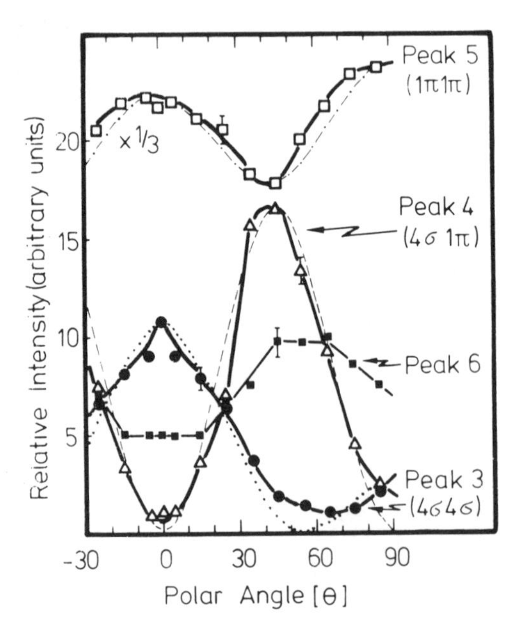

FIGURE 2.8. Variations of relative peak intensities, as obtained by evaluation of the spectra in Figure 2.7 (after Umbach and Hussain[75]).

very similar results) if one assumes that the surface mainly causes the alignment of the adsorbed molecules and that the symmetry of substrate and adsorption site is less important for most of the electronic transitions.

In summary, it is noteworthy that angle-resolved Auger spectra can be utilized to determine the orientation of adsorbed molecules with high accuracy ($< 5\%$) and little ambiguity. The results of Umbach and Hussain confirm what was known from previous studies, namely that CO stands vertically on the Ni(100) surface because they are only compatible with the assumption that the molecular symmetry axis and the surface normal coincide.

Their NO results also show unambiguously that NO is vertically adsorbed on the Ni(100) surface. The angular effects for NO and also for N_2, on Ni(100), are strong, just as they are for the case of CO described above.

2.8. TEMPERATURE EFFECTS, AND COMPARISON BETWEEN THEORY AND EXPERIMENT

In concluding this chapter there is one matter that should be stressed. The theoretical models presented are strictly appropriate at absolute zero of temperature. Therefore, caution is required in comparing experiment with model predictions, because temperature effects are often important in practical situations, to be discussed briefly below by reference to some specific examples.

Thus Grunze *et al.*[77] note that N_2 adsorbs dissociatively on Ni when clean Ni(110) is exposed to N_2 at 190 K. However, they also deduce that at low temperatures (80–130 K) and pressures, N_2 is chemisorbed molecularly on Ni(110).

Turning to the further example of CO on Ni, the discussion of Section 2.4.4 shows that the model there predicts that dissociation of CO on this transition metal is energetically forbidden. It is known that the dissociation of CO on Ni forms the basis of CH_4 formation, but it should be stressed that the catalytic methanation of CO takes place at temperatures above 520 K.[17] There is a considerable body of evidence for the existence of molecular CO on Ni surfaces even at room temperature. In particular, Barber *et al.*[78] claim that their results at room temperature and below indicate little dissociation, though as the temperature is raised the amount of dissociated CO increases. Andersson[79] observed in the energy-loss spectrum of CO adsorbed on Ni(100) at 173 and 293 K vibrational features, which he assigned to bridge and linearly bonded CO. Bertolini and Tardi[80] performed an electron energy loss (EELS) study of CO on clean Ni(111), (100), and (110) surfaces. By analogy between the C–O stretching vibrations they observed with metal carbonyl spectra, they propose that on

Ni(100) and (110) both terminally and bridge bonded CO exist simultaneously. On Ni(111), however, CO occupies twofold sites only.

In the comparison of theory and experiment in Section 2.4.4, the borderline between molecular adsorption and dissociation suggested by Brodén *et al.*[56] seems to be consistently interpreted as a low-temperature borderline. The way the borderline changes with temperature is discussed in the work of Benziger.[81] For diatomic molecules, these considerations of Benziger seem appropriate to discuss the variation with temperature of the borderline considered in Section 2.4.4 in the low-temperature limit.

To summarize, the factors potentially important in determining the energy of chemisorption of atoms on transition and noble metals are:

1. The ideal covalent energy associated with the metal atom-chemisorbed atom bond.
2. The electronegativity difference between the chemisorbed atom and the metal substrate atoms.
3. The breaking of metallic bonds near the surface due to the proximity of the chemisorbed atom. This factor is governed by the average valence of the atoms in the metal.

In this type of treatment, (1) can be altered in an important fashion by (2) and (3), the inclusion of (3) being the new feature beyond Pauling's approach for free space molecules. Using this method, the question as to the dissociation, or binding, of a molecule brought up to a metal surface can be answered in terms of a balance between these three factors. Factor (2) always aids dissociation while (3) opposes it.

Chapter 3

Conformation and Electronic Structure of Polyatomic Molecules

The focus of this chapter will be the geometry and the electronic structure of polyatomic molecules outside metal surfaces. The essential new feature here is obviously molecular conformation. Does benzene remain planar hexagonal on a particular face of tungsten? Is the water H–O–H angle changed from its free-space value when chemisorbed on ruthenium?[82] How does ethene behave on noble and transition metals?[83]

Though there have been quite a number of calculations in this general area as well as a great many experimental studies, it seems difficult at the time of writing to present very many general principles. For free-space polyatomic molecules, one has as important guidelines for predicting shape: Walsh's rules[84] based on the variation of the sum of one-electron energy levels as the geometry is changed,[85] or those based on electron–electron repulsion, pioneered by Sidgwick and Powell,[86] and formalized by Gillespie and Nyholm,[87] all of which demonstrate the vital role of the total number of valence electrons in free-space conformation. Usually, however, when molecules are chemisorbed, one is interested in perturbations of free-space structure and the rules are probably too gross to help with such detail.

One rather general effect is that, in the case of physisorption and perhaps even in weak chemisorption, image effects appropriate to (almost) perfect conductors must be incorporated. These were treated in some detail in the previous chapter on diatoms and will therefore serve as a useful starting point.

3.1. CONFORMATION OF A WATER MOLECULE OUTSIDE A METAL SURFACE

Flores *et al.*[88] considered the effect of a metal surface on the conformation of a water molecule. This theoretical work was motivated by the electron-stimulated desorption-ion angular distribution (ESDIAD) experiment of Madey and Yates[82] concerning the conformation of a water molecule on Ru(001). Their results show that:

1. H_2O is bonded to Ru via the oxygen atom.
2. The angle between the O–H bonds is a few degrees larger than the free-space angle.

From a theoretical standpoint, the work of Mahanty and March[38] and also that of Flores *et al.*,[43] which was covered in Chapter 2, has demonstrated that the chemical bond can be altered as a function of orientation with respect to the surface. However, much less has been done about the effect of surfaces on bond angles. In this section we shall therefore present a summary of the main findings of Flores *et al.*[88] on H_2O outside a metal surface. Later in this chapter we shall consider other polyatomic molecules, both by similar methods[89] and, toward the end of the chapter, by a more sophisticated approach.

In the work of Flores *et al.*,[88] the effect of the metal surface on the conformation of a water molecule is analyzed by considering two independent effects:

1. Screening of the proton–proton repulsion.
2. Interaction of the lone-pair orbitals with the surface.

Both effects (1) and (2) tend to increase the H–O–H angle. However, Flores *et al.*[88] find that the interaction of the lone-pair orbitals with the surface is the dominant effect for a water molecule approaching the surface. In particular, for a chemisorbed state this interaction is responsible for the major part of the deformation. These workers estimate that for H_2O chemisorbed on Ru, the H–O–H angle must increase* from the free-molecule value of 104.5° by $3.1 \pm 0.5°$, consistent with available experimental evidence.

* **Note added in proof.** L. A. Curtiss and J. A. Pople (*J. Chem. Phys.* **82**, 4230, 1985) have demonstrated opening up of the HOH angles for Li and Be atom complexes with water. See also C. W. Bauschlicher (*ibid.* **83**, 3129, 1985) with similar conclusions for H_2O on Ni(100).

3.1.1. Valence Theory of Free-Space Conformation

Briefly, the main features of the water molecule can be understood by assuming that each 1s hydrogen orbital interacts with a p orbital on the oxygen atom, the coordinate system being chosen such that the 2s and $2p_x$ orbitals are doubly occupied while the p_y and p_z orbitals form two electron pair bonds with the two hydrogens. To a first approximation, the H–O–H angle is taken to be 90°, as discussed for example by Goddard and Harding.[90] However, two effects tend to increase this angle to the experimentally observed value of 104.5°:

1. Interaction between the bonds and interaction between the hydrogens.
2. Hybridization of the 2s and p_y, p_z orbitals of oxygen.

Van Vleck and Cross[91] applied the theory of directed valence to the case of a water molecule and concluded that the H–H repulsion increased the H–O–H angle by 10°. As regards the s–p hybridization, Pauling[4] has shown that the effect can account for an increase of approximately 3.5°. When both effects are considered, the H–O–H angle becomes 103.5°, very close to the experimental value. The treatment of Flores *et al.*,[88] summarized below, follows closely these two calculations but modifies them appropriately to account for the presence of the metal surface.

3.1.2. Interaction Energy in a Water Molecule

We first discuss in more detail the model employed by Flores *et al.* for a water molecule in vacuum. First, the theory of directed valence is used to study the H–H repulsion. Then the effect of hybridization is considered, following Pauling's ideas.

Using the directed valence approach, we first neglect the effect of hybridization and write the interaction energy between the two hydrogens and oxygen in the following form:

$$E = E(1\text{H}, 2p_y) + E(1\text{H}, 2p_z) + E(1\text{H}, 2p_x^2) + E(1\text{H}, 2s^2) + E(1\text{H}, 1s^2)$$
$$+ \text{similar terms for the second hydrogen (2H)} + E(1\text{H}, 2\text{H}) \tag{3.1.1}$$

Here, for instance, $E(1\text{H}, 2p_y)$ denotes the interaction energy between the 1s hydrogen orbital and the $2p_y$ oxygen wave function. As already mentioned above, in the theory of directed valence the 1H orbital is paired with the $2p_y$ oxygen orbital while the 2H orbital is similarly paired with the

$2p_z$ wave function. According to this approach, the energy of a paired bond can be written as

$$E(2\text{H}, 2p_z) = E(1\text{H}, 2p_y) = (Q + J)/(1 + S^2) \tag{3.1.2}$$

where

$$Q = \frac{e^2}{R_{\text{OH}}} + \int \frac{e^2}{r_{12}} \psi_{\text{H}}^2(1)\psi_{\text{O}}^2(2) d\mathbf{r}_1 d\mathbf{r}_2 - \int \frac{e^2}{r_{\text{H}2}} \psi_{\text{O}}^2(2) d\mathbf{r}_2 - \int \frac{e^2}{r_{\text{O}1}} \psi_{\text{H}}^2(1) d\mathbf{r}_1 \tag{3.1.3}$$

$$J = e^2 S^2 \frac{1}{R_{\text{OH}}} - S \int \frac{e^2}{r_{\text{O}1}} \psi_{\text{H}}(1)\psi_{\text{O}}(1) d\mathbf{r}_1 - S \int \frac{e^2}{r_{\text{H}2}} \psi_{\text{H}}(2)\psi_{\text{O}}(2) d\mathbf{r}_2$$
$$+ \frac{e^2}{r_{12}} \int \psi_{\text{H}}(1)\psi_{\text{O}}(2)\psi_{\text{O}}(1)\psi_{\text{H}}(2) d\mathbf{r}_1 d\mathbf{r}_2 \tag{3.1.4}$$

In these expressions, ψ_{H} and ψ_{O} refer to the hydrogen and oxygen orbitals while S is the overlap integral. In writing equations (3.1.2)–(3.1.4) it has been assumed that the ionicity of the bond can be neglected. This assumption is justified, for example, by the work of Duncan and Pople,[92] who write for the molecular orbital

$$\psi(\text{OH}) = \lambda\psi_{\text{O}} + \mu\psi(1\text{H}) \tag{3.1.5}$$

with $\lambda/\mu = 1.06$. This is a wave function built from an oxygen wave function with a very small bond ionicity. Although in more sophisticated calculations it has been found that the bond ionicity is somewhat larger than that given above, the Duncan–Pople calculation nevertheless supports the view, adopted in the treatment of Flores *et al.*, that neglect of bond ionicity is a reasonable approximation.

In contrast to the above, the energy between electrons that are not bonded can be written as

$$E(1\text{H}, 2p_x) = \frac{1}{4}\left(\frac{Q + J}{1 + S^2} + 3\,\frac{Q - J}{1 - S^2}\right) \tag{3.1.6}$$

with Q and J given by equations (3.1.3) and (3.1.4). The second term in equation (3.1.6) is associated with the antibonding interaction for electrons with parallel spins. It should be noted that in equations (3.1.2) and (3.1.6) the charge of the oxygen nucleus has been taken as unity: this means that its entire charge has been split into eight different interactions appearing in equation (3.1.2) for each H orbital.

Equations (3.1.1)–(3.1.6) can be used to obtain the potential energy of the water molecule as function of the bond lengths and bond angles:

$$V = f(r_1) + f(r_2) + g(r_{12}) + N \sin^2(\tfrac{1}{2}\theta - \tfrac{1}{4}\pi) \tag{3.1.7}$$

where r_1 and r_2 are the bond lengths and r_{12} is the H–H distance. The angle between the bonds is denoted by θ. For more details the interested reader is referred to the paper of Van Vleck and Cross,[91] who obtained the minimum of equation (3.1.7) at

$$\left.\begin{aligned} r_1^0 = r_2^0 &= 1.00 \text{ Å} \\ \theta^0 &= 100° \end{aligned}\right\} \tag{3.1.8}$$

Moreover, by developing the functions $f(r_1)$ and $g(r_{12})$ around their equilibrium values, they finally expressed $V(r_1, r_2, \theta)$ in the form

$$\begin{aligned} V = {} & V(r_1^0, r_2^0, \theta^0) + \tfrac{1}{2}(k_1\{\Delta r_1\}^2 + k_2\{\Delta r_2\}^2 + 2k_3\{\Delta\theta\}^2 r_1^{02}) \\ & + k_{12}\Delta r_1 \Delta r_2 + 2^{1/2} k_{13} r_1^0(\Delta r_1 \Delta\theta + \Delta r_2 \Delta\theta) + \cdots \end{aligned} \tag{3.1.9}$$

where k_1, k_2, k_3, k_{12}, and k_{13} are constants related to the behavior of f and g. This equation will be used later to discuss the change in the molecular parameters due to the presence of the metal surface.

Having dealt with the hydrogen–hydrogen repulsion, we shall now include the hydridization effects. Following Pauling's argument the bond energy B can be written as a sum of two terms, one of which is proportional to the bond strength [see equation (3.1.16) below] while the other is the promotion energy, namely the energy spent in promoting one electron from an s-level E_s to the p-level E_p.

The wave functions for hydridized orbitals, by reference to Figure 3.1,

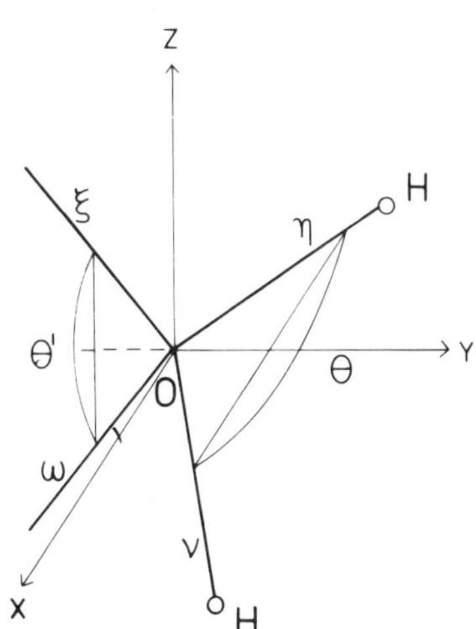

FIGURE 3.1. Notation and nature of wave function for sp^3 hybridized orbitals of the H_2O molecule.

can now by expressed in the form

$$\psi_\eta = \left(\frac{x}{1+x}\right)^{1/2} \left\{ |1s\rangle + \frac{1}{x^{1/2}} |p_\eta\rangle \right\} \tag{3.1.10}$$

$$\psi_\nu = \left(\frac{x}{1+x}\right)^{1/2} \left\{ |1s\rangle + \frac{1}{x^{1/2}} |p_\nu\rangle \right\} \tag{3.1.11}$$

$$\psi_\xi = \left(\frac{x'}{1+x'}\right)^{1/2} \left\{ |1s\rangle + \frac{1}{x'^{1/2}} |p_\xi\rangle \right\} \tag{3.1.12}$$

$$\psi_\omega = \left(\frac{x'}{1+x'}\right)^{1/2} \left\{ |1s\rangle + \frac{1}{x'^{1/2}} |p_\omega\rangle \right\} \tag{3.1.13}$$

where $x = -\cos\theta$, $x' = -\cos\theta'$, $|s\rangle$ is the s-orbital, and $|p_i\rangle$ is a p-orbital along the i-direction. We assume that the $2s$- and $2p$-orbitals have the same radial behavior. Orthogonality conditions for the wave functions (3.1.10)–(3.1.13) then give the following relation between x and x':

$$3xx' + x + x' = 1 \tag{3.1.14}$$

In Figure 3.1, the two hydrogens are bonded to the orbitals η and ν. Following Pauling,[4] the strength of the orbitals η and ν is given by the square of

$$\left(\frac{x}{1+x}\right)^{1/2} \left[1 + \left(\frac{3}{x}\right)^{1/2}\right] \tag{3.1.15}$$

in such a way that one contribution to the bond energy is given by

$$b\frac{x}{1+x}\left[1 + \left(\frac{3}{x}\right)^{1/2}\right]^2 \tag{3.1.16}$$

where b is a parameter adjusted to the experimental data.

On the other hand, if $E_p \neq E_s$, some energy must be expended to form the hybrid orbitals. This promotion energy is taken by Flores *et al.* as

$$\tilde{E}_p \simeq \frac{2x}{1+x}(E_p - E_s) \tag{3.1.17}$$

where, according to equations (3.1.10)–(3.1.13), $2x/(1+x)$ is the number of electrons promoted from an s-state to a p-orbital. However, in the present case $(E_p - E_s)$ turns out to be greater than the bond energy B. This implies that there is an important contribution from interaction between hybrids, tending to lower the promotion energy given by equation (3.1.17). This is taken into account by modifying equation (3.1.17) to read

$$\tilde{E}_p \simeq \frac{2x}{1+x} p(E_p - E_s) \tag{3.1.18}$$

where p is a further parameter to be adjusted to the experimental data.

Equations (3.1.16) and (3.1.18) give the bond energy in the form

$$B = b\frac{x}{1+x}\left[1+\left(\frac{3}{x}\right)^{1/2}\right]^2 + \frac{x}{1+x} p(E_p - E_s) \tag{3.1.19}$$

Equation (3.1.19) allows x to be obtained by minimizing B. Pauling proposed the use of equation (3.1.19) by taking $p = 1$, $(E_p - E_s) = 180$ kcal/mol, and adjusting b in such a way that B is equal to the experimental bond energy. In this manner, the H–O–H angle increase due to hybridization is found to be 3.5°. Flores *et al.* employed a variation of this approach by taking $p(E_p - E_s) = 117$ kcal/mol and $b = 31.7$ kcal/mol. With these values, they fitted the experimental bond energy of 111 kcal/mol and found an increase of 4.5° in the H–O–H angle. If these variations are taken into account, this angle become $(90 + 10 + 4.5) = 104.5°$, which is the experimental free-space value. Below, following Flores *et al.*, this procedure for treating hybridization in conjunction with the model proposed by Van Vleck and Cross for the H–H repulsion will be employed to treat the effect of the surface on the water molecule.

3.1.3. Introduction of the Image Potential

Following Flores *et al.*, whose approach was discussed in Chapter 2, the effect of the surface will again be introduced by means of the image potential. In other words, any interaction potential of the type $1/r_{ij}$ will be replaced by $[r_{ij}^{-1} - (r_{ij}^{\mathrm{I}})^{-1}]$, where r_{ij}^{I} is the distance between the point i and the image of j.

Starting with the Van Vleck and Cross approach, the different interactions can be obtained from equation (3.1.1) by means of equations (3.1.2) and (3.1.6), provided Q is replaced by $Q + \delta Q$ and J by $J + \delta J$, where δQ and δJ are given by

$$\delta Q = -\frac{e^2}{R_{\mathrm{OH}}^{\mathrm{I}}} - \int \frac{e^2}{r_{12}^{\mathrm{I}}} \psi_{\mathrm{H}}^2(1)\psi_{\mathrm{O}}^2(2)d\mathbf{r}_1 d\mathbf{r}_2 + \int \frac{e^2}{r_{\mathrm{H2}}^{\mathrm{I}}} \psi_{\mathrm{O}}^2(2)d\mathbf{r}_2 + \int \frac{e^2}{r_{\mathrm{O1}}^{\mathrm{I}}} \psi_{\mathrm{H}}^2(1)d\mathbf{r}_1 \tag{3.1.20}$$

$$\begin{aligned}\delta J = &-\frac{e^2 S^2}{R_{\mathrm{OH}}^{\mathrm{I}}} + S\int \frac{e^2}{r_{\mathrm{O1}}^{\mathrm{I}}} \psi_{\mathrm{H}}(1)\psi_{\mathrm{O}}(1)d\mathbf{r}_1 + S\int \frac{e^2}{r_{\mathrm{H2}}^{\mathrm{I}}} \psi_{\mathrm{H}}(2)\psi_{\mathrm{O}}(2)d\mathbf{r}_2 \\ &- \int \frac{e^2}{r_{12}^{\mathrm{I}}} \psi_{\mathrm{H}}(1)\psi_{\mathrm{O}}(2)\psi_{\mathrm{O}}(1)\psi_{\mathrm{H}}(2)d\mathbf{r}_1 d\mathbf{r}_2\end{aligned} \tag{3.1.21}$$

In the physisorbed region, it is easy to show that $\delta Q = 0$, a consequence of the spherical symmetry of the $1s$ hydrogen orbital. On the other hand, δJ can be calculated by expanding all the quantities $(r^{\mathrm{I}}_{12})^{-1}$, $(r^{\mathrm{I}}_{\mathrm{H2}})^{-1}$, and $(r^{\mathrm{I}}_{\mathrm{O1}})^{-1}$ around $(R^{\mathrm{I}}_{\mathrm{OH}})^{-1}$, following Flores *et al.*[88] Then to lowest order one can write δJ in the form

$$\delta J = e^2 \int \left[\frac{3(\mathbf{R}^{\mathrm{I}}_{\mathrm{OH}}\cdot\mathbf{r}_1)(\mathbf{R}^{\mathrm{I}}_{\mathrm{OH}}\cdot\mathbf{r}^{\mathrm{I}}_2)}{(R^{\mathrm{I}}_{\mathrm{OH}})^5} - \frac{\mathbf{r}_1\cdot\mathbf{r}^{\mathrm{I}}_2}{(R^{\mathrm{I}}_{\mathrm{OH}})^3}\right]\psi_{\mathrm{H}}(1)\psi_{\mathrm{O}}(1)\psi_{\mathrm{H}}(2)\psi_{\mathrm{O}}(2)\,d\mathbf{r}_1\,d\mathbf{r}_2$$

$$\equiv \frac{3(\mathbf{R}^{\mathrm{I}}_{\mathrm{OH}}\cdot\mathbf{X}_1)(\mathbf{R}^{\mathrm{I}}_{\mathrm{OH}}\cdot\mathbf{X}_2)}{(R^{\mathrm{I}}_{\mathrm{OH}})^5} - \frac{\mathbf{X}_1\cdot\mathbf{X}_2}{(R^{\mathrm{I}}_{\mathrm{OH}})^3} \tag{3.1.22}$$

where we define

$$\mathbf{X}_1 = e\int \mathbf{r}_1\psi_{\mathrm{H}}(1)\psi_{\mathrm{O}}(1)\,d\mathbf{r}_1 \tag{3.1.23}$$

$$\mathbf{X}_2 = e\int \mathbf{r}^{\mathrm{I}}_2\psi_{\mathrm{H}}(2)\psi_{\mathrm{O}}(2)\,d\mathbf{r}_2 \tag{3.1.24}$$

$\mathbf{r}_1$ and $\mathbf{r}_2$ being defined in Figure 3.2. With reference to the $E(1\mathrm{H}, 2\mathrm{H})$ interaction the same arguments can be employed to obtain $\delta E(1\mathrm{H}, 2\mathrm{H})$, the difference from the previous equations being that $R^{\mathrm{I}}_{\mathrm{OH}}$ is replaced by $R^{\mathrm{I}}_{\mathrm{HH}}$ and ψ_{O} by ψ_{H} in equation (3.1.22).

Equations (3.1.20)–(3.1.24) can now be used to calculate the change

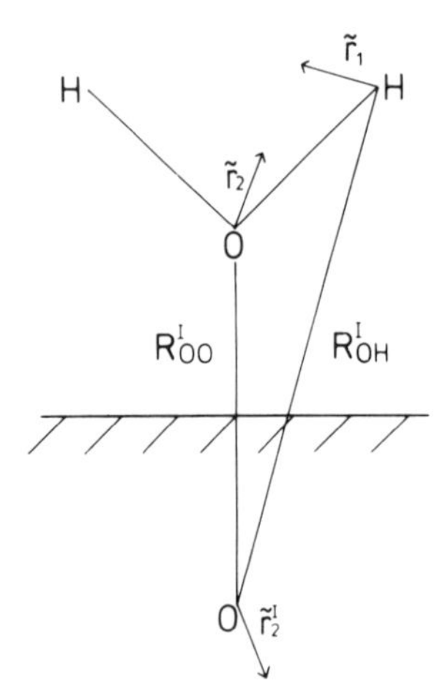

FIGURE 3.2. Notation used in equations (3.1.22)–(3.1.24) for image calculation pertaining to a water molecule outside a metal surface.

δV in V introduced by the presence of the surface:

$$\delta V = E(1\text{H}, 2p_y) + E(1\text{H}, 2p_z) + E(1\text{H}, 2p_x^2) + E(1\text{H}, 2s^2) + E(1\text{H}, 1s) + \text{similar terms with } 2\text{H} + E(1\text{H}, 2\text{H}) \quad (3.1.25)$$

$$\delta E(1\text{H}, 2p_y) = \frac{\delta J(1\text{H}, 2p_y)}{1+S^2} \quad (3.1.26)$$

$$\delta E(1\text{H}, 2p_z) = \frac{1}{4}\left[\frac{\delta J(1\text{H}, 2p_z)}{1+S^2} - \frac{3\delta J(1\text{H}, 2p_z)}{1-S^2}\right] \quad (3.1.27)$$

$$\delta E(1\text{H}, 2p_x) = \frac{1}{4}\left[\frac{\delta J(1\text{H}, 2p_x)}{1+S^2} - \frac{3\delta J(1\text{H}, 2p_x)}{1-S^2}\right] \quad (3.1.28)$$

$$\delta E(1\text{H}, 2s) = \frac{1}{4}\left[\frac{\delta J(1\text{H}, 2s)}{1+S^2} - \frac{3\delta J(1\text{H}, 2s)}{1-S^2}\right] \quad (3.1.29)$$

and similarly for other quantities appearing in equation (3.1.1).

In the actual calculations for δV, we have neglected all interactions originating from the $2p_x$, $2s$, and $1s$ oxygen orbitals in equation (3.1.25) because they are lone-pair orbitals and are not expected to make any important contribution to the change in the proton–proton interactions induced by the surface. Accordingly Flores *et al.* set

$$\delta V \simeq \delta E(1\text{H}, 2p_y) + \delta E(1\text{H}, 2p_z) + \text{similar terms with } 2\text{H} + E(1\text{H}, 2\text{H}) \quad (3.1.30)$$

On the other hand, the H and O orbitals have been approximated by

$$\psi(1\text{H}) = \left(\frac{\pi}{2}\right)^{-1} \exp(-r) \quad (3.1.31)$$

$$\psi_{\text{O}}(2p_y) = (Z^5/32\pi)^{1/2} y \exp(-\tfrac{1}{2}Zr) \quad (3.1.32)$$

with $Z = 4.55$.

Equation (3.1.30) with the other input above gives the change in the interaction potential V due to the presence of the surface. Using δV and equation (3.1.9), the change in the H–O–H angle due to the proton–proton screening can be determined as a function of the distance R^{I}_{OH}.

3.1.4. Effect of the Surface on Hybridization

We shall now consider the effect of the surface on the hybridization contribution. As for the H–H interaction, the surface is introduced through the image potential. In the present case, this potential is responsible for the new interaction between the hybrids given by equation (3.1.20) and their images.

In a valence-bond theory, one can think of the six electrons of the L-shell of O as being distributed as follows. One electron is placed in each of the bond orbitals ψ_η and ψ_ν, and two electrons in each of the lone-pair orbitals ψ_ξ and ψ_ω. As a result of this electronic charge distribution the contributions of the dipole moment from four electrons, one in each hybrid, cancel due to the spatial symmetry. However, there is a net dipole produced by the remaining two electrons in ψ_ξ and ψ_ω. It should be noted that another electric dipole can be created by the O–H bonds. However, since their ionicity is small, as discussed above, this contribution can be neglected.

The above discussion suggests that the effect of the surface on the hybrids is essentially due to the interaction between the electric dipole created by the lone pairs and its image. This electric dipole can be readily obtained using equations (3.1.10) and (3.1.13) and taking for the $|s\rangle$ and $|p\rangle$ orbitals the forms

$$|s\rangle = (Z^5/96\pi)^{1/2} r \exp(-\tfrac{1}{2}Zr), \qquad |p\rangle \equiv \text{equation (3.1.32)} \tag{3.1.33}$$

with $Z = 4.55$.

These wave functions enable us to determine the following dipole moment associated with the lone-pair wave functions:

$$D = 1.27[x'^{1/2}/(1 + x')] \quad \text{au} \tag{3.1.34}$$

and the following dipole–dipole interaction per bond:

$$E_{\text{d-d}} = -\frac{1.62}{(R^{\text{I}}_{\text{OO}})^3} \frac{x(1-x)}{(1+x)^2} \tag{3.1.35}$$

where R^{I}_{OO} is the distance between O and its image. When the interaction of the hybrids with their image has been obtained, it is easy to calculate its effect on hybridization by introducing $E_{\text{d-d}}$ in equation (3.1.19). This procedure yields

$$B = b\frac{x}{(1+x)}\left[1 + \left(\frac{3}{x}\right)^{1/2}\right]^2 - \frac{x}{1+x}\, p(E_p - E_s) + \frac{1.62}{(R^{\text{I}}_{\text{OO}})^3}\frac{x(1-x)}{(1+x)^2} \tag{3.1.36}$$

In this equation, b and $(E_p - E_s)$ are the constants given previously. The effect of the surface on hybridization was then calculated by Flores *et al.* by minimizing B and obtaining x as a function of the distance R^{I}_{OO}.

3.1.5. Results and Discussion

In Figure 3.3, their results are plotted in the form of the change in bond angle, calculated via the directed valence theory, as a function of R^{I}_{OH}. Here, the H–O–H angle tends to increase as the molecule approaches the surface. This is the result of the combined effects of screening of the proton–proton repulsion and bond–bond interaction. This latter effect overcomes that of screening of the proton–proton interaction, which would tend to decrease the H–O–H angle.

In Figure 3.4, the change in bond angle as induced by hybridization is plotted as a function of R^{I}_{OH}. In this case, it can also be seen that the bond angle increases as R^{I}_{OH} decreases. The physical interpretation of this effect is also clear: the dipole–dipole interaction between hybrids and their images tends to decrease the angle between the lone-pair orbitals and, as a consequence, the bond angle increases.

It is of interest that changes in the water molecule conformation, introduced by the dipole–dipole interactions, are greater than those due to the proton–proton and bond–bond interactions. Moreover, the dipole–dipole interaction becomes much more important for smaller values of R^{I}_{OH}.

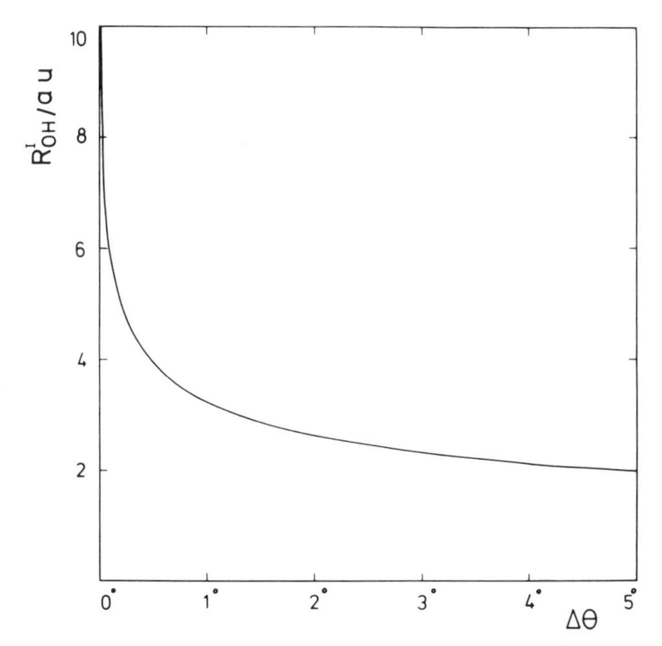

FIGURE 3.3. Results for a change in the bond angle of H_2O, calculated via the directed valence theory, as a function of R^{I}_{OH}, i.e., the image length R_{OH}.

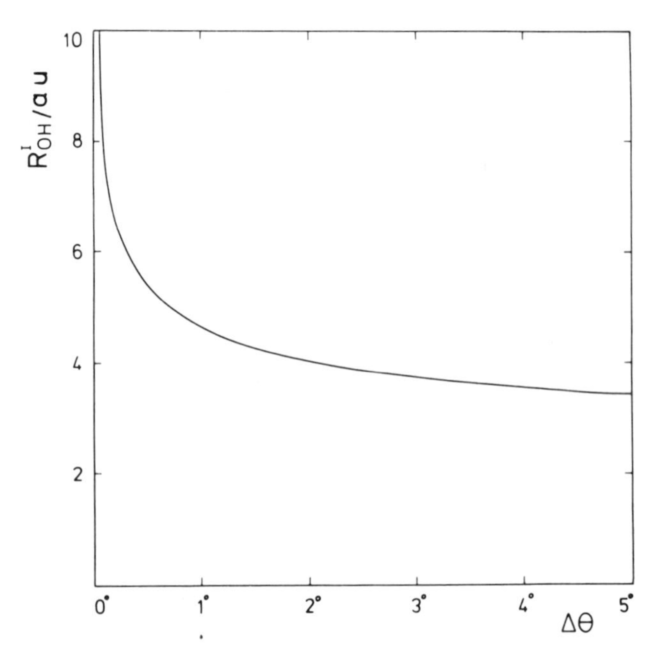

FIGURE 3.4. Change in bond angle calculated from the interaction between lone pairs and their images, plotted as a function of R^{I}_{OH} (cf. Figure 3.3).

It must be stressed, however, that the above analysis gives reliable results for the physisorbed region, which according to estimates of Flores *et al.*[88] is for $R^{I}_{OH} \gtrsim 6$ au. By extrapolating the physisorbed state results somewhat beyond their range of validity, Flores *et al.* were able to estimate what might be expected for a distance $R^{I}_{OH} \simeq 4$ au, appropriate for the chemisorbed situation. In this case they obtained

$$\Delta\theta_{pp} \simeq 0.45° \tag{3.1.37}$$

$$\Delta\theta_{dd} \simeq 2.2° \tag{3.1.38}$$

This demonstrates how interaction between the lone pairs and the surface must dominate in the chemisorbed region. One can now use this result to obtain an estimate of the change in bond angle for the chemisorbed state. The procedure is to use the following equation for the bond energy:

$$B = b\frac{x}{1+x}\left[1+\left(\frac{3}{x}\right)^{1/2}\right]^2 - \frac{x}{1+x}p(E_p - E_s) + \frac{dx(1-x)}{(1+x)^2} \tag{3.1.39}$$

instead of equation (3.1.36), and neglect contributions due to proton–proton interaction. In equation (3.1.39), d replaces $1.62/(R^1_{OO})^3$ in equation (3.1.36) and provides a measure of the interaction between the lone pairs and the metal surface.

Equation (3.1.39) can be employed to obtain x as a function of d by minimizing B. Moreover, d can also be calculated for H_2O on Ru by using the fact that the chemisorption energy is 12.6 kcal/mol in this case. Flores *et al.* thereby obtain an increase in the H–O–H angle given by $\Delta\theta = 2.8°$. This is close to the value given in equation (3.1.38). Their conclusion is therefore that, including the correction given in equation (3.1.37), the increase for the H–O–H angle must be around 3.2°. Madey and Yates reported from ESDIAD studies an increase of $(11.5 \pm 10)°$, the lowest value of this wide range being most likely to occur due to the image effects acting on the ion angular distribution.

To conclude this discussion of H_2O outside a metal surface, some comments are called for on the model used. In the approach of Flores *et al.*, it was assumed that the total H–O–H angle for the free molecule could be accounted for by two effects:

1. Proton–proton repulsion, this being responsible for an increase of 10° from the 90° angle predicted by elementary valence-bond theory.
2. s–p hybridization, this yielding an increase of 4.5°.

These values should not be assumed to be fully quantitative, for some authors have claimed that the proton–proton repulsion cannot increase the H–O–H angle by more than 5°. Were this the case, the s–p hybridization would have to increase the bond angle by 9.5°. Then the effect of the surface on the H–O–H angle through the s–p hybridization would have to be recalculated taking into account the new values for the angle. Flores *et al.* obtain for this case a 3.0° increase, similar to the one obtained above.

To summarize, the effect of the surface on the conformation of a water molecule approaching a metal appears through two contributions to the total energy:

1. Screening of the bond–bond interaction and proton–proton repulsion.
2. Interaction of the lone pairs with the surface.

Both effects tend to increase the bond angle although the second is dominant. In particular, in a chemisorbed state one can expect the interaction between the lone pairs and the surface to be responsible for the major part of the molecular deformation. The results of Flores *et al.*

indicate that for H_2O chemisorbed on Ru, the H–O–H angle must increase from the free-space value of 104.5° by $3.1 \pm 0.5°$.

3.2. CONFORMATION OF NH_3 AND C_2H_4 MOLECULES

Flores *et al.*[89] have used a related semiempirical approach to that discussed in detail above for H_2O to treat the conformation of NH_3 and C_2H_4 molecules outside a metal surface.

Specifically, for ammonia they discuss the shape of the molecule by considering the interaction between the lone pair and its image, while for ethylene (≡ ethene) the equivalent approach is to consider the interaction between the π-orbitals and the surface.

3.2.1. Interaction Between Lone Pair and Image for NH_3

The orientation of an NH_3 molecule approaching a metal surface is illustrated in Figure 3.5. The attractive interaction between the lone-pair orbital and its image changes the hybridization in the molecule and hence the H–N–H angle. In order to treat this effect, Flores *et al.* again follow the work of Pauling on free-space molecules. The four ammonia sp^3 hybrids,

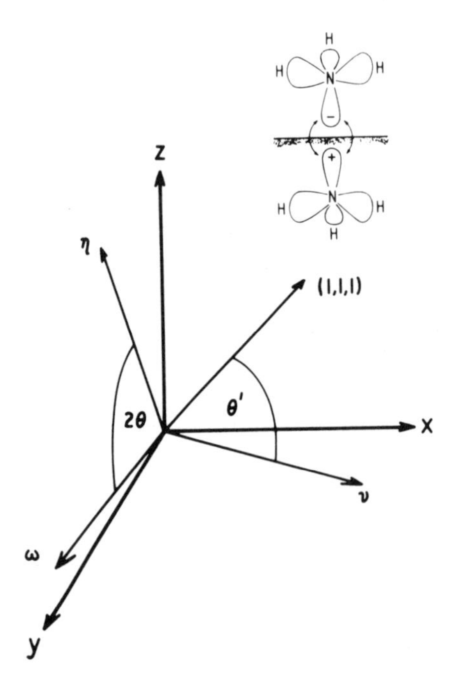

FIGURE 3.5. Orientation of an ammonia molecule approaching a metal surface. Attractive interaction between a lone-pair orbital and its image is also depicted (inset). Four ammonia sp^3 hybrids indicated.

shown also in Figure 3.5, can be expressed as

$$\psi(111) = \frac{x'}{(x'^2 + x)^{1/2}} \left(|s\rangle + \frac{x^{1/2}}{x'} |p_{111}\rangle \right) \tag{3.2.1}$$

$$\psi_\eta = \left(\frac{x}{1+x} \right)^{1/2} \left(|s\rangle + \frac{1}{x^{1/2}} |p_\eta\rangle \right) \tag{3.2.2}$$

and similarly for ψ_ν and ψ_ω, with $x = -\cos 2\theta$ and $x' = -\cos \theta'$ satisfying $3x'^2 = 1 - 2x$. In this notation the lone pair is associated with the 111 hybrid direction and p_i denotes a p wave function along direction i, as before.

According to Pauling, the energy associated with the N–H bond in free space can be written, per bond, in the form

$$E = \frac{bx}{1+x} \left[1 + \left(\frac{3}{x} \right)^{1/2} \right]^2 - \frac{x}{1+x} p'(E_p - E_s) \tag{3.2.3}$$

The first term is Pauling's bond strength, while the second allows for hybridization by virtue of the energy required to promote an electron from an s to a p orbital. The factor p' corrects for the interaction of the hybrids. In the calculation of the energy of Flores *et al.*, they use equation (3.2.3) with $p'(E_p - E_s)$ taken to be the value discussed above for water, scaled by the ratio of the s–p separation in N to that in O, namely about 11/16. By adjusting b in equation (3.2.3) to give an energy minimum equal to the experimental value, they found

$$b = 1.16 \text{ eV}, \qquad 2\theta = 95.6° \tag{3.2.4}$$

The above result for 2θ implies that the difference between the observed H–N–H angle of 107.3° and 90° is partly due to hybridization (5.6) and partly to the H–H repulsion (11.7° if the remaining angle increase is attributed solely to this mechanism).

Turning to the interaction between the ammonia molecule and the metal surface, one can use hybrid orbitals in equations (3.2.1) and (3.2.2) to express the dipole associated with the lone pair, and hence the interaction between the lone pair and its image, the latter taking the form

$$\frac{1.62}{(R_{NN}^{I})^3} \frac{x(1-2x)}{(1+x)^2} \tag{3.2.5}$$

where, as usual, R_{NN}^{I} is the distance between the N atom and its image and the energy is in atomic units. While expression (3.2.5) represents an interaction characteristic of physisorption, it can be generalized to chemisorption in the form $x(1-2x)d/(1+x)^2$ to obtain the energy per bond for the chemisorbed molecule as

$$E_{\text{chem}} = \frac{bx}{1+x}\left[1+\left(\frac{3}{x}\right)^{1/2}\right]^2 + \frac{x}{1+x}p'(E_p - E_s) + \frac{x(1-2x)d}{(1+x)^2} \qquad (3.2.6)$$

where the constant d evidently measures the interaction energy between the lone pair and the surface. Equation (3.2.6) gives the H–N–H angle 2θ as a function of the chemisorption interaction. Thus, if the chemisorption energy is known and by adjusting the parameter d to fit this, it is possible to estimate the change in angle due to the interaction. For NH_3 on Ru(001), $\Delta(2\theta) = 1.2°$. This is less than the increase obtained for water on the same surface, discussed in Section 3.1 above. The difference is due to the different dipole–dipole interactions, which in turn reflect the different geometries of the two molecules.

3.2.2. Adsorption of Ethylene on a Planar Metal Surface

Flores *et al.* have used similar chemical methods to study the adsorption of ethylene on planar metal surfaces, assuming that prior to interaction the molecule is orientated with its molecular plane parallel to the surface. Adopting the conventional picture of bonding in ethylene, the carbon–carbon double bond is described in terms of sp^2 hybrids along the C–C axis, with p_z orbitals perpendicular to the molecular plane.

The treatment of the free-space configuration is standard and we need not therefore go into detail here. Thus we can turn immediately to the deformed ethylene molecule interacting with the surface. The major difference between ethylene and ammonia or water is that the latter molecules possess a net electric dipole moment to which the lone-pair hybrids make a significant contribution. As discussed above, the interaction between the lone pairs and their images is largely responsible for the deformation of these molecules on metal surfaces. Ethylene, although having no net dipole moment, can interact with the surface via its polarizable π-electron assembly. Referring to Figure 3.6, which is reproduced from Flores *et al.* and shows a possible rehybridized distorted ethylene configuration in the inset, the following contributions to the total energy of the molecule are to be taken into account:

1. From the C–H bonds. This is given by

$$4A\frac{x}{1+x}\left[1+\left(\frac{3}{x}\right)^{1/2}\right]^2, \qquad x = -\cos 2\theta \qquad (3.2.7)$$

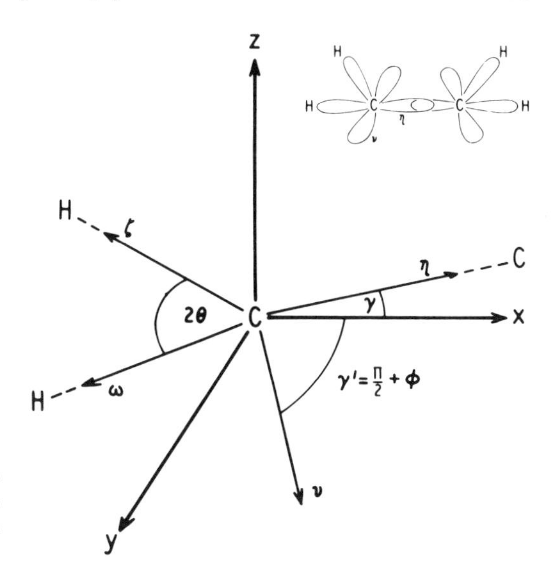

FIGURE 3.6. Possible rehybridized distorted ethylene configuration. Hybrids ψ_η and ψ_ν lie in $x - z$ plane.

2. From the C=C bond. This is given in terms of the ψ_η and ψ_ν hybrids in Figure 3.6 as

$$B(\cos\eta + 3^{1/2}\sin\eta)^2 \tag{3.2.8}$$

and

$$B[\cos\nu - 3^{1/2}\sin\nu\sin(\gamma+\phi)]^2 + C[3^{1/2}\sin\nu\cos(\gamma+\phi)]^2 \tag{3.2.9}$$

respectively. The hybrids ψ_η are taken parallel to the surface. In deriving expression (3.2.9) for the bonding interaction between the two ν hybrids, Flores *et al.*[89] took into account the fact that they are not parallel to each other by projecting the p_ν wave function into components along and perpendicular to the C–C bond direction.

By using these equations and calculating constants A, B, and C from the free-space molecule as $A = 1.08$, $B = 0.92$, and $C = 0.91$, all in eV, Flores *et al.* obtain

$$\begin{aligned} E_{\text{deformed molecule}} = {} & 4.32\frac{x}{1+x}\left[1+\left(\frac{3}{x}\right)^{1/2}\right]^2 + 0.92(\cos\eta + 3^{1/2}\sin\eta)^2 \\ & + 0.92[\cos\nu - 3^{1/2}\sin\nu\sin(\gamma+\phi)]^2 \\ & + 0.91[3^{1/2}\sin\nu\cos(\gamma+\phi)]^2 + b(\cos\nu + 3^{1/2}\sin\nu)^2 \end{aligned} \tag{3.2.10}$$

the final term proportional to b measuring the molecule–surface interaction. The energy in equation (3.2.10) is evidently a function of two angles, say γ and ϕ, and of the parameter b. Table 3.1 summarizes the different angles for the deformed molecule, obtained by minimizing the energy expression (3.2.10) with respect to γ and ϕ for various values of b, which need not be recorded. In this table, E is calculated from equation (3.2.10) while H is taken from experiment for the metals shown.

The metals are seen to have been ordered in increasing H, which one expects to involve increasing distortion γ. For sufficiently large distortions, the bonding character would be so markedly altered that the above model would no longer be appropriate, so that the range of validity of the work discussed here is restricted to H not too large.

According to Table 3.1, the effect of the surface is (1) to tilt the C–H bonds, so that the H atoms lie above the plane containing the C atoms, and (2) to decrease the HCC and HCH angles, the former to a greater extent.

It is of interest here to make contact with the work of Demuth,* whose results listed in Table 3.2 are inferred from photoemission spectra.

Larger deformations of C_2H_4 occur on Pd(111) and Pt(111) suggesting a single-bond character in the C–C linkage. Comparison of Tables 3.1 and 3.2 shows that the model proposed by Flores *et al.* describes the main trends of the observed data.

* Demuth's work is discussed in further detail in Section 3.4.

TABLE 3.1. Chemisorption Energies E and Heat of Chemisorption H for Deformed C_2H_4 Molecule for Different Molecule–Surface Interactions

γ	HCH	HCC	E (eV)	H (eV)	
0	115.9	122	0	Ag	0.37
10	115.9	121.5	0.55	Cu	0.8
20	115.7	120	1.34	Ni	2.51
30	115.4	117.6	3.57	Fe	2.94
40	114.7	114.4	5.15	W, Cr	4.37
50	113.7	110.7	15.56	Ta	5.97

TABLE 3.2. Angles Inferred from Photoemission Spectra (after Demuth[114])

Surface	HCH	HCC
Cu(111)	120	120–117.4
Ni(111)	117.4	120
Pd(111)	106.8–109.5	106.8–119.5
Pt(111)	106.8	106.8–109.5

In summary, ammonia on Ru(001) is expected to exhibit a slight increase in the HNH angle from its free-space value, due to the interaction of the lone-pair electrons with their image in the metal surface. For ethylene, the main effects of chemisorption are (1) tilting upward of the C–H bonds so that the H atoms are not in the plane containing the C atoms, and (2) decrease in the HCC and HCH angles, the HCC angle being the most affected. A more detailed theory of ethylene will be discussed in Section 3.4 below.

3.2.3. Cluster Approach to the Electronic-Level Structure of Ethylene on Ni(100)

In connection with the work of Flores *et al.* described above, a very different, cluster-type approach to the adsorption of ethylene on a Ni(100) substrate has been reported by Howard and Dresselhaus.[93] The Ni(100) surface was simulated by a ten-atom Ni cluster of C_{4v} point group symmetry, to which an adsorbed ethylene molecule was then added. The $Ni_{10}(100)$ cluster was found to have at least some features resembling semiquantitatively those of a Ni(100) substrate. In particular, it shows a high density of states near the Fermi level and a d-bandwidth of about 2.6 eV, compared to the bandwidth of 3.5 eV of bulk Ni. The cluster was found to exhibit a weak magnetization of about 0.2 Bohr magnetons/atom, which is a significant reduction not only from the bulk value of 0.54 μ_B/atom but from the values found for small cubic Ni clusters of comparable size.

Howard and Dresselhaus then calculated the electronic-level structure of the isolated C_2H_4 molecule and compared this with that of a $\{Ni_{10}(100)+C_2H_4\}$ π-bonded cluster. Their calculations indicate that the reported difference in activation energies for ethylene hydrogenation on paramagnetic and ferromagnetic Ni may be due to shifts in the Ni *spd* levels relative to the ethylene $1b_{3u}$ bonding orbital energy level, via the exchange splitting of approximately 0.1 eV.

As to the picture of bonding of C_2H_4 on Ni, the π-bonding mechanism involves charge donation from the C_2H_4 $1b_{3u}(\pi)$ orbital to a $Ni_{10}(100)$ a_1-type *spd* orbital, and corresponding back-donation of charge into the C_2H_4 π^*-orbital. The $Ni_{10}(100)$ levels near the Fermi energy are virtually unaffected.

3.3. MOLECULAR VERSUS DISSOCIATIVE ADSORPTION OF NH_3 ON TRANSITION-METAL SURFACES

At this point, we shall return to the treatment of molecular versus dissociative adsorption given for diatomics in Chapter 2, but now specifi-

TABLE 3.3. Energy $E(\text{M–N}) + 3E(\text{M–H})$ in eV[a]

Cr^D	Mn^D	Fe^D	Co^D	Ni	Cu
16.0	17.3	14.5	13.9	11.6	11.4
Mo^D	Tc	Ru	Rh	Pd	Ag
17.0	—	14.6	12.7	10.9	12.5
W	Re^D	Os	Ir	Pt^M	Au^M
(16.6)	(15.7)	(14.6)	(12.6)	(10.3)	(10.0)

[a] M denotes molecular adsorption while D refers to dissociation. Absence of labeling means no decisive conclusion (errors in energies given by Flores *et al.*)[51]

cally with respect to ammonia on transition-metal surfaces. Here, following Flores *et al.*,[51] one can make use of the treatment of nitrogen and of hydrogen in Section 2.4 of that chapter. The quantity $E(\text{M–N}) + 3E(\text{M–H})$ can be compared with 12.2 eV, i.e., the energy of formation of NH_3, in order to test whether a given metal is sufficiently reactive to cause dissociation of the molecule. The results of Flores *et al.* are collected in Table 3.3.

A study of the available experimental data indicates that the behavior of certain metal substrates toward NH_3 chemisorption is highly sensitive to temperature variations in the 200–400 K range. Although this makes it difficult to define a specific borderline between molecular and dissociative adsorption, there is clear evidence that NH_3 retains its molecular character on Ni, Ir, Pt, and Au[94–97] at room temperature. Both molecular[98] and dissociative[99] adsorption have been reported on Fe surfaces.

The various Mo and W surfaces[100,101] appear to decompose NH_3 at room temperature; however, it has been suggested[102] that NH_3 molecules exist on the W(100) surface.

Experiments on polycrystalline Re[103] indicate that both molecular and dissociative chemisorption occur at room temperature.

The totality of experimental evidence led Flores *et al.* to propose the boundary indicated by the solid line in Table 3.3. Although the experimental uncertainties are considerable for adsorption of NH_3, this evidence indicates that for NH_3 the theoretical model does not yield fully quantitative predictions and a more refined model will eventually be necessary.

3.4. ELECTRONIC STRUCTURE AND CONFORMATION OF ETHENE ON TRANSITION- AND NOBLE-METAL SURFACES

The final part of this chapter will be devoted to the electronic structure and conformation of a specific organic molecule, ethene, when adsorbed on a variety of transition- and noble-metal surfaces. In spite of the very specific nature of this treatment, the methods focused on in treating it theoretically may well turn out to have a reasonably wide range of validity for dealing with electronic structure and conformation of polyatomic molecules on metal surfaces.

As to the conformation, Demuth[104] has developed a method, based on his own experimental work, by which the geometry of molecules may be predicted when they undergo perturbation from their free-space equilibrium conformation due to the proximity of a metal surface. Subsequently, Felter and Weinberg[105] have extended Demuth's approach by relating linearly the shifts in the molecular-energy levels with the distortions of the molecule. In the study of Hiett *et al.*[106] of ethene on transition- and noble-metal surfaces, to be treated below, substantial use is made of the results of these workers.

Before giving details of the approach of Hiett *et al.*, the nature of the bonding of ethene to a transition-metal surface, as well as the change in behavior of the molecule on different surfaces, will be briefly commented on. The bonding of ethene to a transition-metal surface is thought to have similarities with that occurring between π-acids and transition metals in organometallic complexes[107,108], in that it is predominantly the π- and π^*-orbitals that are involved. The π-orbital donates its electrons to vacant d-orbitals of the metal while the π^*-orbital accepts electrons from filled d-orbitals. On the surfaces under study in the work of Hiett *et al.*,[106] ethene adsorbs molecularly at sufficiently low temperatures.

3.4.1. Conformation of Ethene on Metal Surfaces

To the writer's knowledge, experimental work has not yet unambiguously described the geometry of the adsorbed ethene molecule on metal surfaces, or its position relative to a surface. Nevertheless, many predictions based on experiment have been made and a consensus on the major qualitative distortions has emerged,[109–113] as discussed in Section 3.2.2. The conclusions reached are that the C–C bond length (d_{CC}) increases and the C–C–H angle (θ_{CCH}) and the H–C–H angle decrease due to movement of the hydrogen atoms away from the surface. These deductions have been drawn principally from electron energy loss spectroscopy and photoelectron spectroscopy.

In the study of Hiett *et al.*, the above conformation associated with the C–C bond length and the C–C–H and H–C–H angles has been adopted. It must be recognized at the outset that some features may consequently have been omitted and that they will need to be incorporated in subsequent refinements. Thus Demuth proposes that multicenter metal–hydrogen–carbon bonding occurs in some instances of ethene adsorption.

Likewise, by analysis of electron energy loss spectroscopy data, Felter and Weinberg suggest that a twist about the C–C bond may appear with adsorption.

3.4.1.1. Demuth's Model and its Refinement

Demuth has developed a method by which he aims to predict the geometry of adsorbed molecules from ultraviolet photoelectron spectroscopy (UPS). Provided there is a uniform relaxation shift of the energy levels of the molecule, their separation is reflected in the separation of the UPS peaks. The changes in these separations, compared with the free-molecule case, provide information as to the manner in which the molecular framework is distorted when it interacts with the surface. In order to determine this distortion, a free-molecule approximation is made. The changes in the separations of the energy levels of the adsorbed molecule are assumed to arise from the same distortions that lead to identical changes in the separations of the energy levels of the free molecule.

The effect of the distortions on the π-levels was not considered by Demuth, because of the view that these are not only affected by the change in molecular geometry but also by bonding, while only the former affects the σ-orbitals. The π-levels are studied in some detail in Section 3.4.4 of the present work.

The above discussion concerning modeling of the distortions of adsorbed ethene will be completed by extending the above arguments of Demuth. In doing so, we follow closely the work of Felter and Weinberg.[105]

It will be assumed that the effects of each distortion, of say an increase in bond length or an angle change, are purely additive, so that one can set

$$y_i = \sum_j m_{ij} x_j \tag{3.4.1}$$

where x_j are the distortions of the molecule that lead to changes in the energy difference, y_i, between the ith orbital of the free molecule and that of the adsorbed one, and m_{ij} are a set of coupling coefficients.

In order to compare energies of free and adsorbed molecules, one requires a reference energy level, which Felter and Weinberg chose to be the σ_{ss}^*-level, for both free and adsorbed molecules. Demuth[110] gives the experimental values of the ionization potentials of the molecule when adsorbed on various transition-metal surfaces. These energies are given with respect to the σ_{ss}^* level (chosen as zero) in Table 3.4, from which one may therefore determine y_i in equation (3.4.1).

TABLE 3.4. Experimental UPS Ionization Potentials for Various Surfaces[a]

System	Energy levels (eV)			
	σ_{SS}^*	σ_{CH}	σ_{CC}	σ_{CH}^*
Free C_2H_4	0	−3.2	−4.4	−6.4
C_2H_4/Ni(111)	0	−3.6	−4.8	−6.5
C_2H_4/Ni(100)	0	−3.35	−4.8	−6.4
C_2H_4/Ni(110)	0	−3.5	−4.8	−6.45
C_2H_4/Pd(111)	0	−3.6	−5.0	−6.8
C_2H_4/Pt(111)	0	−4.0	−5.4	−6.8

[a] Note that all energies are with respect to σ_{SS}^*.

Coefficients m_{ij} in equation (3.4.1) were found from the self-consistent-field calculation of Demuth[114] on the free molecule. Equation (3.4.1) can then be expressed in the explicit form

$$\begin{bmatrix} y_{\sigma_{CH}}^* \\ y_{\sigma_{CC}} \\ y_{\sigma_{CH}} \end{bmatrix} = \begin{bmatrix} +0.015 & +0.069 & -0.458 \\ +0.018 & -0.018 & +6.916 \\ -0.080 & +0.061 & +5.289 \end{bmatrix} \begin{bmatrix} x_1 \\ x_2 \\ x_3 \end{bmatrix} \tag{3.4.2}$$

where x_1 refers to changes in θ_{CCH}, x_2 to changes in θ_{HCH}, and x_3 to changes in d_{CC}.

Equation (3.4.2) for the adsorbed molecule and values of y_i obtained from Table 3.4 can be used to readily determine the values of x_j given in Table 3.5.

TABLE 3.5. Predicted Geometries for Ethene Adsorbed on Various Surfaces, Obtained from UPS Values of Demuth[110] and Equation (3.4.2)

System	θ_{CCH} (deg)	θ_{HCH} (deg)	d_{CC} (A)
C_2H_4/Ni(111)	119.5	118.0	1.40
C_2H_4/Ni(100)	116.5	119.0	1.41
C_2H_4/Ni(110)	119.0	118.5	1.40
C_2H_4/Pd(111)	113.0	112.0	1.43
C_2H_4/Pt(111)	114.0	112.0	1.49

3.4.2. Tight-Binding Model of σ-Levels of Adsorbed Ethene

The work of Demuth[114] and Felter and Weinberg[105] involved the assumption that the σ-levels of ethene were unaffected by the presence of a surface. In later work, Demuth[110] investigated this by performing self-consistent-field calculations in which an ethene molecule was moved toward a lone Be atom representing a "metal surface." He found that, apart from the σ_{CC} orbital, the above approximation was valid. When the effect of σ_{CC} was taken into account there was a small change in the geometry, as predicted from the UPS experiments.

Hiett *et al.* demonstrate that the findings of Demuth[110] concerning the σ-levels obtained by interaction with a single Be metal atom can be recovered by a rather simple tight-binding model which, however, embodies the entire metal surface. They find, as he did, that the energy of the σ_{CC}-level becomes more negative due to chemisorption. As a consequence, a quantitative argument will be presented below to determine the decrease in the angle θ_{HCH}.

The Hamiltonian matrix of ethene was set up for the valence orbitals using the hydrogen $1s$ and the carbon $2s$ and $2p$ atomic orbitals as a basis set. In principle, this matrix can be constructed by evaluating elements of the form $\langle X | H | Y \rangle$ where X and Y are basis functions and H is a tight-binding Hamiltonian. However, Hiett *et al.* did not attempt first-principles evaluation of the matrix elements, but they rather parametrized them. In Appendix 3.2 we record the explicit matrix, the nonzero elements being obtained by fitting the eigenvalues of the Hamiltonian matrix to the values given by Demuth.[104] The zero matrix elements arise from the neglect of interaction between nonadjacent atoms. In this model, it is assumed that the metal surface distorts the molecule and thereby its symmetry is lowered from D_{2h} to C_{2v}. Group theory enables one to block-diagonalize the 12×12 Hamiltonian matrix into four blocks, the resulting matrix being given explicitly in Appendix 3.2.

3.4.3. Effect of the Surface on σ-Levels

The tight-binding Hamiltonian was adjusted to give (see Appendix 3.2), for the undistorted molecule, the best agreement with the π_{CC}, σ^*_{CH}, σ_{CC}, and σ_{CH} levels as calculated by Demuth.[104] These results are in good agreement with UPS data[115,116] for the free molecule. It is noteworthy that those are the levels expected to be most sensitive to the effects of chemisorption. The fact that other levels are not described to the same accuracy is not important for our present purposes, though we believe that by generalizing the parametrization this could be rectified should it prove of interest later.

Turning to the chemisorbed molecule we note first that when ethene interacts with a transition-metal surface, the energy of the π-orbitals decreases. We shall return to this problem in a later section. From the findings there, it turns out that we can approximately simulate the effect of bonding to the surface by decreasing by a constant amount the energy of the $2p$-orbitals that constitute the π-bond. This constant is shown to be close to the chemisorption energy.

Hiett *et al.* therefore diagonalized the block-diagonal matrices (Appendix 3.2) both without and with the surface present, and the results are presented in Table 3.6. These data show that the interaction with the surface has a noticeable effect only on the π_{CC} and σ_{CC} orbitals. The change in π_{CC} is to be expected, while the change in σ_{CC} is found to be in accord with earlier work.[110]

The conclusion here is that, in order to calculate the molecular geometry in the presence of the surface from the UPS data, it is first necessary to correct the latter for the differences $\Delta\sigma_{CC}$ between σ_{CC} in the presence of and without the surface (Table 3.6). Increasing the value of σ_{CC} by $\Delta\sigma_{CC}$ corresponds to increasing $y_{\sigma_{CC}}$ in equation (3.4.2) by this amount, and the corrected geometries are then to be determined again by solving equation (3.4.2). These distortions are recorded in Table 3.7.

TABLE 3.6. Energies of the Orbitals of Ethene Molecule When Subjected to the Distortions Given in Table 3.2. Both Without and With the Effect of the Surface Included[a]

System	−(Chemisorption energy) (eV)	−Energy (eV)							
		σ_{SS}	σ_{SS}^*	σ_{CH}	σ_{CC}	(σ_{CC})	σ_{CH}^*	π_{CC}	(π_{CC})
C_2H_4/Ni(111)	2.5	30.6	24.0	17.5	15.9	(16.1)	13.8	10.0	(12.4)
C_2H_4/Ni(100)	2.5	30.7	23.9	17.6	15.6	(16.0)	13.9	10.0	(12.0)
C_2H_4/Ni(110)	2.5	30.6	24.0	17.6	15.8	(16.0)	13.8	10.0	(12.2)
C_2H_4/Pd(111)	2.7	30.8	23.9	17.3	15.7	(16.5)	13.6	10.1	(11.7)
C_2H_4/Pt(111)	3.0	30.8	23.9	17.3	15.8	(16.8)	13.6	10.1	(11.9)

[a] Values in parentheses are energies of orbitals of the distorted molecule in the presence of the surface. Only the σ_{CC} and π_{CC} levels change; see text.

TABLE 3.7. Predicted Distortions for Various Surfaces after Applying Corrections to σ_{CC} as Found from Table 3.6

System	θ_{CCH} (deg)	θ_{HCH} (deg)	d_{CC} (A)
C_2H_4/Ni(111)	117.5	117.0	1.42
C_2H_4/Ni(100)	114.5	118.0	1.43
C_2H_4/Ni(100)	117.0	118.5	1.42
C_2H_4/Pd(111)	105.0	110.0	1.53
C_2H_4/Pt(111)	104.0	109.0	1.61

It can be seen that the result of correcting σ_{CC} is to increase the distortions. Unfortunately, at the time of writing, the actual distortions are not known from experiment. As a final point on the conformational studies, one should, strictly speaking, perform a self-consistent calculation because use has been made of Demuth's geometry, which is not quite consistent with the predictions[106] in Table 3.7. Preliminary calculations indicate that the results of iteration support Demuth's geometry rather well.

The main conclusion here is that the corrections to the predicted molecular geometry in the presence of the surface cause angle changes from 2 to 10° for θ_{CCH} with smaller changes for θ_{HCH}, but do not alter the qualitative predictions of an increase in bond length and decrease in bond angles. Calculations for similar molecules, such as ethyne and perhaps also benzene, on a transition-metal surface, would seem to be both feasible and likely to yield a useful description of molecular properties.

3.4.4. π-Orbital Bonding and Charge Transfer Between Molecule and Metal

Evidence was presented in the above discussion of conformation that only the σ-level σ_{CC} and the π-levels are appreciably altered when ethene is chemisorbed on transition metals. It was also found (see Table 3.6) that the shift in σ_{CC} due to chemisorption is rather less important than the change in the π-levels. Because of these findings, one can now set up a simple model to describe the π-orbitals in interaction with the surface.[106] This will give us information on the shift of the π-levels already invoked above and on the direction of the charge transfer, which will in turn relate to the change in work function when chemisorption occurs. This has been done following the method of Newns[117] who, using an Anderson Hamiltonian within the Hartree–Fock approximation, conducted a self-consistent calculation for the chemisorption of hydrogen atoms on transition-metal surfaces. In his calculation, only the d-electrons of the metal were included and the d-band was described by means of a simple model with a semielliptic density of states.

On generalizing to ethene in free space, one takes the basic π-electron model Hamiltonian in the following form:[106]

$$H_m = \sum_{\sigma} (\varepsilon_0 n_{A\sigma} + \varepsilon_0 n_{B\sigma}) + \sum_{\sigma} t(a^{+}_{A\sigma} a_{B\sigma} + a^{+}_{B\sigma} a_{A\sigma})$$
$$+ U[n_{A+}n_{A-} + n_{B+}n_{B-}] + J[n_{A+}n_{B+} + n_{A-}n_{B-}] \qquad (3.4.3)$$

where ε_0 is the isolated $2p$-energy level, t is the hopping parameter between the C $2p$-orbitals on centers A and B. U is the intrasite Coulomb interaction, and J is connected with the exchange interaction between electrons on the different centers. Number operator, and creation and annihilation operators for spin σ are written in the usual notation. Details of how the different parameters are determined are given in Appendix 3.3. Though Hiett *et al.*[106] did not attempt to make the tight-binding results in Appendix 3.2 accord altogether quantitatively with the above parameters, nevertheless it is easy to show that both the $2p$-energy level and the hopping parameter are roughly the same order of magnitude in the two models. Naturally, the Hamiltonian (3.4.3) is a more refined model Hamiltonian than that used as a basis for Appendix 3.2.

One must next couple the free-space π-electron Hamiltonian (3.4.3) to the d-orbitals of the metal surface. This was done following Anderson[118] and Newns[117] by adding terms to H_m of the form

$$H_{\text{metal}} + H_{\text{int}} = \sum_{k\sigma} \varepsilon_k n_{k\sigma} + \sum_{k\sigma} V_{\text{A}k}(a^{+}_{k\sigma} a_{\text{A}\sigma} + a^{+}_{\text{A}\sigma} a_{k\sigma})$$

$$+ \sum_{k\sigma} V_{\text{B}k}(a^{+}_{k\sigma} a_{\text{B}\sigma} + a^{+}_{\text{B}\sigma} a_{k\sigma}) \qquad (3.4.4)$$

where the first term on the right-hand side describes the d-bands of the metal, while $V_{\text{A}k}$ and $V_{\text{B}k}$ represent the $2p$-orbitals on A and B in interaction with the d-electrons. Appendix 3.4 deals with the nonmagnetic Hartree–Fock approximation to the above Hamiltonian $H_m + H_{\text{metal}} + H_{\text{int}}$.

The parameters in the model are interaction parameters between ethene and metal β', which is defined precisely in Appendix 3.4, and those of the metal, namely the work function, the bandwith, and the position of the Fermi level. These parameters for the metal are recorded in Table 3.8 for Ni and Cu, among others.

In order to determine the values of β' for Cu and Ni, use has been made of the experimental chemisorption energies,[119] namely -0.80 eV

TABLE 3.8. Characteristics of Different Metals (see Newns[117] and references therein)

Metal	Width of d-Band (eV)	Energy Difference between E_F and Center of Band (eV)	Work Function (eV)
Ti	8.6	-2.04	3.86
Cr	6.1	0.69	4.56
Ni	3.8	1.76	4.50
Cu	2.7	2.90	4.46

and -2.50 eV, respectively. To this end, Table 3.9 gives a selection of results for Ni that demonstrate both the variation of the chemisorption energy due to π-bonding (denoted by ΔE) with β', and also the way the bonding π-energy level varies with this same parameter. Also recorded are values of the charge transfer c, from the molecule to the metal, as a function of β'.

To find β', one must bear in mind that the above measured values of chemisorption energy include, of course, a contribution ΔE already referred to from π-bonding, and a further term, $\Delta E'$, due to the distortion of the molecule from its free-space geometry; $\Delta E'$ for Cu has been estimated as 0.13 eV[114] and for Ni as 0.35 eV.[114] Hence from Table 3.9, by interpolation, one finds for Ni that $\beta' = 3.1$ eV. Similar results in Table 3.10 for Cu lead to $\beta' = 2.1$ eV.

It can be seen from Table 3.9 that the bonding π-energy level is lowered as a result of the interaction $\beta' = 3.1$ eV by 2.3 eV for Ni. The corresponding lowering for Cu is 1 eV. Both these values are quite close to the chemisorption energies for these metals, so such independent calculations were not conducted for Pt and Pd, merely shifting the appropriate π-level by essentially the chemisorption energy, as anticipated in Section 3.3.

TABLE 3.9. Variation of Energy of the π-Bonding Level (ε_π), Charge Transfer (c), and Chemisorption Energy (ΔE) with β', for Nickel

β' (eV)	ε_π (eV)	c	ΔE (eV)
0.0	−9.05	0.00E0	0.00
1.0	−9.37	1.84E-2	−0.21
2.0	−10.14	8.58E-3	−0.88
3.0	−11.28	−1.99E-1	−2.64
4.0	−12.30	−2.81E-1	−5.33
5.0	−13.26	−3.02E-1	−8.45

TABLE 3.10. As for Table 3.9 but for Copper, so a Different Value of t is Used (see Appendix 3.3)

β' (eV)	ε_π (eV)	c	ΔE (eV)
0.0	−9.55	0.00E0	0.00
1.0	−9.77	2.01E-2	−0.20
2.0	−10.44	−1.58E-2	−0.81
3.0	−11.50	−1.87E-1	−2.72
4.0	−12.45	−2.46E-1	−5.43
5.0	−13.37	−2.73E-1	−8.53

The charge transfer, as already mentioned, takes place from the molecule to the metal for both Ni and Cu. This agrees qualitatively with the experimental fact that when ethene is adsorbed on Ni or Cu, the work function decreases.[114] This corresponds to a dipole oriented perpendicular to the metal surface with its positive end away from the surface, confirming the above prediction[106] of the direction of the charge transfer. The transfer of charge for ethene chemisorbed on Ni is shown in Figure 3.7 as a function of β'.

A further by-product of the calculations is the π-bond order (h). As an example, Figure 3.8 shows this quantity plotted against β' for ethene

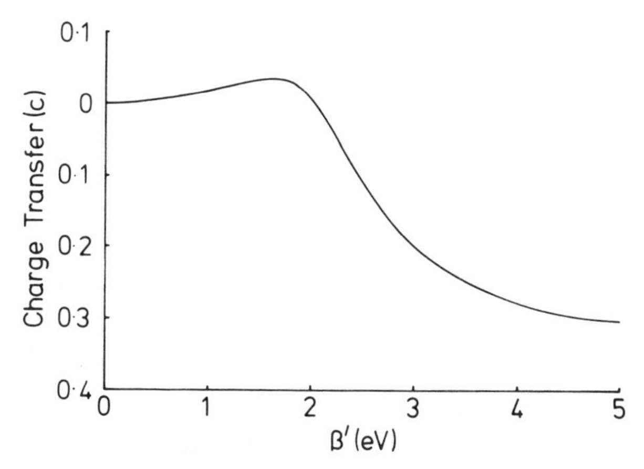

FIGURE 3.7. Transfer of charge for ethene chemisorbed on Ni, as a function of β'.

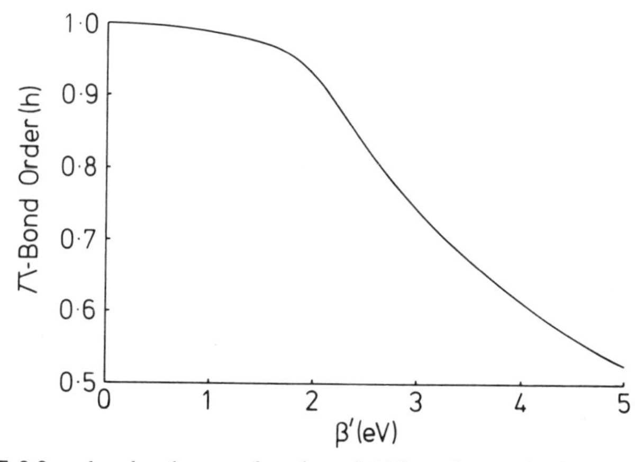

FIGURE 3.8. π-bond order as a function of β' for ethene adsorbed on Ni.

chemisorbed on Ni. The shape of this curve suggests, roughly speaking, two regimes. When $\beta' < 2$ eV the interaction seems to have little influence either on the π-bond order or on the charge transfer. However, for strongly chemisorbed species (corresponding in Figure 3.8 to $\beta' > 2$ eV), the π-bond order decreases quickly with increasing β', and hence at $\beta' = 3.1$ eV for Ni the value is substantially reduced (to 0.72) by the interaction.

In summary, this work confirms the validity of an approach, which considers the ethene molecule as relatively weakly perturbed by its interaction with the metal surface. Hence a relatively simple tight-binding model can be set up and parametrized to fit the free-space σ-levels; it allows one to give a rather simple treatment of the way changes in molecular conformation can be related to UPS ionization potentials. It turns out that with the one modification, that the level σ_{CC} must be changed somewhat by interaction with the surface, the qualitative predictions that d_{CC} increases and that θ_{CCH} and θ_{HCH} decrease clearly emerge. Table 3.7 presents the quantitative results of Hiett *et al.* with the predicted values of θ_{CCH}, θ_{HCH}, and d_{CC} (to be compared with Table 3.5).

A procedure generalizing that introduced by Newns,[117] and based on the Anderson Hamiltonian solved at the Hartree–Fock level,[106] was employed to gain insight into the electronic redistribution involved in the π-levels as interaction takes place with the d-states of the metal. It was thereby shown that as the strength of the metal–molecule interaction increases, i.e., the chemisorption energy becomes more exothermic, the π-bond order decreases and there is a transfer of charge between the metal and molecule. The direction of this charge transfer is found to be sensitive to the electronic structure of the metal and to the strength of the molecule–surface interaction. Using the experimental values of the chemisorption energy, the model predicts that in the cases of Ni and Cu the molecule transfers electrons to the metal. This is in accord with the decrease in the work function that has been experimentally observed in both of these systems.

Subsequent to the work of Hiett *et al.*, related studies on ethene, and to a lesser extent on acetylene, have been carried out by Barnes *et al.* (private communication). This work has proceeded in two directions: (1) to carry the procedure of Hiett *et al.* through to self-consistency, and (2) to utilize a different parametrization of the molecular Hamiltonian, directly related to the Roothaan linear combination of atomic orbitals method.

Though many of the principal conclusions of Hiett *et al.* are left intact, some aspects of the parametrization in their work will eventually require refinement when further experimental data becomes available. The method of Barnes *et al.* has also been used, with some degree of success, to discuss vibrational frequencies of chemisorbed ethene (cf. Chapter 4 for further discussion of vibrational frequencies in relation to infrared spectroscopy and neutron scattering).

3.5. INTERACTION BETWEEN ADSORBATES

A chemical model for H_2O on Ru was discussed earlier in this chapter, so reference will simply be made to an experimental study by Mariani and Horn[120] of the orientation of H_2O adsorbed on Cu(110).

3.5.1. Hydrogen-Bonded Clusters of H_2O

A central question in the study of water adsorption on metal single crystals is the orientation of the water molecules on the surface. Netzer and Madey[121] used the ESDIAD technique already referred to above and concluded that the array of H_2O molecules on Ni(111) contains a distribution of tilt angles with respect to the surface normal. The evidence from vibrational spectroscopy indicates that, even at very low coverages, H_2O forms hydrogen-bonded clusters on Ru(001) and Pt(100) with the hydrogen bonding leading to a large broadening and intensity enhancement of the O–H stretching vibration. (For bonding of water to Pt, cf. Appendix 3.1.)

3.5.2. Some Organic Molecules

The chemisorption of organic molecules such as benzene, naphthalene, and azulene on (111) metal surfaces has been studied experimentally by LEED and related surface techniques. The surface crystallography shows that these molecules appear to obey a close-packing-type principle as adsorbate–adsorbate interactions become more and more important and eventually compete with the adsorbate–surface bond in dictating the chemisorption structure.

Chemisorption and ordering of naphthalene and azulene on Pt has been studied experimentally by Dahlgren and Hemminger.[122] Their previous studies of the ordering and dehydrogenation of naphthalene(II) and azulene(I) on Pt(111) have led to the recognition of the importance of short-range intermolecular interactions for these systems. These two molecules are both essentially planar with dimensions as indicated in Figure 3.9. When the hydrogen Van der Waals radii are taken into account, the linear dimensions of azulene and naphthalene only differ by 10%. Azulene has a dipole moment of magnitude 0.8 D.

When azulene and naphthalene are considered on Pt(111), intact molecular adsorption is observed at room temperature. Complete dehydrogenation with an onset at 200 °C is the eventual fate of these molecules adsorbed on Pt(111).

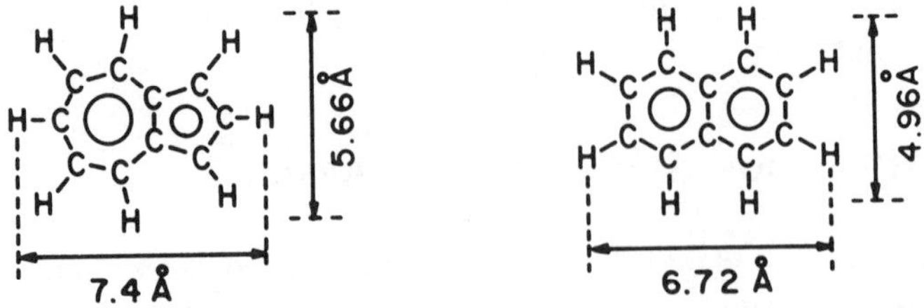

FIGURE 3.9. Dimensions of naphthalene and azulene (after Dahlgren and Hemminger[122]).

Naphthalene and azulene are large enough molecules to cover nine surface Pt atoms in a closest-packed overlayer structure. Thus the bonding of these molecules to Pt(111) is relatively nonsite specific. Site specificity is, however, not totally absent in these overlayers as indicated by the fact that the overlayer unit cell vectors are aligned with those of the substrate.

However, the orientational ordering of neighboring molecules appears to be governed by intermolecular forces rather than the molecule–Pt interaction for both azulene and naphthalene on Pt(111). It is also known that several ordered phases exist for azulene on Pt(111) while only one phase exists for naphthalene.

Gavezzotti and Simonetta[123] have reported molecular force-field calculations on the naphthalene/Pt(111) system. Their calculations indicate that the structure proposed by Dahlgren and Hemminger for the naphthalene/Pt(111) overlayer is a potential-energy minimum for their empirical intermolecular potential function. Dahlgren and Hemminger have also carried out molecular force-field calculations utilizing potentials from the work of Williams[124] on both naphthalene and azulene systems. These calculations are consistent with the observed differences between naphthalene and azulene.

3.6. ELECTRONIC EXCITED STATES OF CHEMISORBED POLYATOMIC MOLECULES

Progress on the study of electronically excited states of chemisorbed diatomic molecules was summarized in Section 2.6 of Chapter 2. The general comments made there are applicable also to polyatomics, so we need only summarize here the main findings regarding a number of specific types of molecule.

Avouris and Demuth[125] reported a systematic study by EELS of the excitations of weakly adsorbed aromatic molecules on metal surfaces at low temperatures. As for diatomics under similar conditions the intrinsic excitations of the adsorbates were left intact with small energy shifts of the order of $\frac{1}{10}$ eV and broadening of this order also relative to the corresponding free (gaseous) adsorbate values.

Avouris *et al.*[126] later used EELS to study the electronic excitations of pyridine on Ni(100). For this system, at low temperatures ($T < 200$ K) and surface coverages $<$ half a monolayer, pyridine assumes a flat adsorption geometry while at higher temperatures or higher coverages it assumes a tilted configuration. Several electronic excitations were observed in the tilted phase and these were assigned by Avouris *et al.* as $\pi \rightarrow \pi^*$ excitations of the aromatic ring. No corresponding excitations were detected for the flat phase, however. Avouris *et al.* suggest that these observations are to be understood in terms of the different nature of the chemisorption bond in the two phases.

Chapter 4

Dynamics of Adparticles and Neutron Inelastic Scattering

Our concern in Chapters 2 and 3 has been with static properties of molecules adsorbed on surfaces. In this chapter, the discussion of dynamic properties will be introduced by considering the vibrational frequencies of molecules adsorbed on metal surfaces. There are a number of techniques available that give useful information on these frequencies and will be briefly referred to, but the main attention will be focused on what can be learned from neutron inelastic scattering about molecular vibrational frequencies.

We shall deal in Section 4.1 below with the principles of neutron scattering from adsorbed molecules.[127,128] A brief comparison with other surface techniques will be presented, followed by a discussion of diffusion and its influence on neutron scattering cross-sections. Electronic and vibration rotation spectra of adsorbed molecules will then be examined.

The next topic to be treated will be adsorbate frequency shifts as influenced by coordination at the surface, followed by a theoretical framework for interpreting neutron inelastic scattering from covered surfaces.

The last major topic of this chapter will be the treatment of adparticle dynamics via Kramers' equation, which will lead naturally into a discussion of molecular desorption in Chapter 5.

4.1. PRINCIPLES OF NEUTRON SCATTERING FROM ADSORBED MOLECULES

Hall and Wright[127] have reviewed the field of neutron scattering from adsorbed molecules. We follow their account and start with a brief outline of the concepts associated with the neutron technique. In any neutron-scattering experiment, the neutrons may undergo changes in their energy and momenta and, as a consequence of the energy changes, both elastic and inelastic scattering events may be observed.

If beams of incident and scattered neutrons have wave vectors $\mathbf{k}_0$ and $\mathbf{k}$ and energies E_0 and E, the resultant momentum transferred on scattering is $\hbar\mathbf{Q}$, where

$$|\mathbf{Q}| = |\mathbf{k}_0 - \mathbf{k}| = (k_0^2 + k^2 - 2kk_0 \cos\theta)^{1/2} \tag{4.1.1}$$

θ being the angle through which the neutron is scattered. The energy transfer ΔE is given by

$$\Delta E = E_0 - E = \frac{\hbar}{2m}(k_0^2 - k^2) = \hbar\omega \tag{4.1.2}$$

where m is the neutron mass.

For elastic scattering, $\Delta E = 0$ and $|\mathbf{k}_0| = |\mathbf{k}|$ so that $|\mathbf{Q}| = 2k_0 \sin(\theta/2)$. Neutron scattering can also be coherent if the scattering amplitudes of the nuclei in equivalent positions in a lattice are identical, or incoherent plus coherent if there are fluctuations in amplitude from site to site.

The wavelengths of thermal neutrons, produced in a reactor under normal operating conditions, are of the order of interatomic separations in the solid state. Consequently they are valuable for structural investigations. At the same time the energies of those thermal neutrons are comparable with the energies associated with atomic and molecular vibrations, so that vibrational excitations in a material can be measured with high accuracy.

It should also be emphasized that one can define neutron energies with precisions of the order of 0.1%, so that one can extend the region of energy transfers which can be explored down toward 10^{-2} cm^{-1}. Consequently the energy range of the inelastic neutron-scattering technique allows investigation of interatomic and intermolecular vibrations, translational and rotational diffusion, and the excitations between quantum levels split by tunneling.

4.2. COMPARISON WITH OTHER SURFACE TECHNIQUES

Let us consider inelastic incoherent neutron scattering. Such experiments can give information of a similar kind to optical,[129] inelastic electron tunneling, and inelastic electron-scattering spectroscopies,[130] and are obviously to be preferred either where a vibration has a particularly low transition moment or where a sample or adsorbate is highly absorbing.

It has been pointed out for optical spectroscopy that this first group may contain all those vibrations for which the dipole-moment change lies parallel to the surface of a "metal." It is argued that metal films, clusters, and particles approximate infinitely polarizable media sufficiently well so that vibrations leading to dipole-moment changes parallel to a metal surface induce an equal image moment within the metal, resulting in no net dipole-moment change and therefore no absorption intensity. For inelastic electron scattering the selection rule leads to the same conclusion; namely that a net perpendicular component of a static dipole moment or induced dipole moment perpendicular to the surface is necessary for observation of a particular vibration. This is only true to a first approximation in the case of tunneling spectroscopy.

Two advantages that stem from the fundamental nature of the neutron-scattering process are the ability to exactly calculate the intensities expected from a neutron spectrum and the ability to measure scattering from all contributions to the vibrational density of states. The first property contrasts with the position in optical spectroscopy, where to make a similar prediction requires knowledge of the changes in the electric dipole moment as a consequence of a change in nuclear positions. This advantage has been found valuable in assigning surface vibrations.

The second property becomes valuable when interaction force constants between adsorbed atoms are important. The neutron technique is the only one of those referred to above where the momentum transfers in the scattering processes are of similar magnitude to reciprocal lattice vectors, and so it becomes possible to excite all collective excitations whatever the phase difference in amplitude between neighboring atoms.

4.3. DIFFUSION MEASUREMENTS

Here, neutron scattering gives unique information on individual nuclear displacements, in contrast with the molecular and nuclear dipole–dipole correlation functions obtained from infrared and nuclear-magnetic-resonance measurements.

In cases where precise geometric information is available on the structure of an adsorbate, nuclear spin relaxation is probably to be preferred as an investigative tool for exploring molecular motion because the spin-lattice relaxation times can then be accurately calculated and compared with measured values. In other cases where intersorbate distances are not known, the neutron technique is favorable.

The differing measurement times of the two techniques also lead to complementary information. The relatively long time of the relaxation technique leads to a time-averaged diffusion constant, which may be insensitive to microscopic inhomogeneities. On the other hand, the shorter time-scale number-averaged neutron technique will be sensitive to the diffusion process in which most molecules are participating at any one time.

4.4. THEORY

A brief review of the theory will now be given. For detailed accounts, the reader should consult the book by Marshall and Lovesey[131] and the articles by Stiller,[132] Windsor,[133] and Wright and Sayers.[134]

In principle, the intensity of inelastic neutron scattering from a sample can be calculated exactly, in contrast to the situation familiar in optical spectroscopy. For incoherent phonon scattering, the cross-section $d^2\sigma_{\text{inc}}/d\Omega dE$ for neutron energy loss is given by*

$$\frac{d^2\sigma_{\text{inc}}}{d\Omega dE} = \sum_{\mathbf{q},j} \frac{k}{k_0} \delta[\hbar\omega - \hbar\omega_j(\mathbf{q})] \frac{\hbar[n_j(\mathbf{q})+1]}{2\omega_j(\mathbf{q})} \sum_{\rho} \frac{\sigma_{\text{inc}}}{4\pi M_\rho} \exp(-2W_\rho) |\mathbf{Q}\cdot\mathbf{U}_\rho^j(\mathbf{q})|^2 \tag{4.4.1}$$

where $\hbar\omega_j(\mathbf{q})$ is the characteristic frequency of the phonon mode j of atom ρ in the unit cell, having amplitude $\mathbf{U}_\rho^j$ and wave vector $\mathbf{q}$; $n_j(\mathbf{q})$ denotes the Bose occupation probability, and $\exp(-2W_\rho)$ the Debye–Waller factor.[131] It is not always necessary to employ this equation in interpreting neutron spectra and, where possible, there is the advantage of considerable simplicity in approximating the dynamics of an isolated adparticle by that of a simple harmonic oscillator. The cross-section for the isotropic oscillator is

$$\frac{d^2\sigma_{\text{inc}}}{d\Omega dE} = \frac{k}{k_0}\frac{\sigma_{\text{inc}}}{4\pi} \exp\left(-\frac{\hbar Q^2}{2M\omega_0}\coth\frac{\hbar\omega_0}{2k_B T}\right) \exp\left(\frac{n\hbar\omega_0}{2k_B T}\right) \delta(\hbar\omega - n\hbar\omega_0)$$
$$\times I_n\left(\frac{\hbar Q^2}{2M\omega_0}\operatorname{cosech}\frac{\hbar\omega_0}{2k_B T}\right) \tag{4.4.2}$$

where n is the number of units of energy $\hbar\omega_0$ lost or gained by the neutron while I_n is the modified Bessel function of the first kind.

For the anisotropic case of the atom adsorbed at a surface, Wright[135] has shown that the cross-section involves similar terms in both ω_A and ω_E, the vibrations perpendicular and parallel to the surface.

* The incoherent cross-section σ_{inc} in equation (4.4.1) is defined as $\sigma_{\text{inc}} = 4\pi(\overline{b^2} - \bar{b}^2)$, the scattering length b being given through $V(r) = (2\pi\hbar^2/m)b\delta(r)$, where $V(r)$ is the Fermi pseudopotential of the nucleus. Only the average scattering lengths of atoms, $\bar{b}$ in equivalent crystallographic positions will contribute to the coherent scattering, whereas departure of an atom's scattering length from the mean gives rise to incoherent scattering.

For coherent phonon scattering, the cross-section is given by

$$\frac{d^2\sigma_{\text{coh}}}{d\Omega dE} = \frac{(2\pi)^3}{V} \sum_{j,q} \frac{k}{k_0} \delta(\hbar\omega - \hbar\omega_j(q)) \sum_{\tau} \delta(\mathbf{Q} - \mathbf{q} - \boldsymbol{\tau}) \frac{\hbar[n_j(\mathbf{q}) + 1]}{2\omega_j(\mathbf{q})}$$

$$\times \exp(i\mathbf{Q}\cdot\boldsymbol{\rho}) \sum_{\rho} \frac{\langle b\rangle_\rho}{M} [\mathbf{Q}\cdot\mathbf{U}^j_\rho(\mathbf{q})]^2 \exp(-2W_\rho) \tag{4.4.3}$$

where τ denotes a reciprocal lattice vector while b is the appropriate neutron-scattering length. It can be seen that, due to the presence of the product $\mathbf{Q}\cdot\mathbf{U}$ in the cross-section, considerable selectivity can be obtained using a partially aligned sample.

4.4.1. Influence of Diffusion on Scattering Cross-Section

When the particle from which scattering occurs is undergoing diffusion on the time scale of the neutron–nucleus interaction, the beam of elastically scattered neutrons has its energy distribution broadened. The variation of the broadening with momentum transfer depends on the nature of the diffusion process involved, but continuous, jump, and rotational diffusion all lead to recognizably different scattering laws $S(Q, \omega)$, given by

$$S(Q, \omega) = (d^2\sigma/d\Omega dE) \frac{k_0 4\pi}{k\sigma} \tag{4.4.4}$$

After performing a polycrystalline average one obtains:

1. For three-dimensional macroscopic diffusion

$$S(Q, \omega) = \frac{DQ^2}{\omega^2 + (DQ^2)^2} \tag{4.4.5}$$

   where D is the diffusion coefficient.
2. For two-dimensional diffusion[135]

$$S(Q, \omega) = \frac{1}{2}\int_0^\pi d\theta \sin\theta \frac{1}{\pi} \frac{B\sin^2\theta}{\omega^2 + (B\sin^2\theta)^2} \tag{4.4.6}$$

   where $B = \tau^{-1}[1 - J_0(Q\rho)]$ and τ is the mean time between jumps over a distance ρ.
3. For rotation of an atom about a single axis at distance a in jumps of $360°/N$

$$S(Q, \omega) = B_0(Qa)\delta(\omega) + \frac{1}{N} \sum_{n=1}^{N-1} B_n(Qa) \frac{\tau_n}{1 + (\omega\tau_n)^2} \tag{4.4.7}$$

where

$$B_n(Qa) = \frac{1}{N}\sum_{P=1} j_0\left(2Qa \sin\frac{\pi P}{N}\right)\cos n\left(\frac{2\pi P}{N}\right) \tag{4.4.8}$$

$$\tau_n = \frac{\tau}{1-(\cos 2\pi/N)}\frac{\sin^2 \pi/N}{\sin^2 n\pi/N} \tag{4.4.9}$$

and j_0 is the spherical Bessel function of order zero. The theory of surface diffusion is discussed in Section 4.8 below.

4.5. ELECTRONIC AND VIBRATION–ROTATION SPECTRA OF ADSORBED MOLECULES

Considerable theoretical and experimental effort has been devoted to the spectroscopy and dynamics of molecules adsorbed on surfaces. Spectroscopic investigations of this field employing photons,[129] electrons,[130] and atoms as incident particles are fundamental in studying the geometric arrangement of adsorbed species, discussed in earlier chapters, and the energy levels of adsorbed systems. They also provide a key to understanding various chemical reactions, which will be examined in more detail in Chapter 6, and relaxation processes at surfaces.

However, there are a number of points to be considered in analyzing spectra of adsorbed systems. Following Kono *et al.*,[136] we note especially the following ones:

1. Overlayer structures and multiple trapping sites of adsorbed species.
2. Coupling of the nuclear motion of adsorbed species with electron–hole pairs in the solid (see, for example, Ref. 325, p. 519).
3. Effects of bulk and surface phonons, reviewed by Wallis.[137]
4. Localized vibrational modes whose amplitudes are well localized at the trapping sites of adsorbed species, following Grimley,[138] Mahanty *et al.*[139] (see also Stockmeyer[140]), Goodman,[141] and Black[142] (see also Section 4.6 below).
5. Hindered rotation of admolecules.

Points (3) to (5) are considered in the work of Kono *et al.*, to be summarized below, but they note that the five effects are not independent of one another.

One of the most important aspects in the theoretical treatment of adsorbed systems will be to deal with the large number of degrees of freedom in the adsorbed systems. Kono *et al.* apply the adiabatic approximation specifically to an adsorbed diatomic molecule on a surface. In particular, the band shape of electronic spectra for this system is studied.

In the above approximation, the Schrödinger equation for the total system is separated into those for electronic motion, intramolecular vibration, and low-frequency motions, such as hindered molecular rotation and lattice vibration. They then apply the theory to an admolecule on the (001) surface of face-centered-cubic monatomic crystals and determine the line shape of a vibronic transition for both chemisorption and physisorption cases. In chemisorption, a side band due to low-frequency motions can appear in the line shape of a vibronic transition. On the other hand, in physisorption cases the line shape is a single peak.

The above treatment of Kono *et al.*[136] is applicable not only to electronic spectra, but also to vibration–rotation spectra (infrared spectra) of a molecule adsorbed on a surface. However, detailed knowledge is required of the perturbed lattice vibration, which is one of the important factors that determine the line shape in the infrared region.

A comment is worth making here on relaxation processes of the admolecule. The processes of vibrational relaxation and desorption are known to play a key role in chemical reactions, as reviewed, for example, by Zhdanov and Zamarev.[143] Although numerous experimental and theoretical studies have been conducted in this field, most of them cover only some specific aspects of the relaxation processes on surfaces. Since the theory of Kono *et al.* is based on the adiabatic approximation, the relaxation phenomena can be formulated as nonadiabatic processes. The kinetic-energy operator of the intramolecular vibration will induce the electronic relaxation and the kinetic-energy operators of the molecular rotation and perturbed lattice vibrations will induce the intramolecular vibrational relaxation. This may therefore offer a route to further progress in this area.

4.6. ADSORBATE FREQUENCY SHIFTS

4.6.1. Influence of *n*-Fold Coordination

In the work of Sayers, it is shown that the adsorbate stretching and bending modes for H on MoS_2 are shifted to higher frequency by interaction with the lattice modes of the substrate.[144,145] This is, in fact, a general result and was studied in detail by Grimley[138] for adsorption on the ⟨100⟩ surface of a simple cubic lattic.

Following Grimley's work,[138,134] let us consider a simple cubic lattice of atoms of mass m and stretching and bending force constants γ_1 and γ_2. The adsorbate is taken to bond to a single surface atom on the ⟨100⟩ surface, the origin being taken at this atom. The x-axis is chosen perpendicular to the surface. To account for the free surface, the stretching-force constant between metal atoms in the surface and first subsurface planes is changed to γ_1'. The adatom has mass M_A and is bonded to the metal atom at the origin with stretching and bending force constants μ_1 and μ_2. Let $\Omega_A = (\mu_1/M_A)^{1/2}$ be the stretching frequency of the adatom with the metal atoms held fixed. When the interaction with the lattice modes is incorporated, this frequency is shifted to a value ω_A given by

$$\omega_A^2 = \frac{\Omega_A^2}{1 + \lambda\Omega_A^2 G(0,0)} \tag{4.6.1}$$

where $\lambda = M_A/m$ and $G(0,0)$ is the site diagonal Green function at site 0.

The upper limit of the simple cubic lattice phonon spectrum occurs at $\omega = \Omega_L$ with $m\Omega_L^2 = 4(\gamma_1 + 2\gamma_2)$ and therefore any solution of equation (4.6.1) that represents a local mode must evidently satisfy $\omega_A > \Omega_L$. For $\omega > \Omega_L$, the site diagonal Green function $G(0,0)$ is negative and decreases in magnitude with increasing ω. Let us denote the value of $G(0,0)$ at $\omega = \Omega_L$ by G_L. Then it follows from equation (4.6.1) that if $1 + \lambda\Omega_A^2 G_L \leq 0$, a local mode will exist at frequency $\omega_A > \Omega_A$, but if $1 + \lambda\Omega_A^2 G_L > 0$, this solution only exists if

$$\Omega_A \geq \frac{\Omega_L}{(1 - \lambda\Omega_L^2 G_L)^{1/2}} \tag{4.6.2}$$

and for a given value of λ, there is a minimum value of Ω_A below which a local mode does not persist when interaction with the vibrational modes of the substrate is incorporated. If λ is small ($M_A \ll m$) there is a local mode only if $\Omega_A > \Omega_L$, but as λ increases a local mode can be found for Ω_A in the band.

For large ω

$$G(0,0) \underset{\omega\to\infty}{\longrightarrow} -\frac{1}{\omega^2}\left[1 + \frac{1}{2}\left(\frac{\Omega_L}{\omega}\right)^2 + \frac{(z-2)\gamma_1}{m\omega^2}\right] \tag{4.6.3}$$

where $z = \gamma_1'/\gamma_1$. Thus if $\Omega_A \gg \Omega_L$, the solution of equation (4.6.1) is

$$\omega_A^2 = \Omega_A^2(1 + \lambda) \tag{4.6.4}$$

We shall return briefly to this result below.

Mahanty *et al.*[139] have considered the bonding of an adatom on a two-dimensional square lattice. The change in the vibrational density of states due to coupling of the adatom to the lattice in the on-top position is given by (see Section 4.7.2 below)

$$g(\omega) = -\frac{2\omega}{\pi}\,\mathrm{Im}\,\frac{d}{d\omega^2}\{\ln[\eta - B^T G_s(\omega^2)B - M_A\omega^2]\} \tag{4.6.5}$$

where η is the adsorbate–metal force constant. Details are given in Section 4.7 below[139]: for $\omega \gg \omega_L$, with ω_L the maximum frequency of the lattice phonons, one finds (with ω_0 the 'local mode frequency'; see below)

$$\frac{\omega^2}{\omega_0^2} = \frac{1 + (1 + 4\lambda)^{1/2}}{2} \tag{4.6.6}$$

which agrees with equation (4.6.4) for small λ.

If the adatom is coupled to two surface atoms by equal force constants η, then, as shown in Section 4.7.2 below,

$$B^T G B = \eta^2 [G(11) + G(22) + G(21) + G(12)] \tag{4.6.7}$$

where $G(11) = G(22) = g_0$ is given by $G_s(0,0)$ in Mahanty *et al.*,[139] while $G(21) = G(12) = g_1$, say, is given by

$$g_1 = \frac{1}{mV^*} \int \frac{d^2k \cos k}{\omega^2(k) - \omega^2 - i\varepsilon} \tag{4.6.8}$$

with V^* the area of the two-dimensional Brillouin zone. The Green functions g_0 and g_1 satisfy the sum rule

$$g_1 = \left(1 - \frac{m\omega^2}{4\gamma}\right) g_0 = \frac{1}{4\gamma} \tag{4.6.9}$$

where γ is related to the maximum frequency ω_L by $\omega_L^2 = 8\gamma/M$. The position of the local mode is then found to lie at ω_0, given by

$$\frac{\omega_0^2}{\omega_L^2} - \frac{\omega^2}{\omega_L^2} + \frac{\lambda\omega_0^4}{\omega_L^4} - \frac{\lambda\omega_0^4}{\omega_L^4} R\left(1 - \frac{\omega^2}{\omega_L^2}\right) = 0 \tag{4.6.10}$$

where R is the real part of the Green function, for solutions outside the band. This quantity is given explicitly in Mahanty *et al.*[139]: one finds

$$\left(\frac{\omega}{\omega_0}\right)^2 = \frac{1 + (1 + 4\lambda)^{1/2}}{2} \tag{4.6.11}$$

as before.

For adsorption at a threefold and fourfold coordinated site, Black[146] has noted that the shifted frequencies are given by

$$\omega_{3\text{-fold}} = \omega_0 \left(1 + \frac{M_A}{3m \sin^2 \theta}\right)^{1/2} \tag{4.6.12}$$

and

$$\omega_{4\text{-fold}} = \omega_0 \left(1 + \frac{M_A}{4m \sin^2 \theta}\right)^{1/2} \tag{4.6.13}$$

respectively, where θ is the angle between the substrate plane and the adatom bond.

4.6.2. Intensity in Local Mode at Adsorbate Site

From the work of Wright and Sayers[134] for the case of adsorption of H on MoS_2 at low coverages, it was demonstrated that the vibrational density of states of the H atom for frequencies lying within the substrate density of states strongly reflects the motion of the sulfur atom to which the hydrogen is bound. It is of interest to consider this phenomenon in a little more detail using the one-dimensional model of Mahanty *et al.*[139,134]

Let us consider a one-dimensional lattice of N matrix atoms of mass m satisfying periodic boundary conditions. The adatom, of mass M_A, is assumed bound to atom 1 with force constant η. If γ denotes the force constant between nearest-neighbor matrix atoms, then

$$M_A G(\omega^2) = \frac{1}{\omega^2 - (\eta/M_A) + (\eta^2/M_A) G(1, 1, \omega^2)} \tag{4.6.14}$$

where

$$\begin{aligned} G(1, 1, \omega^2) &= \frac{i}{m[\omega^2(\omega_L^2 - \omega^2)]^{1/2}}, \qquad \omega < \omega_L \\ G(1, 1, \omega^2) &= \frac{-1}{m[\omega^2(\omega^2 - \omega_L^2)]^{1/2}}, \qquad \omega > \omega_L \end{aligned} \tag{4.6.15}$$

$\omega_L = (4\gamma/M_A)^{1/2}$ being the upper limit of the substrate density of phonon states. For $\omega < \omega_L$, the local density of vibrational states on the adatom is obtained from equation (4.6.14) in the form

$$n(\omega) = \frac{2M_A}{\pi m} \frac{\omega[\omega^2(\omega_L^2 - \omega^2)]^{1/2}}{(\omega^2/\omega_0^2 - 1)^2 \omega^2(\omega_L^2 - \omega^2) + \eta^2/m^2} \tag{4.6.16}$$

which, in the limit ω_0/ω_L tends to infinity, yields

$$n(\omega) = \xrightarrow[\omega_0/\omega_L \to \infty]{} \frac{2M_A}{\pi m} \frac{\omega[\omega^2(\omega_L^2 - \omega^2)]^{1/2}}{\omega^2(\omega_L^2 - \omega^2) + \eta^2/m^2} \tag{4.6.17}$$

It then follows that in this limit[147]

$$\int_0^{\omega_L} n(\omega)d\omega = M_A/m \tag{4.6.18}$$

independently of the value of η^2/m^2.

Scattering from an adsorbate with a local mode above the high-frequency cutoff of the substrate density of states therefore always has intensity in the substrate phonon region with integral intensity M_A/m in this region. It is found from Figure 4.1[148] that this is the case for hydrogen and deuterium adsorbed on Raney nickel.

For $\omega > \omega_L$

$$M_A G = \left[\omega^2 - \frac{\eta}{M_A}\left(1 + \frac{\eta}{m[\omega^2(\omega^2 - \omega_L^2)]^{1/2}}\right)\right]^{-1} \tag{4.6.19}$$

Vanishing of the denominator yields the position of the local mode. This

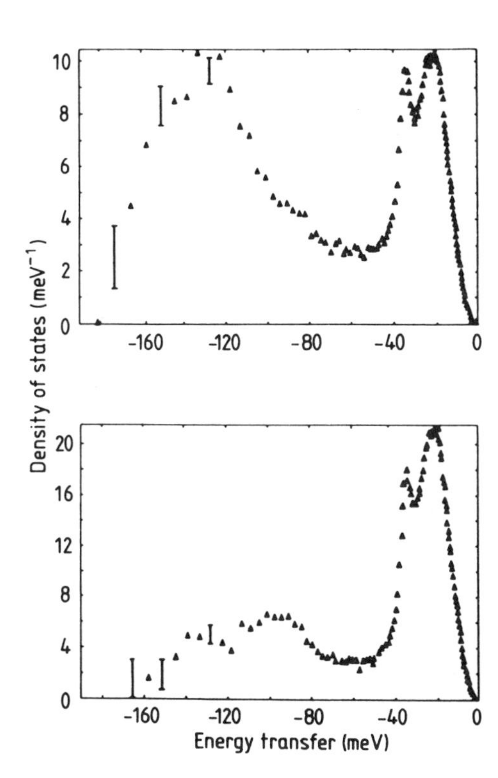

FIGURE 4.1. Frequency distribution for adsorbed hydrogen (top) and deuterium (bottom) on Raney nickel (after Wright and Sayers[134]).

occurs at

$$\omega^2 = \omega_0^2 \left(1 + \frac{M_A \omega_0^2}{m[\omega^2(\omega^2 - \omega_L^2)]^{1/2}}\right) \tag{4.6.20}$$

to first order in M_A/m, if ω_0 is not too near the band edge.

For $\omega_0^2 \gg \omega_L^2$, equation (4.6.20) gives $\omega^2 \simeq \omega_0^2(1 + M_A/m)$, in agreement with equation (4.6.4).

The dispersion of adsorbate modes, based especially on the work of Alldredge *et al.* as well as that of Armand and Theeten, is reviewed in detail by Wright and Sayers[(134)] and the interested reader is referred to this account.

4.7. THEORETICAL FRAMEWORK FOR INTERPRETING NEUTRON INELASTIC SCATTERING FROM COVERED SURFACES

In relation to the above discussion, a framework for interpreting neutron inelastic scattering from covered surfaces was provided by Mahanty *et al.*[(139)] with particular reference to the theory of hydrogen on platinum.

Their starting point was the experiment of Wright *et al.*,[149] where the neutron inelastic scattering from adsorbed hydrogen on the surface of powdered metallic Pt was measured, the data being reproduced in Figure 4.2. These data were obtained by subtracting the scattering from pure Pt, so that they represent only the effect of the adsorbed H on the Pt surface. Peaks in the neutron scattered intensity are found to lie at energy transfers of 9, 19, and 50 meV in addition to the elastic scattering peak. The positions of the two low-energy peaks correlate almost precisely with peaks in the bulk vibrational spectrum of pure Pt, as given by Dutton and Brockhouse.[(150)] The high-energy peak can therefore be associated with a local vibrational mode of hydrogen coupled to Pt.

The experimental data represent a weighted density of phonon states of the H–Pt system at the surface and can be expected to reveal the peaks and other structural features of the phonon density of states. To facilitate interpretation of the main structural features of these neutron data, Mahanty *et al.* have examined the phonon density of states of the simplest possible model incorporating most of the essential features of the H–Pt system. The model is based on near-neighbor forces. Eventually, one must expect that, for a Pt substrate, the long-range interactions characteristic of metals, discussed in the 3*d* transition series, for example, by Matthai *et al.*,[(151)] will have to be incorporated. However, Mahanty *et al.* have qualitatively described the corrections to their model expected for long-

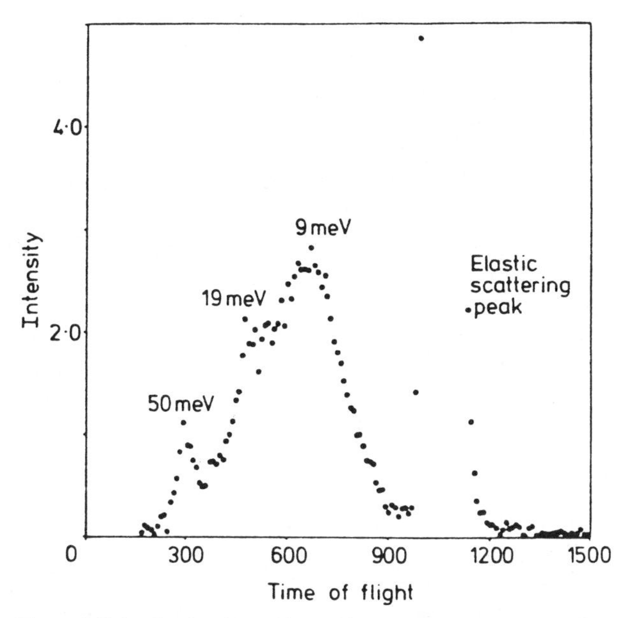

FIGURE 4.2. Time-of-flight distribution of inelastically scattered neutrons from an experiment of Wright *et al.* for adsorbed hydrogen on the surface of powdered metallic platinum (after Mahanty *et al.*[139]).

range forces, and these go in the right direction to account for some of the discrepancies between the predictions of the model and the experiment of Wright *et al.*[149]

It should also be stressed that even though a specific crystal surface will be treated below, the neutron experiment of Wright *et al.* was carried out on a powder. One must anticipate that this will lead to less pronounced structure in the experiment than would be observed if the neutron inelastic scattering could be measured from a single-crystal face.

In addition to the specific use of the model described below to interpret neutron inelastic data for H on Pt, it is also worth reiterating its utility to explore (1) the way the position of the local mode of a light impurity relative to the high-frequency band-edge of the matrix phonon spectrum will affect the change in the phonon density of states due to the adsorbed species, and (2) the effect of coupling the light impurity to more than one atom of the matrix, as in Section 4.6.1 above.

4.7.1. Description of the Model

Following Mahanty *et al.*, and in order to be quite specific below, the vibrations of the Pt–H system will be analyzed in the framework of a "scalar model." However, though the assumption of a "scalar model" in which each atom has only one degree of freedom may seem somewhat unrealistic, it is well known (see, for example, Maradudin *et al.*[152]) that the nature of the singularities in the phonon spectrum are revealed even in such a scalar model, although their exact locations are not. Unfortunately, because of the simplicity of the model of Mahanty *et al.*, to be elaborated below, precise evaluation of relative intensities of the peaks in the side band will not be possible. However, the qualitative features of the structure can be exhibited in terms of the singularities of the surface-mode spectrum and the local mode itself. The effect of the anisotropy in the interatomic coupling along different directions can also be studied in this model. It will be shown, for instance, that anisotropy can lead to broadening of the peaks in the side band.

Following Mahanty *et al.*,[139] the vibrational degrees of the entire system will be divided as follows:

1. An internal degree of freedom of the Pt–H system. The scalar displacement of the adsorbed H atom will be denoted by v.
2. A set of degrees of freedom represented by a vector $\mathbf{U}_\nu$, whose components are the displacements $\mathbf{u}_m$ of the atoms in the interior of the crystal, $m = 1, 2, \ldots, N_\nu$, where N_ν is the total number of atoms in the interior. Again, the "scalar model" is assumed.

The equation of motion of the system can be written formally in matrix form, where η is the force constant of the H–Pt coupling,[153,154]

$$\left[\begin{array}{c|c|c} A_\text{v} - m\omega^2 & 0 & 0 \\ \hline C^\text{T} & A_\text{s} - m\omega^2 & B \\ \hline 0 & B^\text{T} & \eta - M\omega^2 \end{array}\right] \begin{bmatrix} \mathbf{U}_\text{v} \\ \mathbf{U}_\text{s} \\ v \end{bmatrix} = 0 \qquad (4.7.1)$$

Here, the full dynamic matrix has been partitioned to explicitly indicate coupling between the H atom of mass M, the surface Pt atoms, and the Pt atoms (mass m) in the interior of the crystal. The division between the surface and volume atoms is, to some extent, arbitrary; for instance, the surface need not be a monolayer. The superscript T in equation (4.7.1) denotes the transpose. This equation is equivalent to

$$\left.\begin{aligned} (A_\text{v} - m\omega^2)\mathbf{U}_\text{v} + C\mathbf{U}_\text{s} &= 0 \\ C^\text{T}\mathbf{U}_\nu + (A_\text{s} - m\omega^2)\mathbf{U}_\text{s} + Bv &= 0 \\ B^\text{T}\mathbf{U}_\text{s} + (\eta - M\omega^2)v &= 0 \end{aligned}\right\} \qquad (4.7.2)$$

If one eliminates $\mathbf{U}_\text{v}$ and $\mathbf{U}_\text{s}$, the last equation of (4.7.2) becomes

$$D(\omega)v = \{\eta - B^\text{T}[A_\text{s} - C^\text{T}(A_\text{v} - m\omega^2)^{-1}C - m\omega^2]^{-1}B - M\omega^2\}v = 0 \quad (4.7.3)$$

The frequency-distribution function obtained from equation (4.7.3) by means of the relation

$$g(\omega) = -\frac{2\omega}{\pi} \operatorname{Im} \frac{d}{d(\omega^2)} \ln D(\omega) \tag{4.7.4}$$

is what will be relevant to the interpretation of the data of Wright *et al.*,[149] since the neutron cross-section for H is much larger than that for Pt and it is this frequency-distribution function that will govern the motion of the adsorbed hydrogen.

Initially, Mahanty *et al.* ignore the term $C^T(A_\nu - m\omega^2)^{-1}C$ in $D(\omega)$ and assume that many of the observed features arise from the surface modes, given by $(A_s - m\omega^2)$. This is a somewhat drastic approximation since the long-range forces between Pt atoms, that determine the contribution from the neglected term, are rather important. Furthermore, the term omitted would give nonvanishing contributions right up to the bulk band-edge, while the surface modes would be cut off somewhat earlier.

4.7.2. Coupling of an Adparticle to a Two-Dimensional Square Lattice

Mahanty *et al.*[139] then consider a very specific model in which one of the surface Pt atoms is bonded with a hydrogen atom. For simplicity, they regard the Pt surface phonon mode as due to a two-dimensional square lattice with nearest-neighbor interactions. Furthermore, the vibrational displacements of the surface atoms are treated as scalars.

Equation (4.7.2) then takes the form

$$[\eta - B^T G_s(\omega^2) B - M\omega^2] v = 0 \tag{4.7.5}$$

where G_s is the Green-function matrix defined as

$$G_s(\omega^2) = (A_s - m\omega^2)^{-1} \tag{4.7.6}$$

Equation (4.7.5) yields the change in the density of frequencies due to coupling of H to Pt, in the form

$$g(\omega) = -\frac{2\omega}{\pi} \operatorname{Im} \frac{d}{d(\omega^2)} \{\ln[\eta - B^T G_s(\omega^2) B - M\omega^2]\} \big|_{\omega^2 + i\varepsilon} \tag{4.7.7}$$

In the above model

$$B^T G_s B \equiv \eta^2 G_s(1, 1; \omega^2) \tag{4.7.8}$$

where

$$G_S(1,1;\omega^2) = \frac{1}{mV^*}\int \frac{d^2K}{\omega^2(\mathbf{K}) - \omega^2 - i\varepsilon}$$
$$= \frac{1}{m\omega_L^2}[R(1,1;\omega^2) + iI(1,1;\omega^2)] \qquad (4.7.9)$$

Here, R and I are the real and imaginary parts of the Green function, a discussion of which is given, for example, by Maradudin *et al.*[152] The quantity V^* in equation (4.7.9) is the volume of the Brillouin zone.

By this approach one reaches the basic equation for this model, namely that the change in the phonon density of states due to adsorbed hydrogen is given by

$$g(\omega) = \frac{\frac{2\omega}{\pi}\left[\left(1 + \frac{\lambda\omega_0^4}{\omega_L^2}R'\right)\frac{I}{\omega_L^2} + \left(\omega_0^2 - \frac{\lambda\omega_0^4}{\omega_L^2}R - \omega^2\right)\frac{I'}{\omega_L^2}\right]\lambda\omega_0^4}{\left(\omega_0^2 - \frac{\lambda\omega_0^4}{\omega_L^2}R - \omega^2\right)^2 + \frac{\lambda^2\omega_0^8}{\omega_L^4}I^2} \qquad (4.7.10)$$

ω_L being the maximum frequency of the surface phonons. In addition $\lambda = M/m$, the ratio of the impurity to surface atom masses, $\omega_0 = (\eta/M)^{1/2}$ and will be referred to as the local-mode frequency, and R' and I' are the derivatives with respect to ω^2 of R and I. Moreover, if $\omega_0 > \omega_L$, there is a δ-function contribution to $g(\omega)$ at the local-mode frequency, given in general by the vanishing of the first term in the denominator of equation (4.7.10), outside the band.

Mahanty *et al.*[139] also consider the case where the hydrogen atom is coupled to more than one Pt atom. This is necessary to provide criteria for deciding whether or not the bond between Pt and H is metallic-like, in which case H can be expected to be bonded to more than one Pt atom. The secular equation for this case is obtained by a straightforward generalization of the above arguments. If, for example, the H atom is coupled to two neighboring Pt atoms (say 1 and 2) by equal force constants η, then

$$B^T G B = [\eta\eta]\begin{bmatrix} G(11) & G(12) \\ G(21) & G(22) \end{bmatrix}\begin{bmatrix} \eta \\ \eta \end{bmatrix}$$
$$= \eta^2[G(11) + G(22) + G(21) + G(12)] \qquad (4.7.11)$$

where $G(11) = G(22) = g_0$, say, is given by $G_s(11)$ in equation (4.7.9), while $G(21) = G(12) = g_1$, say, is given by equation (4.6.8) in Section 4.6.1 above.

The generalization of equation (4.7.5) then becomes

$$[2\eta - 2\eta^2(g_0 + g_1) - M\omega^2] = 0 \tag{4.7.12}$$

The Green functions g_0 and g_1 are readily shown to satisfy the sum rule quoted in equation (4.6.9) of Section 4.6.1 above, the quantity γ being related to the maximum frequency ω_L by $\omega_L^2 = 8\gamma/m$. This result can be used to express equation (4.7.12) in the form

$$\left[\omega_0^2 - \frac{\lambda\omega_0^4}{\omega_L^2}\left(1 - \frac{\omega^2}{\omega_L^2}\right)(R + iI) + \frac{\lambda\omega_0^4}{\omega_L^2} - \omega^2\right] v = 0 \tag{4.7.13}$$

where $\omega_0^2 = 2\eta/M$; equation (4.7.13) constituting a generalization of equation (4.6.10) of Section 4.6.1 above. Hence one arrives again at the desired change in the phonon density of states:

$$g(\omega) = \frac{2\omega}{\pi}\frac{\lambda\omega_0^4}{\omega_L^2}\frac{P}{Q} \tag{4.7.14}$$

where

$$P = \left(1 - \frac{\omega^2}{\omega_L^2}\right) I\left[1 - \frac{\lambda\omega_0^4}{\omega_L^4} R + \frac{\lambda\omega_0^4}{\omega_L^4}\left(1 - \frac{\omega^2}{\omega_L^2}\right) R'\right]$$
$$+ \left[\left(\frac{\omega_0^2}{\omega_L^2} - \frac{\omega^2}{\omega_L^2}\right) + \frac{\lambda\omega_0^4}{\omega_L^4} - \frac{\lambda\omega_0^4}{\omega_L^4}\left(1 - \frac{\omega^2}{\omega_L^2}\right) R\right]\left[\left(1 - \frac{\omega^2}{\omega_L^2}\right) I' - I\right] \tag{4.7.15}$$

and

$$Q = \left[\frac{\omega_0^2}{\omega_L^2} - \frac{\omega^2}{\omega_L^2} + \frac{\lambda\omega_0^4}{\omega_L^4} - \frac{\lambda\omega_0^4}{\omega_L^4} R\left(1 - \frac{\omega^2}{\omega_L^2}\right)\right]^2$$
$$+ \frac{\lambda^2\omega_0^8}{\omega_L^8}\left(1 - \frac{\omega^2}{\omega_L^2}\right)^2 I^2 \tag{4.7.16}$$

If $\omega_L^2 \to \infty$ in equation (4.7.13), one recovers the previous result for H interacting with one Pt atom, because $(R + iI)/\omega_L^2$ remains finite while terms such as ω^2/ω_L^2 and ω_0^4/ω_L^2 tend to zero.

4.7.3. Interpretation of Neutron Intensity for Hydrogen on Platinum

Having discussed rather fully the consequences of the model of Mahanty *et al.*, the final step is to return to the interpretation of the experiment of Wright *et al.*[149] for H on Pt.

Mahanty *et al.* conclude that they are observing the local model associated with H coupled to Pt, at 50 meV. It is therefore natural to interpret their peaks at 9 meV and 19 meV as a side band. Indeed, the positions of these peaks do correlate with the prominent peaks of the matrix vibrational spectrum.

Furthermore, from the above discussion, the substantial intensity observed in the side band relative to the local mode suggests, by comparison with the model calculations, that more than one Pt atom is interacting strongly with hydrogen.

There remain two significant discrepancies between the predictions of the theoretical model and the measured intensity of Wright *et al.*:

1. Their peaks at 9, 19, and 50 meV are very broad, compared with the much more spiky structure of the model.
2. Their neutron measurements indicate a change in the density of states that is always positive, while the model predicts regions in which the density of states is decreased.

These points will now be discussed in turn.

4.7.3.1. Broadening of Peaks in Observed Neutron Intensity

The experiment was carried out on a powdered specimen, but the model calculations are for a specific surface. Because a powder was used, one expects a distribution of local-mode frequencies, due to hydrogen in somewhat different environments on the different surfaces involved. This, together with instrumental resolution, seem to be the prime reason for broadening of the observed local mode.

Similarly, for the principal peaks in the side band, one must again average over different faces of the face-centered cubic structure, and one expects substantial broadening.

4.7.3.2. Effects of Long-Range Forces

With regard to the regions of decreased density of states given by the present model, it should be borne in mind that the interionic potential in metallic Pt is long-range,[151] while the model calculation was based on near-neighbor forces.

4.8. THEORY OF SURFACE DIFFUSION

Thomas[155] has considered the specific example of the motion of an atom physisorbed on a (100) surface of a simple cubic solid of static atoms. He employed numerical integration of its classical equations of motion. It is found that over short times the atom moves as a free particle, but that over long times it behaves like a diffusing particle executing a random walk. Thomas demonstrates that his numerical results can be fitted to a "random" walk, which has some correlation between successive steps.

This work is relevant here because one of the important processes in the growth of surfaces, and in catalytic reactions treated in Chapter 6, is the motion of atoms adsorbed on surfaces. Prior to Thomas' study, theoretical work had concentrated on the related problem of the scattering of incident atoms by surfaces. Due to the difficulty of performing full formal calculations of atom–surface interactions, computer simulation of the problem has been employed. Thomas' work extends these calculations to the motion of atoms along surfaces. Unlike the scattering situation, the motion of the atom can only be described if the long-range periodic influence of the surface potential is taken into account. In the work of Thomas, the surface atoms are assumed fixed at their equilibrium positions. His calculations are based on classical dynamics with a two-body central force interaction. This gives a reasonable description of the interaction of many physisorbed atoms with crystalline solids.

Without going into details, for which the interested reader must consult the original paper, the computer results are nicely fitted by the following analytical expression for the mean-square distance $\overline{R^2(t)}$ as a function of time t:

$$\frac{\overline{R^2(t)}}{t} = \frac{2k}{\lambda}\left\{1 - \frac{1}{\lambda k t}\left[1 - \lambda \exp(-\lambda k t)\right]\right\} \tag{4.8.1}$$

When $t \ll 1/\lambda k$, this equation reduces to the free-particle behavior $\overline{R^2(t)}/t \simeq k^2 t$, and when $t \gg 1/\lambda k$, $\overline{R^2(t)}/t \simeq 2k/\lambda$. Thus small and very large t behavior as found in the computer experiment is subsumed into equation (4.8.1). Thomas[155] gives a small table of λ and k data for three different values of the energy E of the adsorbed atom starting from a position near the origin. In these computations, the adsorbed atom is assumed to interact via a Lennard-Jones 6-12 potential with each of the surface atoms, so the detailed results should not perhaps be overstressed in the context of metal surfaces. (For later work, see Notes in Proof, p. 262.)

4.9. VIBRATION EXCITATION IN MOLECULE–SURFACE COLLISIONS DUE TO TEMPORARY NEGATIVE MOLECULAR-ION FORMATION

With regard to inelastic electron scattering, Gadzuk[156] has considered vibrational excitation in molecule–surface collisions due to temporary negative molecular-ion formation. He draws particular attention to some interesting features associated with vibrational excitation of diatomic molecules when they form temporary negative ion resonance states. He cites as an example the N_2^- shape resonance seen in inelastic electron scattering from gaseous and condensed states. When a free-space N_2 molecule, initially in its ground state (i.e., vibrational ground state), is hit by an electron with about 2 eV kinetic energy, there is a significant probability that the electron will be trapped within the molecule, forming a temporary negative molecular ion.

Gadzuk[(156)] proposes that, due to the image-potential lowering of the electron-affinity level of a diatomic molecule in interaction with a metal surface, somewhere outside the surface an incident molecule would find its affinity level degenerate with or lower than the substrate Fermi level, at which point a substrate electron could hop onto the molecule. A negative molecular ion is thereby formed and remains until the molecular ion reflects from the surface and the affinity level rises above the Fermi level, thus permitting reverse electron-hopping back into the metal.

The lifetime of the molecular ion can be controlled by varying both the kinetic energy of the incident molecule and also the substrate work function. By analogy with the electron-scattering events, greatly enhanced vibrational excitation of overtones is expected in the molecules of the scattered beam. Induced fluorescence probing of the vibrational state distribution should then yield useful information about the dynamics of charge-transfer reactions and nonadiabatic effects in molecule–surface interactions. For a theory of this phenomenon the reader is referred to Gadzuk's paper,[(156)] where the numerical consequences for a model system designed to simulate N_2 or NO scattering from metal surfaces are also presented.

4.10. ADPARTICLE DYNAMICS: KRAMERS' EQUATION IN A METALLIC MEDIUM

The primary concern in this section will be with the dynamics of an adatom in a metallic medium. Nevertheless, it will be helpful first to consider the range of validity of a Fokker–Planck equation in a classical medium bearing in mind subsequent extension to a degenerate electron system.

Some insight can be gained by comparing a derivation based on the phenomenological Langevin equation with an *ab initio* expansion in powers of the square root of the ratio of medium mass to Brownian mass, $\mu = (m/M)^{1/2}$ (cf. Lebowitz and Rubin[(157)]). A simple extension of both treatments to an inhomogeneous medium is possible, the Langevin equation, with $\mathbf{r}$ and $\mathbf{v}$ adatom position and velocity, being

$$M\frac{d\mathbf{v}}{dt} = -\frac{\partial V_s(\mathbf{r})}{\partial \mathbf{r}} - \boldsymbol{\eta}(\mathbf{r})\mathbf{v} + \mathscr{F}(\mathbf{r}, t) \tag{4.10.1}$$

where $V_s(\mathbf{r})$ is the surface potential acting on the adatom of mass M, $\boldsymbol{\eta}(\mathbf{r})$ is a friction tensor experienced by the adatom moving through the electron spillout region (generalizing the friction constant η), while $\mathscr{F}(\mathbf{r}, t)$ is the force exerted by the medium particles (eventually including specifically electrons) as a result of equilibrium fluctuations of the adatom. In a fully realistic treatment of the metal-surface problem, $V_s(\mathbf{r})$ will, of course, have to incorporate contributions from the lattice, while the phonons will be subsumed into $\boldsymbol{\eta}(\mathbf{r})$ and $\mathscr{F}$.

Although the Langevin equation (4.10.1) is entirely plausible on physical grounds, it should be noted that it may also be obtained rigorously to lowest order in μ by extending the arguments of Lebowitz and Rubin to the inhomogeneous case and the Langevin representation. The advantage gained by working with the Langevin equation rather than the Fokker–Planck equation for the phase-space density of adatoms in this case is that it allows one to expose the additional assumptions about the strength of the fluctuations in $\mathscr{F}$ and of the potential gradient $\partial V_s/\partial \mathbf{r}$ made in an analysis such as that in the work of Lebowitz and Rubin, apart from $\mu \ll 1$ (see also Appendix 4.2.)

In order to derive from equation (4.10.1) the Fokker–Planck equation for the phase-space density function $f(\mathbf{r}, \mathbf{v}, t)$ of the adatom, namely

$$\frac{\partial f(\mathbf{r}, \mathbf{v}, t)}{\partial t} - \mathbf{v}\cdot\frac{\partial f}{\partial \mathbf{r}} - M^{-1}\frac{\partial V_s}{\partial \mathbf{r}}\cdot\frac{\partial f}{\partial \mathbf{v}} = \frac{\partial}{\partial \mathbf{v}}\mathbf{D}(\mathbf{r})\left(\frac{M}{k_B T}\mathbf{v} + \frac{\partial}{\partial \mathbf{v}}\right) f \tag{4.10.2}$$

and

$$\mathbf{D}(\mathbf{r}) = \frac{1}{m^2}\int_0^\infty d\tau \langle FF(-\tau)\rangle = \frac{k_B T}{m}\boldsymbol{\eta}(\mathbf{r}) \tag{4.10.3}$$

one requires only two properties of the medium (determining the statistical properties of the fluctuating force $\mathscr{F}$): (1) it drives the adatom toward thermal equilibrium, and (2) the velocity and position of the adatom are constant over the correlation time, τ_F say, of the force $\mathscr{F}$. The implication of the latter requirement is that the potential $V_s(\mathbf{r})$ and also $\boldsymbol{\eta}(\mathbf{r})$ do not change appreciably over a characteristic distance.

It was quite clear from the pioneering work of Kramers that stochastic equations, such as those discussed above, could be used to treat the dynamics of adatoms near surfaces. He was concerned primarily with the rate constants of chemical reactions in viscous media (see also Chapters 5 and 6). Anticipating what is to be discussed more fully later, his basic idea was to treat the chemical reaction as a Brownian-motion process in the presence of a potential barrier. He obtained an approximate solution to the Smoluchowski equation and derived the following expression for the rate constant k:

$$k = (A/\eta_s)\exp(-E_b/k_B T) \tag{4.10.4}$$

where A is a constant, usually determined empirically, η_s is the solvent viscosity, while E_b is an effective potential barrier height.

While the so-called Kramers' rule (4.10.4) has been confirmed experimentally for reactions in polymer solutions[158] as well as for enzyme-catalyzed reactions,[159] and also has been generalized experimentally[160] and theoretically[161] to treat ligand migration in proteins, the main interest here is in the application of Kramers' equation to the treatment of the dynamics of adatoms near metal surfaces. Contributions in this latter area have been made by Suhl and co-workers, and by Schaich.

There is a necessity to pay attention to the role of itinerant electron contributions to adatom dynamics. Therefore, one should note here that Kramers' description of adatoms[162,163] on metals requires a treatment of the friction constant $\eta(\mathbf{r})$ in the inhomogeneous electron gas at a planar metal surface. This can be derived using inverse transport theory, also invoked in the pioneering work of Suhl and his colleagues[164,165] on the electronic contribution to the friction constant. After order-of-magnitude estimates of the electronic contribution to the friction constant in a bulk electron gas, results of an approximate evaluation of $\eta(\mathbf{r})$ in simple metals will be summarized, the details being recorded in Appendix 4.4. In order to carry out calculations of $\eta(\mathbf{r})$ analytically, the following fairly drastic simplifications are called for:

1. The density matrix of the inhomogeneous electron gas at the metal surface is replaced by that of the infinite-barrier model, set out in Appendix 2.1.
2. The screened potential around an adatom in the presence of the surface has had to be approximated, as explained in Appendix 4.4.

4.10.1. Electronic Contribution to the Surface-Friction Constant

One has in mind specifically atomic and ionic motion in metals. This motion might well be considered in either solid or liquid phases, though the main applications in this book are to solids. Naturally, such motions are never entirely free but may be described by the introduction of a potential energy, $U(\mathbf{r})$ say, and a frictional drag, denoted by $\eta(\mathbf{r})$ above, at least as a first step toward including dynamic interactions.

In Appendix 4.4, some consideration is given to the validity of Kramers' equation for single-atom (or ion) motion in a metal. In addition, the interested reader is referred to the work of Brenig,[166] who discusses more general kinetic equations, of Brako and Newns,[167] and of Brivio and Grimley,[168] where quantization of the adatom motion in weakly coupled modes is important. In the present context, it will merely be noted that the source of the potential energy $U(\mathbf{r})$ for atomic motion also provides an electronic potential $V(\mathbf{r})$ in addition to that of a single atom in the electron gas. This has consequences for (1) finite concentration effects in the electrical resistivity of metals, (2) configurational dependence in the phenomenon of electromigration, and (3) cooperative and hydrodynamic interactions in atomic motions.

What is of prime concern here then is the evaluation of the frictional drag term entering Kramers' equation (4.10.2). The frictional force $\mathcal{F}$ is written proportional to the velocity $\mathbf{v}$ in the form

$$\mathcal{F} = -\eta \mathbf{v} \tag{4.10.5}$$

and, as recognized by Suhl *et al.*,[164] the frictional constant is given in terms of the screened potential $V_{sc}(\mathbf{r})$ created by a charge in the electron gas. By using inverse transport theory[169–171] (see equation (4.10.7) below and also Appendix 6.2), with the Dirac density matrix replaced by the free-electron density matrix, one obtains the bulk result

$$\eta = \frac{m^2}{12\hbar^3\pi^2}\int_0^{2k_f} dk\, k^3 |\tilde{V}_{sc}(k)|^2 \tag{4.10.6}$$

This is the Born approximation to scattering from a weak screened potential energy $V_{sc}(\mathbf{r})$, its Fourier transform being denoted by $\tilde{V}_{sc}(k)$.

4.10.2. Introduction of Inhomogeneity at a Metal Surface

The inverse transport theory yields[169,171] the frictional term

$$\eta = 2\pi\hbar \int d\mathbf{r}_1 d\mathbf{r}_2 \frac{\partial V(\mathbf{r}_1)}{\partial \mathbf{r}_1}\frac{\partial V(\mathbf{r}_2)}{\partial \mathbf{r}_2} |\sigma(\mathbf{r}_1\mathbf{r}_2 E_f)|^2 \tag{4.10.7}$$

where σ is the energy derivative of the Dirac density matrix. One must be careful in establishing a suitable screened potential to insert in this equation, if the independent electron assumption involved in its derivation is to give meaningful results. In Appendix 4.4 therefore, there are developed two models providing inhomogeneous surface screened potentials for adatoms near the surface, following McCaskill *et al.*[172] The first of these gives good results well inside the surface where gradient corrections are small, and the second is reliable well outside the metal surface.

4.10.2.1. Order of Magnitude of the Friction Constant

As a preliminary to treating the inhomogeneity in the metal electron density near a planar surface, the electronic contribution to the frictional drag on an adatom, whose dynamics involves activated diffusion over potential barriers (provided by the surface lattice structure of the metal host), may be evaluated from the friction constant for the homogeneous electron gas. For a singly charged screened ion, this has a value 2×10^{-15} $\mathrm{kg\,s^{-1}}$ or about one proton mass per picosecond. (A hydrogen atom in uniform motion at a room-temperature thermal velocity would be stopped in a nanosecond over a micron scale if reexcitation did not occur.)

When the adatoms are constrained by the lattice structure of the surface of the metal to diffuse between different sites by activation over a barrier of height E_b, the significance of the friction constant requires further consideration. According to Kramers, the transition-state rate equation

$$r = \omega \exp(-E_b/k_B T) \tag{4.10.8}$$

where ω is the vibrational frequency of the potential well, applies independently of any friction constant whenever $\eta/M \ll \omega$, where M is the atomic mass. In fact, if the barrier heights are of the order of room-temperature thermal energy or less, this inequality is not well satisfied for first-row adatoms. The frictional behavior then modifies the hopping diffusion rate (see also Appendix 4.1).

The remainder of this section is concerned with a more detailed evaluation of the friction constant in the inhomogeneous electron gas.

4.10.2.2. Friction-Constant Evaluation with Screened Potentials

Apart from constants, the inverse transport expression evaluated by McCaskill and March[172] for the friction η on adatom motion is [cf. equation (4.10.7)]

$$\eta = C \int \frac{\partial V}{\partial \mathbf{r}_1} \cdot \frac{\partial V}{\partial \mathbf{r}_2} |\sigma_{IB}(\mathbf{r}_1\mathbf{r}_2E_f)|^2 \, d\mathbf{r}_1 d\mathbf{r}_2 \tag{4.10.9}$$

where σ_{IB} is the energy derivative of the density matrix for the infinite barrier. This is directly related to the retarded Green function G^- yielding, with the constant C normalized to $4m^2/h^3$,

$$\sigma_{IB} = \partial\rho/\partial E = (1/\pi)\mathrm{Im}\, G^-$$
$$= \frac{\sin k_f|\mathbf{r}_1 - \mathbf{r}_2|}{|\mathbf{r}_1 - \mathbf{r}_2|} - \frac{\sin k_f|\mathbf{r}_1 - \mathbf{r}_2^{im}|}{|\mathbf{r}_1 - \mathbf{r}_2^{im}|} \qquad (4.10.10)$$

where the last line follows from the infinite-barrier boundary condition, which provides an image-theory expression for the Green function. Here

$$\mathbf{r}_2^{im} = \mathbf{r}_2 - 2z_2\hat{\mathbf{z}} \qquad (4.10.11)$$

where $\hat{\mathbf{z}}$ is the unit vector perpendicular to the surface.

Two possible methods of tackling this equation with a general potential are now developed following McCaskill and March.[172] The first of these is particularly suited to potentials expressed in spherical harmonics. The second exploits the cylindrical symmetry about $\hat{\mathbf{z}}$.

Let us first consider the interference term in equation (4.10.9):

$$\eta_{int} = 2C \int \frac{\partial V}{\partial \mathbf{r}_1} \cdot \frac{\partial V}{\partial \mathbf{r}_2} |\sigma(\mathbf{r}_1\mathbf{r}_2^{im}k_f)\sigma(r_2 r_1 k_f)|\, d\mathbf{r}_1 d\mathbf{r}_2, \qquad z_1, z_2 < 0 \qquad (4.10.12)$$

The aim is then to expand $\sigma(\mathbf{r}_1\mathbf{r}_2^{im}k_f)$ in terms of products of functions of $r_{12} = |\mathbf{r}_1 - \mathbf{r}_2|$, z_2, and $\cos\theta_{12}$, where θ_{12} is the angle $\mathbf{r}_1 - \mathbf{r}_2$ makes with the z-axis. The other terms are readily handled by Fourier transforms, as in the homogeneous problem. The interference term reduces to

$$\eta_{int} = \frac{2C}{8\pi} \sum_{l=0}^{\infty} (2l+1) \int_0^{2k_f} k\,dk \int d\Omega_{\mathbf{k}}\, i\mathbf{k}\, \tilde{V}_{l<}(\mathbf{k})\mathbf{V}'_{l<}(\mathbf{k}) P_l(k/2k_f) P_l(\cos\theta_k) \qquad (4.10.13)$$

where the $P_l(x)$ are Legendre polynomials and the angular integration $\int d\Omega_{\mathbf{k}}$ and l summation represent the essential generalization of the weak-scattering theory in a bulk electron gas to the inhomogeneous case. The Fourier transform of the potential gradient is decomposed into harmonics:

$$\mathbf{V}'_{l<}(\mathbf{k}) = \int_{z<0} \exp(-i\mathbf{k}\cdot\mathbf{r}) j_l(2k_f z) \frac{\partial V}{\partial \mathbf{r}}\, d\mathbf{r} \qquad (4.10.14)$$

The interference term must be added to the direct and image terms:

$$\eta_{dir} = C \int_{z_1, z_2 < 0} \frac{\partial V}{\partial \mathbf{r}_1} \frac{\partial V}{\partial \mathbf{r}_2} |\sigma_0(\mathbf{r}_1\mathbf{r}_2 k_f)|^2\, d\mathbf{r}_1 d\mathbf{r}_2$$
$$= \frac{C}{8\pi} \int_0^{2k_f} k\,|\tilde{V}_<(\mathbf{k})|^2 \hat{\mathbf{k}}\hat{\mathbf{k}}\, d\mathbf{k} \qquad (4.10.15)$$

$$\eta_{\text{im}} = C\int_{z_1,z_2<0} \frac{\partial V}{\partial \mathbf{r}_1}\cdot\frac{\partial V}{\partial \mathbf{r}_2}|\sigma_0(\mathbf{r}_1\mathbf{r}_2^{\text{im}}k_{\text{f}})|^2\,d\mathbf{r}_1 d\mathbf{r}_2$$

$$= \frac{C}{8\pi}\int_0^{2k_{\text{f}}} d\mathbf{k}kV_<(-\mathbf{k}^{\text{im}})V_<(k)\hat{\mathbf{k}}^{\text{im}}\hat{\mathbf{k}} \tag{4.10.16}$$

where $\mathbf{k}^{\text{im}} = \mathbf{k} - 2k_z\hat{\mathbf{z}}$.

As an example of calculations up to this point, McCaskill and March demonstrate that the above theory leads to the η_{xx}-component for the specific case of a delta-function potential, evaluated directly by Kumar and Sorbello.[173]

It is naturally of some importance to assess the reliability of this result, in which the potential is not self-consistently screened, by evaluating the influence of the nonspherical potential correction, which arises from the inhomogeneous screening equation (A4.4.1) for $k \ll q$, this being proportional to $(\partial/\partial z)\delta(\mathbf{r})$. The details will be omitted, but the results may be summarized as follows. The homogeneous screening term is found to be

$$\eta_{xx}^{(0)}(z_0') = C'\left(1 - \frac{\sin z_0'}{z_0'}\right)\left(1 + \frac{3\cos z_0'}{z_0'^2} - \frac{3\sin z_0'}{z_0'^3}\right) \tag{4.10.17}$$

and the linear inhomogeneous screening correction is

$$\eta_{xx}^{(1)}(z_0') = -C'\frac{48\pi Ze^2}{q_0^3}k_{\text{f}}^2\left(\frac{\cos z_0'}{z_0'} - \frac{\sin z_0'}{z_0'^2}\right)\left(\frac{1}{3} + \frac{1+\cos z_0'}{z_0'^2} - \frac{2\sin z_0'}{z_0'^3}\right) \tag{4.10.18}$$

where $z_0' = 2k_{\text{f}}z_0$, q_0^{-1} is the bulk screening length, and

$$C' = \tfrac{2}{3}C\lambda^2k_f^4, \qquad \lambda = 4\pi Ze^2/q_0^2 \tag{4.10.19}$$

Hence though of similar magnitude to equation (4.10.17) near the surface, the correction to the friction constant from the inhomogeneous screening becomes less significant as the adatom moves into the bulk metal, falling off as z_0^{-2}, i.e., faster than the homogeneous screening surface-friction constant, which decays as z_0^{-1}.

4.10.2.3. Charge Outside a Metal Surface

The approximations to the inhomogeneous screening in the second model of McCaskill and March,[172] that employing classical image theory as a boundary condition, enable one to discuss the problem of the friction constant for a charge outside the surface.

Because one has basically subsumed the electron density surface profile into a step function, it is consistent to apply the procedure to the density matrix as well:

$$\sigma_{\mathrm{IB}} \rightarrow \sigma_0(\mathbf{r}_1\mathbf{r}_2E_{\mathrm{f}})\theta_-(z) \tag{4.10.20}$$

in which case calculation of the friction constant with the cylindrically symmetric potential of equation (A4.4.5) becomes tractable.

In particular, the first four terms in the asymptotic expansion for large distances z_0 from the surface, for the friction constant for motion parallel to the surface, are

$$\eta_{xx}(z_0) = \frac{C(\pi Ze^2)^2}{2(2q_0z_0)^4}\left[6\ln(2q_0z_0) - C_1 - \frac{90\ln(2q_0z_0)}{(2q_0z_0)^2} - \frac{C_2}{(2q_0z_0)^2} + \cdots\right] \tag{4.10.21}$$

where C_1 and C_2 depend on the Fermi energy but are approximately 8 and $40 + 15q_0^2/k_{\mathrm{f}}^2$, respectively. Thus q_0^{-1}, the bulk screening length, determines the length scale for the range of the electronic contribution to the friction constant, which falls off as $(\ln x)/x^4$ when the adatom moves further away from the surface.

The two model calculations discussed above for adatom motion in and outside the surface electron density profile display complementary features. It is clear that vanishing of the friction constant as the adatom approaches the infinite barrier from inside it is a peculiarity of the localized potential of the first model. On the other hand, the asymptotic result for adatoms well inside the surface density profile seems independent of the detailed approximations made and is unaffected by the inhomogeneous screening discussed in the first model. Likewise, matching of an image-theory potential at the infinite barrier with bulk screening into the metal, in the second model, while a doubtful approximation when the adatom is very close, nevertheless can be anticipated to predict the general asymptotic behavior.

This leads to the final topic of this section, the range of dynamic interaction between adatoms.

4.10.3. Long-Range Dynamic Interaction Between Adatoms

In Chapter 2, considerable attention was paid to the range of the indirect interaction between static adatoms outside planar metal surfaces. One important conclusion was that, while the bulk interaction between a pair of impurities at separation a in a degenerate electron assembly with Fermi wave number k_{f} fell off asymptotically at large a as $(\cos 2k_{\mathrm{f}}a)/a^3$, for the adatom pair parallel to a planar metal surface the interaction was of substantially shorter range, decaying with distance as $(\cos 2k_{\mathrm{f}}a)/a^5$. This, of course, has consequences for the structure of adsorbed atoms on metal surfaces, such as H on Pt* discussed in Section 4.6.2.

McCaskill and March[172] have discussed, in addition, the range of dynamic interactions associated with adatom motion and their main findings will be summarized below.

What these workers do is to study the excess friction constant for motion of an adatom pair parallel to the surface when the pair is aligned in the following three different directions with respect to this motion and the surface:

1. The pair is parallel to the surface and motion is perpendicular to the pair axis.
2. The pair is parallel to the surface and motion is along the pair axis.
3. The pair is perpendicular to the surface.

The excess friction constant calculated by McCaskill and March is the difference between the friction constant of the pair and the sum of those of the individual atoms.

The argument below relies on localized potentials, which obey linear superposition, plus use of the unperturbed density matrix which, as noted earlier, constitutes a Born approximation to the scattering. Thus the localized potential due to the proximity of two adatoms is chosen as

$$V = \lambda\delta(\mathbf{r}-\mathbf{r}_0) + \lambda\delta(\mathbf{r}-\mathbf{r}_0') \tag{4.10.22}$$

while the density matrix is given by the image equation (4.10.10). The interaction term in the friction constant is

$$\eta_{\text{int}}(\mathbf{r}_0\mathbf{r}_0') = 2\lambda^2 C \frac{\partial}{\partial \mathbf{r}_1}\frac{\partial}{\partial \mathbf{r}_2}(\sigma_{\text{IB}})^2\Big|_{\substack{\mathbf{r}_1=\mathbf{r}_0\\ \mathbf{r}_2=\mathbf{r}_0'}} \tag{4.10.23}$$

and the three possible configurations considered are:

$$\left.\begin{array}{ll}\text{(i)} & \mathbf{r}_0 = (0, y_0, z_0),\ \mathbf{r}_0' = (0, -y_0, z_0)\\ \text{(ii)} & \mathbf{r}_0 = (x_0, 0, z_0),\ \mathbf{r}_0' = (-x_0, 0, z_0)\\ \text{(iii)} & \mathbf{r}_0 = (0, 0, z_0),\ \mathbf{r}_0' = (0, 0, z_0')\end{array}\right\} \tag{4.10.24}$$

for which η_{xx} is of prime interest. After some detailed manipulation, with distances now being measured in units of $(2k_f)^{-1}$ and setting $C'' = 4Ck_f^4$, McCaskill and March[172] find

$$\eta_{(i)} = C''\lambda^2[j_0(a) - j_0(\sqrt{a^2+z^2})][j_1(a)/a - j_1(\sqrt{a^2+z^2})/\sqrt{a^2+z^2}] \tag{4.10.25}$$

* See the review of R. Burch: Notes added in Proof, p. 261.

$$\eta_{(ii)} = C''\lambda^2\left[\left(j_1(a) - \frac{q}{\sqrt{a^2+z^2}}\, j_1(\sqrt{a^2+z^2})\right)^2 + j_0(a) - j_0(\sqrt{a^2+z^2})\right]$$

$$\times\left[4\left(j_1(a)/a - \frac{j_1(\sqrt{a^2+z^2})}{\sqrt{a^2+z^2}}\right) + j_0(a) - j_0(\sqrt{a^2+z^2})\right] \quad (4.10.26)$$

$$\eta_{(iii)} = C''\lambda^2[j_0(a) - j_0(2z)][j_1(a)/a - j_1(2z)/2z] \quad (4.10.27)$$

We note that the bulk behavior differs in the two configurations (i) and (iii) of alignment relative to motion,

$$\eta_{(i)\,\mathrm{bulk}} \sim (\sin 2a)/a^3 \quad (4.10.28)$$

$$\eta_{(ii)\,\mathrm{bulk}} \sim (1 - \cos 2a)/a^2 \quad (4.10.29)$$

and that at the surface the asymptotic behavior is ($a \to \infty$, $z \to 0$)

$$\eta_{(i)} \sim z^4(\sin 2a)/a^5 \quad (4.10.30)$$

$$\eta_{(ii)} \sim z^4(1 + \cos 2a)/a^4 \quad (4.10.31)$$

which confirms a previous conjecture that the leading term behaves as a^{-4} but has the Friedel oscillation of twice the period of the static interaction terms. With one adatom near the surface, one obtains in the perpendicular configuration (iii),

$$\eta_{(iii)} \sim z_s^2(\sin 2a)/a^2 \quad (4.10.32)$$

where z_s is the distance from the surface of the atom situated closer to it.

We conclude this discussion of Kramers' equation, and the friction coefficient, by referring to the work of Jack and Kreuzer.[174] They start from a master equation with transition probabilities calculated quantum mechanically for one-phonon cascades and derive via a Kramers–Moyal expansion the classical equation of Kramers and also van Kampen's[175] linear Fokker–Planck equation for physisorption. They show further that the friction coefficient calculated in terms of the microscopic parameters agrees with the result from a classical model. The macroscopic laws of friction for an adsorbing or desorbing gas particle are derived and solved, establishing time scales for the intermediate quasi-equilibrium on and for the overall adsorption and desorption processes.

Chapter 5

Molecular Desorption

In this chapter, the processes underlying molecular desorption from metal surfaces will be discussed in some detail. Three areas will be dealt with:

1. Thermal desorption.
2. Flash desorption.
3. Electronically induced desorption.

The principal aim in this chapter, and throughout the book, is to treat chemical bonds, not atoms, in interaction with metal surfaces. Such an approach proves possible in this chapter to a considerable extent, in particular in statistical treatments of (1) and to a lesser degree (2). However, when one reaches the level of the microscopic theory of (1), it proves necessary at present to simplify and treat simple closed-shell atoms. This is true in Section 5.4, where a microscopic theory of physisorption is given.

Another general comment is that, while substantial theory is presented for (1) especially, for (3) the development of the theory at the time of writing has been mainly concerned with the interpretation of the experimental results for photon- and electron-induced desorption. Though, in principle, one stands to learn a lot about the adsorbate–surface bond from studies in the area of (3), this has not yet quite developed to this point. However, since the conformational studies reported in Chapter 3 were, at least partly, motivated by the ESDIAD technique of studying, say, the HOH angle for water on Ru, it was obviously important to discuss this technique at a first-principles level, insofar as is presently feasible.

5.1. KRAMERS' THEORY GENERALIZED TO DESORPTION FROM SOLID SURFACES

The initial part of this chapter deals with the first of the basic processes (1)–(3) listed above, namely thermal desorption of adsorbed species from solid surfaces. Accurate experimental techniques have been developed and include molecular beam relaxation spectroscopy (MBRS),[176] temperature-programmed desorption (TPD),[177] LEED,[178] AES,[179] and thermal desorption mass spectrometry (see, for example, Schmidt[180]). This development has provided a considerable literature related to adsorption–desorption experiments for a large variety of systems. The need for a basic understanding of the nature of such processes has stimulated the proposal of many theoretical methods and models.[181–187]

The most commonly used framework for the discussion of thermal desorption is a statistical thermodynamic approach based on transition-state theory,[181,182] presented in Section 5.2 below. Although this method accounts correctly for the observed magnitude of atomic desorption rates, quantitative predictions are difficult since it requires knowledge of the transition state of the desorbing species. This state is not well defined, leading to ambiguities in the resulting description of the system. These problems are particularly serious for molecules, the main topic of interest here, where various choices of parameters can lead to a change of several orders of magnitude in the desorption rates. Although transition-state theory does not provide a prediction of specific desorption rates, it does afford a justification of the Arrhenius form

$$R = A \exp(-E_d/k_B T) \tag{5.1.1}$$

usually employed in characterizing experimental desorption rates in the limit of low coverage, i.e., in the regime of negligible interaction between the adparticles.

A number of authors[183–185] have combined transition-state theory with stochastic trajectory calculations for the evaluation of desorption rates. The theories which employ stochastic trajectory calculations are computationally involved and hence difficult to apply directly to time scales longer than 10^{-10}–10^{-11} s. This prevents their use for desorption, where time scales of 1 s may be involved.

There are also microscopic theories[186,187] in which explicit coupling between the adparticle and the surface phonon modes is introduced to calculate the energy flow between the surface and the adparticle. At present, there remain discrepancies between experimental and calculated desorption rates for physisorbed atoms. Moreover, there appear to be difficulties in applying these approaches to molecular desorption.

The experimental and theoretical studies undertaken over a few decades have deepened our understanding of the desorption mechanisms. They have, however, simultaneously raised some additional basic questions (see the work of Zeiri *et al.*[188] which is followed quite closely in Section 5.1). One puzzling result pertains to the measurements[176,189–191] of both A and E_d in equation (5.1.1) for the desorption of CO from different metal surfaces. Although the observed values for the preexponential factor (in the limit of low coverage) vary over three orders of magnitude, all the experimental results indicate that the values of A are at least two orders of magnitude larger than the corresponding values observed for atomic systems ($A \sim 10^{13}\ s^{-1}$). In addition, several experiments lead to different functional relationships between A and E_d and the surface coverage.

A puzzling theoretical result is due to Grimmelmann *et al.*[184] and Adams and Doll,[192] who calculated the rate of desorption as a function of surface temperature for Ar and Xe on Pt(111) and He on Xe(111), respectively. These calculations showed the existence of two temperature regimes for desorption, in which different values of A and E_d were observed. They found that the preexponential factor and the activation energy in the low-temperature range are larger than the corresponding values for the high-temperature regime. It is difficult to obtain a clear explanation of this effect from within the theory.

These examples illustrate that the detailed processes governing desorption rates are not yet fully understood. One of the basic reasons is that one needs explicit expressions for the rate of desorption in terms of the microscopic properties, such as vibrational frequencies and bond energy, and dynamics of the adsorbate–surface system. Moreover, with the exception of transition-state theory,[181] to be considered in some detail in Section 5.2, most existing treatments have been applied only to the case of atomic desorption.

Zeiri *et al.*[188] have developed a theoretical description for desorption that yields a simple rate expression given in terms of such microscopic properties as mentioned above. The rate of desorption takes the form

$$R = \frac{\Omega_0}{2\pi} f(T) \exp(-D_e/k_B T) \tag{5.1.2}$$

where D_e is the bond enthalpy and Ω_0 is the surface–adsorbate vibrational frequency. For atoms the factor $f(T)$ is unity, but for molecules $f(T)$ depends on the parameters for the frustrated rotations at the surface. This factor for molecules can be 10^2 to 10^3, leading to the much larger desorption rates observed in molecules.

The rate expression (5.1.2) is derived below by extending the treatment of Kramers[162] (given in Chapter 4) in a manner appropriate for desorption. Results obtained in this way will then be summarized for the desorption of atoms and molecules from solid surfaces. Comparison with available experimental data is excellent.

5.1.1. Derivation of the Rate Equation

The starting point is consideration of the relative motion of the adparticle with respect to the surface atoms. In doing so, the ideas introduced by Adelman and Doll[193–195] are embodied. The motion of a few surface atoms, which interact strongly with the adsorbed particle, is considered while the rest of the crystal is assumed to act as a heat bath. Without loss of generality, one can consider the motion of only one surface atom.

First, the adparticle is assumed to be an atom and the results will then be generalized to molecules. A further simplification is introduced by assuming a one-dimensional system in which both the adatom and surface atom are restricted to move in a direction normal to the surface. Thus the motion of the adatom will be described by

$$m\ddot{z} = -\frac{\partial V(z - \zeta)}{\partial z} \tag{5.1.3}$$

where z and ζ are the positions of the adatom and surface atom, respectively, m is the mass of the adatom, and $V(z - \zeta)$ is the interaction potential between the adatom and surface.

The motion of the surface atom will be described by a generalized Langevin equation, following Adelman and Doll[193,194]:

$$m_s\ddot{\zeta} = -\frac{\partial V(z - \zeta)}{\partial \zeta} - m_s\omega_s^2\zeta - m_s\int_0^t \theta(t - t')\zeta(t')dt' + \tilde{f}(t) \tag{5.1.4}$$

where m_s represents the mass of the surface atom, ω_s is the characteristic frequency of the solid, and $\theta(t - t')$ and $\tilde{f}(t)$ respectively correspond to a memory kernel and random force, which incorporate the influence of the heat bath on the motion of the surface atom. These functions $\theta(t)$ and $\tilde{f}(t)$ are related by the second fluctuation–dissipation theorem, as discussed by Adelman and Doll.[193,194]

If the memory kernel in equation (5.1.4) is integrated by parts and the Markovian limit taken, following Adelman and Doll[193,194,196] one obtains a Langevin equation of motion for the surface atom in the form

$$m_s\ddot{\zeta} = -\frac{\partial V(z - \zeta)}{\partial \zeta} - m_s\Omega_s^2\zeta - m_s\eta_s\dot{\zeta} + f(t) \tag{5.1.5}$$

where Ω_s is the effective frequency, η_s is the friction constant, and $f(t)$ is a random force with a Gaussian distribution, following Kramers,[162] Chandrasekhar,[163] and also Adelman and Doll.[193,194] The interaction potential $V(z-\zeta)$ between the adatom and surface is frequently described in terms of Morse or Lennard-Jones type potentials, though Zeiri *et al.*[188] use parabolic splines to represent it.

After some manipulation and a number of approximations, Zeiri *et al.* reduce the equation of motion of the desorbing particle to that of a Brownian particle moving under the influence of a force, which they express in terms of microscopic quantities. The next step then consists in calculating the rate of desorption of the adparticles via Kramers' method, discussed in Chapter 4 and suitably modified for desorption.

5.1.2. Rate of Desorption of Atoms

In a desorption process, one is interested in time scales much longer than the characteristic time scales of the molecular motion. One can therefore follow the derivation of Kramers[162] in obtaining an expression for the desorption rate. The generalized Liouville equation yields

$$\frac{\partial W}{\partial t}+u\frac{\partial W}{\partial z}+K\frac{\partial W}{\partial u}=\beta u\frac{\partial W}{\partial u}+\beta W+q\frac{\partial^2 W}{\partial u^2} \tag{5.1.6}$$

where $W(z, u, t)$ is the probability of finding a particle at position z with velocity u at time t and $q = k_B T\eta/m$. Here η is the friction constant and K represents the acceleration caused by the interaction potential $\tilde{V}$ coupling the adatom to the surface. The work of Zeiri *et al.*[188] leads to an approximation for this quantity in the form

$$K=-(\tilde{B}+\tilde{\Omega}^2 z)=-\frac{1}{m}\frac{\partial\tilde{V}(z)}{\partial z} \tag{5.1.7}$$

Kramers[162] noted that equation (5.1.6) is satisfied by the Maxwell–Boltzmann distribution

$$W_{\mathrm{MB}}(z,u)=C_0\exp\{-[mu^2+2\tilde{V}(z)]/2k_B T\} \tag{5.1.8}$$

where C_0 is a normalization constant. One expects the distribution (5.1.8) to be valid near the bottom of the potential $\tilde{V}$ so that, in the case of desorption, one will be seeking solutions to equation (5.1.6) of the Kramers form

$$W(z,u)=C_0 F(z,u)\exp[-(mu^2+2\tilde{V})/2k_B T] \tag{5.1.9}$$

where $F(z, u)$ is nearly unity in the neighborhood of $z = 0$ and $F(z, u)$ tends to zero as z tends to infinity. One readily obtains the equation for $F(z, u)$ as

$$u \frac{\partial F}{\partial z} + K \frac{\partial F}{\partial u} = q \frac{\partial^2 F}{\partial u^2} - \eta u \frac{\partial F}{\partial u} \tag{5.1.10}$$

Following Kramers, the substitution $\xi = u - az - b$ enables one to integrate and obtain

$$F(z, u) = \left(\frac{a - \eta}{2\pi q}\right)^{1/2} \int_0^{\xi} \exp\left[-\frac{(a - \eta)\xi'^2}{2q}\right] d\xi' \tag{5.1.11}$$

where $a = \frac{1}{2}[\eta + (\eta^2 - 4\tilde{\Omega}^2)^{1/2}]$ and $b = -2\tilde{B}/[-\eta + (\eta^2 - 4\tilde{\Omega}^2)^{1/2}]$.

An expression for the rate of desorption is derived by next calculating the flux of desorbing particles at a given distance z_0 from the surface. In principle, this distance should be taken as infinite so that z_0 corresponds to an atom–surface separation for which the interaction potential $\tilde{V}$ vanishes. If one chooses z_0 to be some finite distance, the flux one should calculate must take into account only those particles possessing sufficient kinetic energy to escape. Thus the desired flux at a given value of z_0 is

$$j(z_0, u_0) = \int_{u_0}^{\infty} W(z_0, u) u \, du \tag{5.1.12}$$

where u_0 is the smallest positive velocity for which a particle at z_0 will desorb, i.e.,

$$D_e = \tilde{V}(z_0) + \tfrac{1}{2} m u_0^2 \tag{5.1.13}$$

Substitution of equations (5.1.9) and (5.1.11) into (5.1.12) gives

$$j(z_0, u_0) = C_0 \left(\frac{a - \eta}{2\pi q}\right)^{1/2} \exp\left(-\frac{\tilde{V}(z_0)}{k_B T}\right)$$
$$\times \int_{u_0}^{\infty} \exp(-mu^2/2k_B T) \int_0^{\xi} \exp[-(a - \eta)\xi'^2/2q] d\xi' u \, du \tag{5.1.14}$$

After some calculation, the final expression for the flux is

$$j(z_0, u_0) = C_0 \frac{k_B T}{m} \mathcal{T}(T) \exp(-D_e/k_B T) \tag{5.1.15}$$

where, with $\phi(\alpha)$ denoting the error function,

$$\mathcal{T}(T) = \phi(\alpha_1) + \exp(-\eta^*)\left(\frac{\pi\eta k_B T}{2ma}\right)^{1/2}[1 - \phi(\alpha_2)] \tag{5.1.16}$$

with

$$\left.\begin{aligned} \alpha_1 &= \left[\frac{m(a-\eta)}{2k_B T\eta}\right]^{1/2}[u_0 - (az_0 + b)], \\ \alpha_2 &= \left(\frac{ma}{2k_B T\eta}\right)^{1/2}[-\theta(az_0 + b) + u_0] \\ \theta &= 1 - \beta/a, \qquad \eta^* = \theta\eta(az_0 + b)^2/2q \end{aligned}\right\} \tag{5.1.17}$$

The rate for a desorption process, again following Kramers,[162] is given by the ratio of the flux at z_0 to the number N_0, of particles at the surface. After a short calculation, this yields the approximate form for the desorption rate:

$$\left.\begin{aligned} R &= \frac{\Omega_0}{2\pi}\mathcal{T}(T)\exp(-D_e/k_B T) \\ N_0 &= C_0(2\pi k_B T/m\Omega_0) \end{aligned}\right\} \tag{5.1.18}$$

where Ω_0 is the value of $\tilde{\Omega}$ at the bottom of the potential well. A related expression, but without the $\mathcal{T}(T)$ term, has been derived from transition-state theory by Garrison and Adelman.[197]

5.1.3. Rate of Desorption of Molecules

The main focus of this chapter is on the case of a molecule desorbing from a surface. For such a system, in addition to translational motion there also exist frustrated motions, such as frustrated rotations and translations. Let us consider, for example, the diatomic molecule CO. The most important frustrated motion corresponds to the bending mode of the O atom about the surface-C bond. For the free molecule there is no direct coupling between the pure-translation and pure-rotational motions, hence coupling between the two motions is through the surface. Thus the total interaction potential between the molecule and solid surface can be written in the form

$$V_{\text{total}} = \tilde{V}(z) + V_{\text{rot}}(\gamma) \tag{5.1.19}$$

where $\tilde{V}(z)$ and $V_{\text{rot}}(\gamma)$ represent respectively the potentials due to pure translation and frustrated rotation (i.e., the bending mode). In such a case the molecules, which are in equilibrium with the surface, will have a Maxwell–Boltzmann distribution of frustrated rotational energy. The total distribution function, obtained as a solution of the generalized Liouville equation, will have the form

$$W(z, u, \gamma, \dot{\gamma}) = W(z, u)\exp\left(\frac{-\mu l^2 \Omega_r^2 \gamma^2 + \mu l^2 \dot{\gamma}^2}{2k_B T}\right) \quad (5.1.20)$$

where $W(z, u)$ is defined in equation (5.1.9) and the harmonic approximation is used for the frustrated rotational motions (cf. Andersson[198]). In equation (5.1.20), μ is the reduced mass for the frustrated rotational motion, l is the corresponding length, and Ω_r is the rotational frequency.

Due to coupling of the translational and rotational motions to the surface, Zeiri *et al.*[188] assume that the energy stored in the frustrated rotation can be converted to translational kinetic energy of the desorbing molecule. As a result, one must modify equation (5.1.13) to read

$$D_e = \tilde{V}(z_0) + \tfrac{1}{2}mu_0^2 + \tfrac{1}{2}\mu l^2 \dot{\gamma}^2 + \tfrac{1}{2}\mu l^2 \Omega_r^2 \gamma^2 \quad (5.1.21)$$

By substituting distribution function (5.1.20) and equation (5.1.21) into equation (5.1.12) and averaging over the rotational motion, one obtains

$$j(z_0, u_0) = \frac{C_0 k_B T}{m} \exp(-D_e/k_B T)\mathcal{T}(T) \int_{-\gamma_0}^{\gamma_0} \int_{-\dot{\gamma}_0}^{\dot{\gamma}_0} d\gamma\, d\dot{\gamma} \quad (5.1.22)$$

where γ_0 is the maximum bending angle for the molecule and $\dot{\gamma}_0$ the corresponding maximum angular velocity.

Similarly, the number of particles at the bottom of the interaction potential is calculated to yield the general expression for the desorption rate of molecules adsorbed in solid surfaces. Hence

$$R = \left(\frac{k_B T}{2\pi m}\right)^{1/2} \left(\frac{2\mu l^2 \Omega_r^2 \gamma_0^2}{\pi k_B T}\right) \mathcal{T}(T) \left[\int_{-\infty}^{\infty} \exp\left(-\frac{\tilde{V}(z)}{k_B T}\right) dz\right]^{-1} \times \exp\left(-\frac{D_e}{k_B T}\right) \quad (5.1.23)$$

If $\tilde{V}$ is approximated by a harmonic potential, one obtains

$$R = \frac{\Omega_0}{2\pi} \left(\frac{2\mu l^2 \Omega_r^2 \gamma_0^2}{\pi k_B T}\right) \mathcal{T}(T) \exp\left(-\frac{D_e}{k_B T}\right) \quad (5.1.24)$$

As a final step in the derivation, Zeiri *et al.*[188] consider the temperature dependence of the function $\mathcal{T}(T)$ in equation (5.1.16). Examination of this function shows that for most physical systems, including both physisorption and chemisorption of atoms and molecules, the parameters entering the right-hand side of equation (5.1.16) are such that $\mathcal{T}(T)$ is equal to unity (independent of the choice of z_0). In particular, for systems such as Xe and K on W(111) and CO on Ni(110), this function is unity to one part in 10^6 up to $T \sim 10^6$ K. Hence one obtains

$$R_{\text{atom}} = \frac{\Omega_0}{2\pi} \exp(-D_e/k_B T) \tag{5.1.25}$$

and

$$R_{\text{molecule}} = \frac{\Omega_0}{2\pi} \left(\frac{2\mu l^2 \Omega_r^2 \gamma_0^2}{\pi k_B T} \right) \exp(-D_e/k_B T) \tag{5.1.26}$$

5.1.4. Results: Atomic and Molecular Desorption

5.1.4.1. Atomic Desorption

Zeiri *et al.*[188] consider first the application of the above treatment to desorption rates for a chemisorbed atomic system, K on W(111), and for a physisorbed atomic system, Xe on W(111). Specifically, Figure 5.1(a) shows the experimental and calculated rates for desorption of K on W(111). Here, the bond energy is 2.64 eV. The results of Zeiri *et al.* deviate from experiment[199–201] by 0.2 orders of magnitude, which is within the experimental uncertainty.

The results for the desorption of physisorbed Xe from W(111) are plotted in Figure 5.1(b) along with the experimental data.[201] The bond energy here is 0.22 eV. It can be seen that equation (5.1.25) provides a quantitative description of the rate of desorption of atoms from solid surfaces. The simple relation between the rate of desorption R and the characteristic microscopic parameters of the system, Ω_0 and D_e, makes it possible to predict one of the quantities from experimental determinations of the other two. In general, the measurement of frequencies is more accurate than that of absolute rates of desorption; thus equation (5.1.25) should be used as a tool to predict desorption rates.

Zeiri *et al.*[188] stress that equation (5.1.25) was obtained from equation (5.1.23) by assuming that the potential $\tilde{V}(z)$ could be approximated by a harmonic potential with frequency Ω_0. This assumption is justifiable whenever $k_B T \ll D_e$, which turns out to be the case in the range of temperatures covered by the experimental results shown in Figure 5.1(a) and (b). However, when this condition is not fulfilled the rate of desorption should be calculated using equation (5.1.23).

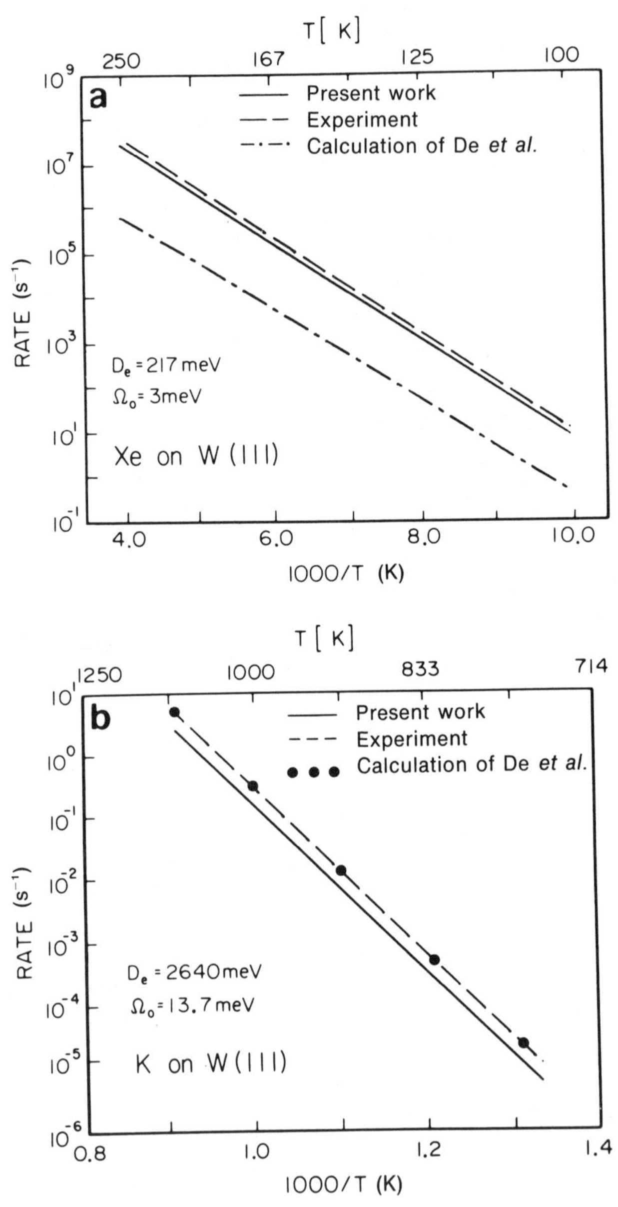

FIGURE 5.1. (a) Experimental and calculated rates of desorption for K on W(111). (b) Desorption for physisorbed Xe on W(111). Calculations and experimental data after Zeiri *et al.*[188] See also G. S. De *et al.*[186]

5.1.4.2. Molecular Desorption

The main differences between molecular and atomic desorption are due to two characteristic features of molecular adsorbates:

1. Molecules have internal degrees of freedom that are absent in atomic systems.
2. Molecules exhibit frustrated rotational modes (surface–adsorbate bending modes) strongly coupled to the surface, but which become pure rotational modes in an activated state.

These differences are of importance in the ensuing discussion, both in this section and in Section 5.2, where transition-state theory is applied to desorption from solid surfaces.

To date, several molecular-desorption systems have been studied experimentally. As mentioned above, a common feature is the large preexponential factor, namely several orders of magnitude above the A of about 10^{13} expected for atomic desorption.

As an example, Zeiri *et al.*[188] have considered CO chemisorbed on Ni(100). In Figure 5.2, the results of their calculations are compared with experiment.[176] Also shown is the rate of desorption one would obtain if

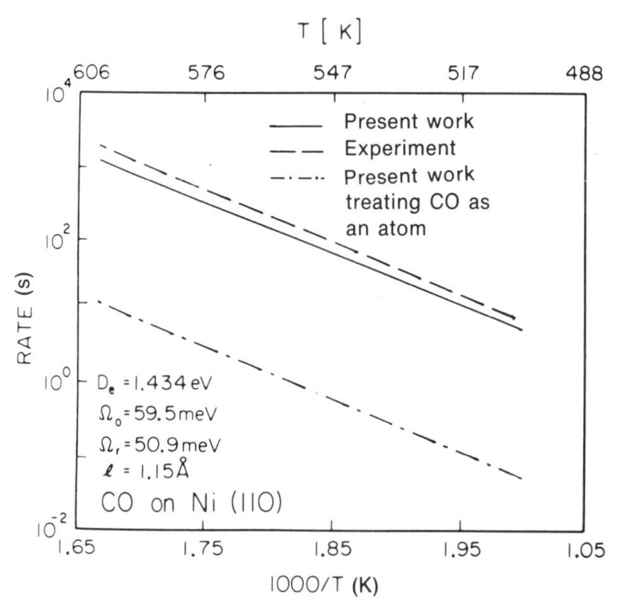

FIGURE 5.2. Rate of desorption for CO chemisorbed on Ni(100). Calculations compared with experimental data (after Zeiri *et al.*[188]).

the CO molecule were treated as an atom. Zeiri *et al.* suggest that the excellent agreement between theory and experiment means that the frustrated rotational motion is responsible for the two orders of magnitude increase in the preexponential factor.

To summarize, the theoretical models going back to Kramers, refined and extended by Adelman and Doll and others, have been shown by Zeiri *et al.* to lead to a relatively simple expression for the rate of desorption, namely equation (5.1.26) involving only microscopic parameters (vibrational frequencies and bond energies) of the adsorbed system that are accessible experimentally. The theoretical results are in excellent agreement with experimental data. Since the theoretical rate involves only parameters that can be independently determined from experiment and theory, this method may well be useful as a tool for the accurate prediction of desorption rates for a variety of atomic and molecular systems. The extension of this treatment to higher coverages will be discussed later in this chapter.

5.2. TRANSITION-STATE THEORY APPLIED TO DESORPTION FROM SOLID SURFACES: AMMONIA ON Ni(111)

In spite of some difficulties with the foundations of transition-state theory, referred to in the previous section [see equation (5.1.1)], the theory has been applied by Redondo *et al.*[202] to desorption of molecules from metal surfaces. These workers have pointed out that, although transition-state theory (TST) has been applied to rate processes for numerous gas-phase reactions, there are few examples to date in which it has been used to treat desorption from solid surfaces. The TST[181] rate of desorption is given by

$$R = \frac{k_B T}{h} \frac{f^*}{f} \exp(-E/k_B T) \tag{5.2.1}$$

where f^* and f are the partition functions for the activated complex and the adsorbed particle, respectively, while E is the activation energy for desorption.

It has already been mentioned in Section 5.1 that the main difficulty with TST is concerned with the definition of the transition state for the desorbing species. The most common choice is to assume that the transition state is located at an infinite adsorbate–surface separation. For example, Garrison and Adelman[197] have used the TST approach to derive an expression for the one-dimensional desorption rate of an atom from a solid surface (where the motion of the adparticle is restricted to the direction normal to the surface). They assumed an activated complex located at an infinite distance from the substrate. In this approach, the partition function for this configuration corresponds to a free atom moving in a direction normal to the surface. The rate expression they obtained is

$$R = \frac{\Omega_0}{2\pi} \exp\left(-\frac{D_e - \frac{1}{2}\Omega_0}{k_B T}\right) \tag{5.2.2}$$

where Ω_0 and D_e are the stretching frequency and binding energy associated with the adsorbate–surface interaction potential.

5.2.1. Rate of Molecular Desorption

5.2.1.1. Case of Diatomic Molecule Desorption

The partition function of the activated complex for a diatomic molecule desorbing from a solid surface is given by

$$f^* = \int dx\, dy\, dr\, d\theta\, d\phi \frac{dp_x}{h} \frac{dp_y}{h} \frac{dp_z}{h} \frac{dp_\theta}{h} \frac{dp_\phi}{h} \times \exp(-P^2/2k_B T)\exp[-U(r)/k_B T] \tag{5.2.3}$$

where

$$P^2 = \frac{p_x^2 + p_y^2}{M} + \frac{p_r^2}{\mu} + \frac{p_\theta^2}{I} + \frac{p_\phi^2}{I \sin^2\theta} \tag{5.2.4}$$

Here M denotes the total mass of the molecule, μ is the reduced mass, while I is the moment of inertia. The coordinates x and y refer to the position of the center of mass, while r, θ, and ϕ are the spherical polar (internal) coordinates defined with respect to the center of mass. Since the activated complex corresponds to an infinite surface–molecule separation, the total potential reduces to the internal potential, represented by $U(r)$. When calculating f^* in equation (5.2.3), Redondo *et al.* assume that the internal potential $U(r)$ is harmonic, with associated frequency ω_i. Within this approximation, the partition function at the transition state reduces to

$$f^* = \frac{A}{h^5}\left[\frac{32\pi^4 M (k_B T)^3 I}{\omega_i}\right] \tag{5.2.5}$$

The partition function for the adsorbed state is

$$f = \int dq \frac{dp}{h^6} \exp\left[-\frac{(P^2 + p_z^2)/M}{2k_B T} - \frac{V(q)}{k_B T}\right] \tag{5.2.6}$$

where q and p represent the coordinates and momenta both of the center of mass and of the internal degrees of freedom, while $V(q)$ is the three-dimensional potential felt by the molecule near the surface. Integration over momenta yields

$$f = \frac{(2\pi k_B T)^3 M^{3/2} \mu^{1/2}}{h^6} I_r \int dq \sin\theta \exp\left[-\frac{V(q)}{k_B T}\right] \tag{5.2.7}$$

where I_r is the moment of inertia associated with the frustrated rotation of the surface. As for atomic desorption, too, the main contribution to the partition function comes from configurations near the equilibrium position. Hence Redondo *et al.*[202] express $V(q)$ as (with $E(x, y)$ the corrugation)

$$V(q) = \tfrac{1}{2}\{M\Omega_0^2(x, y)[z - z_0(x, y)]^2 + I_r\Omega_r^2(x, y)\theta^2 + \mu\bar{\omega}_i^2(x, y)r^2\} + E(x, y) \tag{5.2.8}$$

where Ω_r is the frequency associated with the bending (angle θ) motion and $\bar{\omega}_i$ is the frequency of the internal vibration at the surface. Equation (5.2.8) has been written on the assumption that the total potential is independent of azimuthal angle ϕ, which Redondo *et al.*[202] argue is generally a good approximation for binding sites on low-Miller-index surfaces but would not necessarily be valid for a molecule binding at a step or at a surface defect.

The quantities that depend on x and y and have the same periodicity as the surface can be expanded in appropriate Fourier series. Integration over z, r, θ, and ϕ yields

$$f = M\frac{(2\pi k_B T)^5}{h^8} \int dx\, dy\, \Omega_0^{-1}(x, y)\Omega_r^{-2}(x, y)\bar{\omega}_i^{-1}(x, y)\exp\left[-\frac{E(x, y)}{k_B T}\right] \tag{5.2.9}$$

where the evaluation of the θ integral involves the assumption $I_r\Omega_r^2(x, y)\pi^2 \gg 2k_B T$, which is valid under typical experimental conditions.

When the Fourier-series expansions referred to above are employed, and in the limit of a flat surface, the rate of desorption reduces to

$$R = \frac{\Omega_0}{2\pi}\left(\frac{2I\Omega_r^2}{k_B T}\right)\left(\frac{\bar{\omega}_i}{\omega_i}\right)\exp(-D_0/k_B T) \tag{5.2.10}$$

Except for a factor of $\pi/4$, this expression agrees with that obtained from classical stochastic diffusion theory.[203,188]

Redondo *et al.* generalize the above argument to polyatomic molecules and apply their results to the desorption of ammonia from the

Ni(111) surface. They find D_0 in equation (5.2.10) generalized to polyatomic molecules as a function of coverage θ, the form of $D_0(\theta)$ being given by

$$D_0(\theta) = D_0(0) + \beta\theta \tag{5.2.11}$$

where they estimate $D_0(0) = 21$ kcal/mol and $\beta = -3.7$ kcal/mol.

In the course of this work Redondo *et al.* conducted Hartree–Fock calculations for ammonia adsorbed on the three-coordinate site of the Ni(111) surface modeled with a Ni_{13} cluster. They found an optimum H–N–H angle of 104.8°. This seems different from the arguments of Flores *et al.* in Chapter 3, but the spirit of the two approaches is so different that it is difficult to isolate the specific cause of the discrepancy. Certainly the arguments of Flores *et al.* rely heavily on image theory applied to a perfect conductor, while it is difficult to see how such arguments would be reproduced by small cluster treatments such as that of Redondo *et al.*

In determining $D_0(0)$ and β, Redondo *et al.*[202] fitted published temperature programmed desorption spectra and in so doing invoked the Redhead equation discussed in Chapter 1, namely

$$\frac{d\theta}{dt} = -\frac{\nu_0\theta}{T}\exp\left[-\frac{D_0(\theta)}{k_B T}\right] \tag{5.2.12}$$

with ν_0 approximately equal to 10^{18} K/s.

5.3. MICROSCOPIC THEORIES

Reyes *et al.*[204] have worked out the predictions of a model within the one-phonon approximation for the angle-resolved thermal and flash desorption of atoms from surfaces. They emphasize that, while their model as presented is restricted to desorption of structureless particles, with few exceptions the experimental data are on desorption of the three stable diatomic hydrogenic molecules. With such a model, it is clear that one cannot interpret any effect due specifically to the polyatomic nature of the desorbing particles.

However, classical mechanical models of desorption of atoms, such as those used in the studies of Goodman and Garcia,[205–207] indicate that at least some of the interesting features observed experimentally for desorbing molecules may be properties of desorbing systems in general. This also has some experimental support in the work of Taborek[208] on the nonequilibrium desorption of He.

5.3.1. One-Phonon Approximation

Reyes *et al.* emphasize that the one-phonon approximation restricts the range of validity to systems involving physisorption at low temperatures. Therefore, to illustrate their model predictions, they have carried out calculations with parameters chosen to simulate the He–W system at temperatures below 10 K.

It is important to note here that two different types of desorption are of interest:

1. Steady-state desorption.
2. Flash desorption.

In both these types, the gas–surface system is assumed to be in complete equilibrium for all times $t < 0$. In steady-state desorption, the removal of the gas phase for times $t \geq 0$ is the process under consideration, while in flash desorption the process to be treated is due to an increase in the surface temperature for times $t \geq 0$. While the model of Reyes *et al.* is primarily suited to type (1), they have also undertaken calculations for type (2). Though this extension lacks complete justification, it may serve as a guideline for future calculations.

5.3.1.1. Definitions and Experimentally Measured Properties

The quantities to be calculated below in the model of Reyes *et al.*[204] are those measured in the experiments on desorption.[209–212] The polar angle is denoted by θ ($\theta = 0$ is the surface normal) and the azimuthal angle by ϕ, and it is assumed that no desorption property depends on ϕ.

In this situation, the most detailed information is contained in the function $\eta(\varepsilon, \theta)$, which is defined such that $\eta(\varepsilon, \theta) d\varepsilon\, d\Omega$ is the fractional number rate of atoms desorbed in the energy increment $d\varepsilon$ at ε and in the solid angle increment $d\Omega$ ($\equiv \sin\theta\, d\theta\, d\phi$) at (θ, ϕ).

In a steady-state desorption experiment, it is more usual to measure one or more of the three quantities $n(\theta)$, $\langle\varepsilon(\theta)\rangle$, and $s(\theta)$, which are defined as follows:

1. $n(\theta) d\Omega$ is the fractional number rate of atoms desorbed in the solid-angle increment $d\Omega$ at (θ, ϕ):

$$n(\theta) = \int_0^\infty \eta(\varepsilon, \theta) d\varepsilon \tag{5.3.1}$$

2. $\langle\varepsilon(\theta)\rangle$ is the average value of ε for these atoms:

$$\langle\varepsilon(\theta)\rangle = [n(\theta)]^{-1} \int_0^\infty \eta(\varepsilon, \theta)\varepsilon\, d\varepsilon \tag{5.3.2}$$

3. $s(\theta)$ is the so-called speed ratio, a nonnegative quantity defined by

$$s^2(\theta) = \langle \varepsilon(\theta) \rangle / \langle \varepsilon^{1/2}(\theta) \rangle^2 - 1 \tag{5.3.3}$$

where

$$\langle \varepsilon^{1/2}(\theta) \rangle = [n(\theta)]^{-1} \int_0^\infty \eta(\varepsilon, \theta) \varepsilon^{1/2} d\varepsilon \tag{5.3.4}$$

It can be seen that s is a measure of the velocity dispersion of the desorbing atoms, and equals zero for a so-called monoenergetic stream and unity for a Maxwellian stream (see, for example, Goodman and Wachman[(213)]).

Let us now consider a flash-desorption experiment. One important experimental quantity is $B(t, \theta)$, a detector-bolometer signal for the arrival at time t of the desorbing atoms. If the detector bolometer is assumed to be a linear detector of energy flux, then $B(t, \theta)$ is proportional to the product of the arrival rate $N(t, \theta)$ of atoms and the energy $\Delta\varepsilon(t)$ deposited on the bolometer surface by the atoms:

$$B(t, \theta) \propto N(t, \theta) \Delta\varepsilon(t) \tag{5.3.5}$$

The arrival rate $N(t, \theta)$ is related to $\eta(\varepsilon, \theta)$ by

$$N(t, \theta) = \eta(\varepsilon, \theta) d\varepsilon / dt \tag{5.3.6}$$

where

$$\varepsilon = ml^2 / 2t^2 \tag{5.3.7}$$

m being the mass of a desorbing atom and l the relevant flight path length. It follows from equations (5.3.6) and (5.3.7) that

$$N(t, \theta) \propto \varepsilon^{3/2}(t) \eta(\varepsilon, \theta) \tag{5.3.8}$$

and from equations (5.3.8) and (5.3.5) that

$$B(t, \theta) \propto \varepsilon^{3/2}(t) \eta(\varepsilon, \theta) \Delta\varepsilon(t) \tag{5.3.9}$$

This equation provides a direct connection between steady-state and flash-desorption measurements. In the present context, Kenney *et al.*[(214)] clearly describe the method of interpretation of the experimental time-of-flight spectra in desorption in terms of moments of the speed distribution.

5.3.1.2. Results from Equilibrium Theory

Goodman and Wachman[(213)] reviewed the ordinary equilibrium theory and though it is clear from comparison with measured data that there are fundamental differences, yet it is often useful to have the explicit results for the above quantities for comparison. Thus the equilbrium-theory approximations are

$$\left.\begin{aligned} \eta(\varepsilon,\theta) &= \pi^{-1}(k_B T)^{-2}\varepsilon\exp(-\varepsilon/k_B T)\cos\theta \\ n(\theta) &= \pi^{-1}\cos\theta \\ \langle\varepsilon(\theta)\rangle &= 2k_B T \\ s(\theta) &= 1 \\ B(t,\theta) &\propto \varepsilon^{5/2}(t)\Delta\varepsilon(t)\exp(-\varepsilon/k_B T)\cos\theta \end{aligned}\right\} \quad (5.3.10)$$

It is useful also to consider the maximum value of $B(t,\theta)$, namely $B_{max}(\theta)$; $t_{max}(\theta)$, which is the time at which $B = B_{max}$; and $\Delta(\theta)$, the full width at the half maximum of the curve $B(t,\theta)$. Again ordinary equilibrium theory gives

$$\left.\begin{aligned} B_{max}(\theta) &= B_0\cos\theta \\ t_{max}(\theta) &= t_0 \\ \Delta(\theta) &= \Delta_0 \end{aligned}\right\} \quad (5.3.11)$$

where t_0 and Δ_0 are independent of θ.

5.3.1.3. Other Models

Several other models have been proposed:

1. The activated adsorption model of Van Willigen.[215]
2. Its generalization, allowing for a nonzero fraction of molecules to be adsorbed with negligible activation energy, due to Comsa and David.[216]
3. The generalization of (1) to include frictional effects.[217,218]

The question of whether nonequilibrium characteristics are exclusive to polyatomic as opposed to monatomic molecules, or are general properties of all desorbing systems, was treated by Goodman and Garcia.[205,206]

5.3.1.4. Debye-like Model

The surface of the solid is assumed to behave as a bulk continuum model, where the solid possesses temperature T and occupies the region $z \leq z_0$, and the surface is flat and defined by the plane $z = z_0$. The precise value of z_0 plays no role in the theory and, in the absence of thermal vibrations of the surface, may be set equal to zero. The atom position is denoted by $\mathbf{r} \equiv (\mathbf{R}, z)$, with[219] $\mathbf{R} = (x, y)$. Its motions parallel and perpendicular to the surface are not coupled, and appropriate quantum numbers are $\mathbf{K}$ and ε_z, where $\mathbf{k} = (\mathbf{K}, k_z)$ is the atom wave vector and ε_z is the energy associated with motion perpendicular to the surface. The total energy ε of the atom is given by

$$\varepsilon = \hbar^2 K^2/2m + \varepsilon_z \tag{5.3.12}$$

For bound states of the atom at the surface, $\varepsilon_z = -\varepsilon_n < 0$, $n = 0, 1, 2, \ldots$, while for continuum states, $\varepsilon_z > 0$ and is not restricted (other than by $\varepsilon_z \le \varepsilon$). The first part of this section implies that only bulk modes of the solid will be considered, with frequency ω and wave vector $q = (\mathbf{Q}, q_z)$. For both longitudinal and transverse modes, Reyes *et al.* employ a Debye model

$$\omega(\mathbf{q}) = v_s q = v_s (Q^2 + q_z^2)^{1/2} \tag{5.3.13}$$

where v_s is a (constant) velocity of sound. The restrictions on ω and q may be expressed as $\omega < \omega_D$ and $q < q_D$, where ω_D and q_D are the Debye frequency and wave vector of the model. The phonons induce transitions among the quantum levels of the atom–surface potential, although the model is restricted to zero-phonon and one-phonon (annihilation and creation) processes. Under the conditions explored by Reyes *et al.*,[204] just three different types of process need to be considered:

1. Bound-state activation transitions (lower level to higher level) via phonon annihilation.
2. Bound-state deactivation transitions via phonon creation.
3. Bound to continuum transitions via phonon annihilation.

The last type (3) of process is desorption. Other, more complicated types of process are possible if many-phonon processes are permitted, but these will be neglected here.

5.3.2. Atom–Phonon Interaction

The so-called atom–phonon interaction may be thought of as a perturbation of the noninteracting atom–solid Hamiltonian. Following, for example, Goodman[220] this perturbation may be written as

$$H_p = (dv/dz)\delta z_0(\mathbf{R}, t) \tag{5.3.14}$$

where δz_0 is the fluctuation induced by the phonon on the surface equilibrium position z_0 and $v(z - z_0)$ is the atom–solid interaction potential. The fluctuation δz_0 is a function of $\mathbf{R}$ and may therefore be expanded in terms of Fourier components of a well-defined wave vector. The amplitudes of the phonon processes mentioned above are obtained by taking matrix elements of this perturbation between wave functions of the noninteracting atom–solid system. The well-known (one-phonon) conservation rules

$$\varepsilon_f = \varepsilon_i \pm \hbar\omega \tag{5.3.15}$$

$$\mathbf{K}_f = \mathbf{K}_i \pm \mathbf{Q} \tag{5.3.16}$$

follow, where subscripts f and i denote final and initial states and the plus and minus signs refer to phonon annihilation and creation, respectively.

Reyes *et al.* introduce a two-dimensional band picture, which encompasses the descriptions both of the noninteracting atom and phonon states and of the transitions induced by the atom–phonon interaction. When plotted in a coordinate system (K_x, K_y, ε), equation (5.3.12) gives a set of paraboloids of revolution around the ε-axis, each with its bottom at the point $(0, 0, \varepsilon_z)$. There is one paraboloid for each bound state and a continuum of paraboloids for the continuum state, as depicted in Figure 5.3. A similar procedure is adopted for the phonons by plotting their energies $\varepsilon = \hbar\omega$ in a coordinate system (Q_x, Q_y, ε). Equation (5.3.13) results in a set of hyperboloids of revolution (one for each q_z) around the ε-axis, each with its bottom at $(0, 0, \hbar v_s q_z)$ or, effectively, a continuum of phonon states limited by the two surfaces $\varepsilon = \hbar v_s (Q_x^2 + Q_y^2)^{1/2}$ and $\varepsilon =$

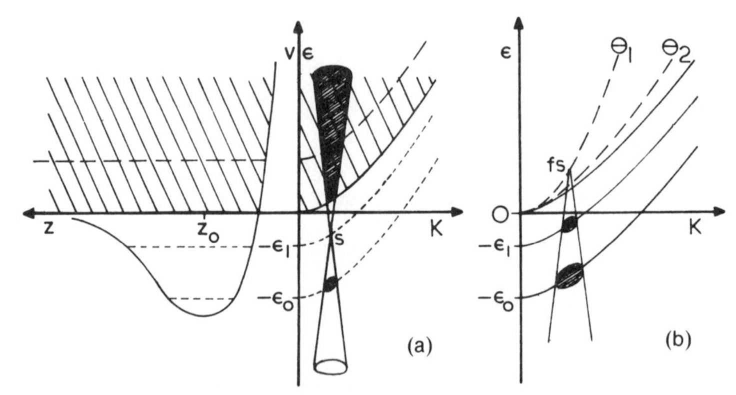

FIGURE 5.3. (a) Left-hand part shows schematically the atom–surface potential $v(z - z_0)$; two bound states having energies $-\varepsilon_0$ and $-\varepsilon_1$ are drawn. The positive energies $\varepsilon > 0$ indicated by the shaded region form a continuum of states (in the limit of infinite box length L). The right-hand part of (a) shows a 'vertical" cut through the paraboloids derived from the bound states and the continuum states. Also shown heavily shaded are the phonon cones for upward and downward transitions from a state (s) located at their common vertex. (b) Paraboloids of the final states corresponding to atoms desorbed at angles $\theta_1 = 30°$ and $\theta_2 = 60°$ are depicted as broken lines; the initial states contributing to a given final state (fs) are enclosed in shaded regions (after Reyes *et al.*[204]).

$\hbar\omega_D$, this continuum being referred to by Reyes *et al.* as the "phonon cone."

It is now easy to envisage transitions, induced by a phonon process, between atom states. Consider, for example, the system shown in Figure 5.3 where, for simplicity, only two bound states (of energies $-\varepsilon_0$ and $-\varepsilon_1$) are included. An atom in a state $(\mathbf{K}_i, -\varepsilon_1)$ belonging to the upper band may make a downward transition to all those states $(\mathbf{K}_f, -\varepsilon_0)$ which lie on the intersection of this band and a phonon cone drawn downward, with its vertex at $(\mathbf{K}_i, -\varepsilon_1)$; a phonon of frequency

$$\omega = \frac{\hbar(K_i^2 - K_f^2)}{2m} + \frac{(\varepsilon_0 - \varepsilon_1)}{\hbar}$$

and wave vector $\mathbf{Q} = \mathbf{K}_i - \mathbf{K}_f$ is created in this process. Allowed transitions in which a phonon is annihilated are those to atom states that lie on the intersection of the continuum of atom states with a phonon cone drawn upward. It is important to note that the slope of a paraboloid is $\hbar^2 K/m$, while that of a phonon cone is $\hbar v_s$. Since the ratio $\hbar K/m v_s$ of these slopes must be small, the phonon cone is "thin" or, in other words, the atom bands are "flat." This purely geometric fact lends support to the one-dimensional models of Efrima *et al.*[187] and Gortel *et al.*[221]; however, the conservation of momentum parallel to the surface, as expressed by equation (5.3.16), remains an essential ingredient of the work of Reyes and co-workers.

These workers note that this two-dimensional band picture is able to encompass desorption from localized sites, provided the paraboloid bands of initial states are replaced by flat bands. Also diffraction, and its important consequence, namely the appearance of band gaps, are readily introduced.

Omitting the detailed analysis given by Reyes *et al.* for the function $n(\theta)$ in equation (5.3.1), the subsequent summary follows their discussion of the factors which determine the transition rates and hence contribute to the qualitative nature of $n(\theta)$. They isolate (1) the allowed final states, (2) the occupancy of initial states, (3) the phonon occupancy, and (4) the matrix element, and are then led via $n(\theta)$ to the expression for $\eta(\varepsilon_f, \theta)$, given by,* with H the Heaviside step function and Γ the gamma function:

$$\eta(\varepsilon_f, \theta) \propto [\varepsilon_f \cos\theta \exp(-\varepsilon_f/T_g)] \sum_n H[\varepsilon_{f,n}^{\max}(\theta) - \varepsilon_f](k_{fz}^2 + k_n^2)^2$$
$$\times \frac{\sinh(2\pi k_{fz})}{\cosh(2\pi k_{fz}) + \cos(2\pi d)} |\Gamma(d + \tfrac{1}{2} + ik_{fz})|^2$$
$$\times \int d\phi \int Q\, dQ\, q_{z0}^{-1} n_\omega \exp(\omega/T_g) \qquad (5.3.17)$$

* For a given initial band, say m, the allowed final states have energy $\varepsilon_f \leq \varepsilon_{f,m}^{\max}(\theta)$, where $\varepsilon_{f,m}^{(\max)}(\theta) \simeq (\hbar\omega_D - \varepsilon_m)\sec^2\theta$, while $d = (2mA)^{1/2}$, where A is the well depth (the classical adsorption energy).

Reyes *et al.* note that terms in square brackets in equation (5.3.17) give the equilibrium-theory result [cf. equation (5.3.10)], the origins of these terms being a factor k_f from the density of final states, a factor $\exp(-\varepsilon_i/T_g) = \exp(\omega/T_g - \varepsilon_f/T_g)$ from the distribution of initial states, and a factor $k_{f_z} = k_f \cos\theta$ from the square of a (suitably reduced; see Reyes *et al.*) matrix element.

The above discussion, leading to equation (5.3.17) for $\eta(\varepsilon_f, \theta)$, is the basis for the calculations of Reyes *et al.* However, they find that the integrations can often be usefully approximated analytically with errors of about 10%. Quantities $n(\theta)$, ε, and s are plotted in Figure 5.4 for steady-state desorption with parameters chosen to represent He desorbing from a W surface. For flash desorption, $B_{max}(\theta)$, $t_{max}(\theta)$, and $\Delta(\theta)$ were calculated for temperatures $T_s = 4T_g = 8$ K. The equilibrium-theory results are presented in Figure 5.5 for comparison with B_{max} and n.

While the paper of Reyes *et al.* must be consulted for a detailed discussion of the above results, it is worthwhile here briefly to consider the relevance of their results to the interpretation of available experimental data. While one-phonon processes have been recognized as dominant in scattering by surfaces held at well below room temperature, as discussed, for example, by Horne and Miller,[222] there seems to be no experimental work reported to date in which He desorbs from a low-temperature surface in a steady-state mode. One could try to compare data on desorption of H_2 and D_2 but, with the only exception to date of the work of Bradley *et al.*,[223] they all show far greater nonequilibrium effects than are displayed by the results of Reyes *et al.* Though the reasons for this are not yet clear, it would

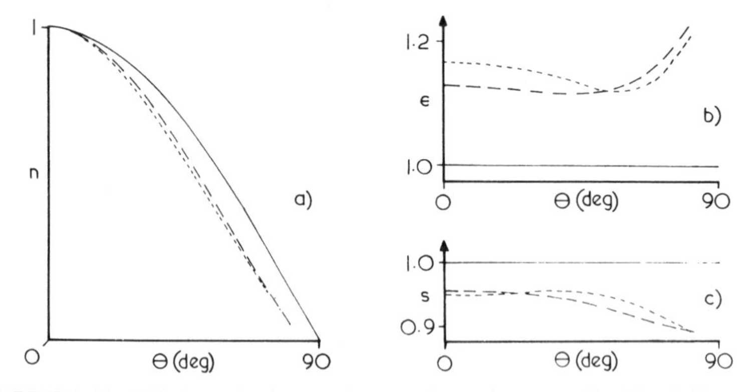

FIGURE 5.4. Numerical results for steady-state desorption; ---, $T_s = T_g = 2$ K; ———, $T_s = T_g = 4$ K; ———, equilibrium theory. (a) Number rate of desorption, n denoting $n(\theta)/n(0)$; (b) average energy per atom, ε denoting $\langle\varepsilon(\theta)\rangle/2bT_{es}$; (c) speed ratio $s(\theta)$ (after Reyes *et al.*[204]).

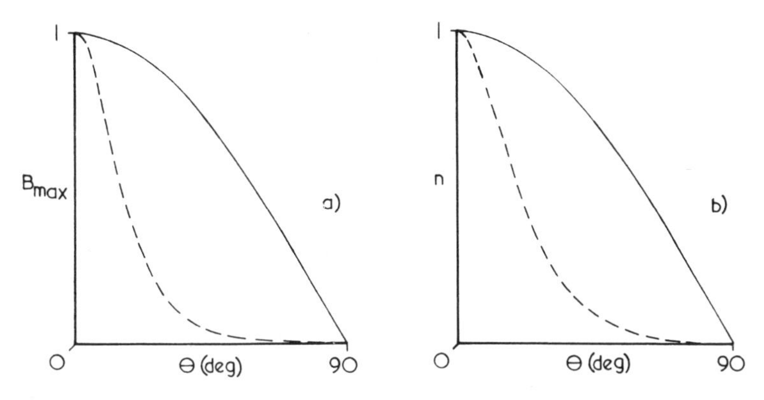

FIGURE 5.5. Numerical results for flash desorption; ———, $T_s = 4T_g = 8$ K; ———, equilibrium theory. (a) Maximum in the bolometer signal; B_{max} denoting $B_{max}(\theta)/B_{max}(0)$; (b) number rate of desorption, n denoting $n(\theta)/n(0)$ (after Reyes *et al.*[204]).

seem that a nonequilibrium statistical distribution introduced in the work of Reyes *et al.* instead of the adopted Maxwellian form could change the features of their results.

In the case of flash desorption, there are data obtained by Taborek.[208] In connection with this work it is relevant to refer to the paper of Gortel *et al.*,[224] who have developed a theory for the time of flight (TOF) spectra in thermal description of helium. Later work by Gortel *et al.*[225] presents TOF spectra for the isothermal (at 8 K) and flash (2–8 K) desorption of He from nichrome. The difference in the angular dependence is striking. The main cause is the following. The studies of isothermal description within a cascade model[226] indicate that thermal equilibrium among all physisorbed bound states is maintained during the whole desorption process. Similarly, in flash desorption from a highly mobile adsorbate, the motion of atoms perpendicular to the surface thermalizes quickly (within 0.2 desorption times) to the final flash temperature (8 K), as demonstrated by Gortel *et al.*[225] The motion of adatoms along the surface remains characterized by the initial temperature of 2 K even after 2 desorption times. This nonequilibrium between the motions parallel and perpendicular to the surface suffices to produce observed strong angular dependence of the TOF spectra.

Surprisingly, the calculated desorption times for full three-dimensional theories of mobile adsorption are the same as for one-dimensional ones, the latter neglecting entirely the motion of atoms along the surface. Also, as observed much earlier, the flash (2–8 K) and isothermal (at 8 K) desorption times are very close to each other.

5.4. QUANTUM-STATISTICAL THEORY OF PHYSISORPTION

This section presents a summary of the work of Gortel and Kreuzer[227,228] on phonon-mediated physisorption. They developed a quantum-statistical theory by setting up the initial-value problem within nonequilibrium statistical mechanics and calculating the desorption times corresponding to the various experimental procedures, namely virgin adsorption, isothermal desorption, and flash desorption. They were thus able to delineate the temperature range over which the Arrhenius–Frenkel parametrization of the desorption time $t_d = t_d^0 \exp(Q/k_B T)$ is acceptable. They calculate desorption times for the specific case of He on a particular Cu–Ni alloy, constantan, which develops a bound state at an energy $E_0/k_B = -25$ K, and they show by comparison with experiment that the rather long flash desorption times (t_d^0 approximately 10^{-7} s at temperatures between 4 and 20 K) are the result of weak coupling (range of surface potential about 2.5 Å) between the gas and phonons of the solid. They correlated isothermal and flash desorption times and advanced a proposal concerning the temperature range where differences in these times might be detected. A fourth-order calculation permits them to delineate the range of validity of the one-phonon (second-order) theory and of the relaxation-time approach to desorption.

In earlier work, Gortel *et al.*[227,228] formulated and worked out the details of a quantum-statistical theory of adsorption and desorption of a gas at the surface of a solid in systems that exhibit localized physisorption, i.e., in which no chemical reactions take place during the adsorption process and in which lateral diffusion in the adsorbate atoms along the surface can be neglected.

Though the kinetics of adsorption and desorption are controlled by the same molecular interactions between the particles making up the gas and solid phases, no matter what the experimental procedure, it is nevertheless necessary to analyze the various experiments by setting up the appropriate initial-value problems of nonequilibrium statistical mechanics to calculate and correlate the various characteristic times measured. Gortel *et al.*[228] have done this for adsorption, flash desorption, and isothermal desorption in simple gas–solid systems within which localized physisorption dominates.

Following Gortel and Kreuzer,[227] let us briefly summarize the main results of such a treatment for the adsorption of a gas at a virgin surface. They have shown that adsorption proceeds in two stages with very different time scales. The first stage is one of fast transients in which the gas particles take notice of the solid surface by readjusting their wave functions to the presence of the surface potential. These transients, which Gortel *et al.* have calculated in an exactly soluble quantum-statistical model, typically evolve in atomic times of the order of 10^{-13} s and will prepare the gas–solid system in the appropriate state from which the second phase starts. The latter, the kinetic regime, is controlled by energy-dissipating mechanisms and will lead to the build-up of the gas adsorbate on the surface, i.e., to filling of the surface-bound states. Gortel *et al.* have calculated the characteristic relaxation time for phonon-mediated adsorption for He adsorbing on constantan, since this system was studied in flash-desorption experiments by Cohen and King.[229] Gortel *et al.* found that in this system, at 20 K, the time of adsorption t_a, about 10^{-7} s, is larger than the atomic time, about 10^{-13} s, by approximately six orders of magnitude, slowing down to about 3 ms at $T = 3$ K. Unfortunately, at the time of writing there seem to be no measurements of these adsorption times in the He/constantan system. A connection with sticking coefficients has been established by Kreuzer and Obermair.[230]

Following Gortel and Kreuzer, a typical flash-desorption experiment will now be described briefly after which the appropriate initial-value problem in their quantum-statistical framework will be set up to calculate the flash-desorption time in second-order time-dependent perturbation theory, which agrees well with experiments on the He/constantan system. Isothermal desorption will then be considered next.

The main conclusions may be summarized as follows. In a fourth-order perturbation theory, it is essential to calculate one-phonon and two-phonon processes as they arise from the phonon-mediated gas–solid interaction, including terms up to third order in the derivative of the surface potential. A relaxation-time description of desorption from a bound state at energy E_0 is possible as long as $\hbar/E_0 \ll t_d$, where t_d is the isothermal desorption time. Gortel *et al.* find that a second-order calculation can be trusted as long as $|E_0| \lesssim k_B T \ll \hbar\omega_D$, where ω_D is the Debye frequency of the solid. Fourth-order contributions become important for $|E_0| < \hbar\omega_D$ and $k_B T \gtrsim \hbar\omega_D$. Moreover, for the range $\hbar\omega_D \lesssim |E_0| \lesssim 2\hbar\omega_D$ fourth-order terms are essential, because second-order contributions are zero in this region of bound-state energies.

5.4.1. Flash Desorption

Following Gortel and Kreuzer, one wishes to formulate explicitly the initial-value problem appropriate for a flash-desorption experiment. Using a one-dimensional treatment, these workers consider a model system in which the surface potential develops only one bound state.*

In developing a quantum-statistical theory of flash desorption, it is

* The potential form factor was of the so-called Wheeler–Yamaguchi form $\exp(-\gamma x)$; the range of the surface potential γ^{-1} being a parameter in their treatment.

useful to recall that such an experiment starts from an initial state in which gas and solid are in equilibrium with each other at temperature T_g. At time $t = 0$ the temperature of the solid is suddenly raised to a value $T_s > T_g$ and the resulting nonequilibrium time evolution, in which surplus adsorbate particles are desorbed, is measured and, assuming exponential decay, characterized by a desorption time $t_d(T_g, T_s)$.

To formulate the corresponding initial-value problem, one can assume that the initial equilibrium occupation for $t < 0$ is described adequately by the static part of the Hamiltonian H of the gas–solid system. The macroscopic time evolution in the system is then started at time $t = 0$ by raising the temperature of the phonon bath to T_s. This produces a nonequilibrium state of the gas and solid that can only be equilibrated by the dynamic part of H. Therefore, following Gortel and Kreuzer, the interaction is switched on at time $t = 0$ by setting

$$H = H_s + H_{st} + H_{dyn}\theta(t) \tag{5.4.1}$$

where $\theta(t) = 0$ for $t < 0$ and $\theta(t) = 1$ for $t > 0$. The physical quantity to be calculated is the time-dependent occupation of the bound state. In equation (5.4.1), H_s describes the dynamics of the solid in the absence of the gas, H_{st} the static gas Hamiltonian,[228] while the dynamic term H_{dyn} is switched on at time $t = 0$.

If this occupation is denoted by $n_0(t)$, Gortel and Kreuzer argue that the linear time dependence arising from Fermi's Golden Rule is to be interpreted as the lowest-order term in the expression

$$n_0(t) = n_0(\infty) + [n_0(0) - n_0(\infty)]\exp(-t/t_{d(f)}) \tag{5.4.2}$$

Gortel and Kreuzer then derive the flash-desorption time from such a second-order time-dependent perturbation calculation of the form

$$t_{d(f)}^{-1} = -48\omega_d\sqrt{\varepsilon_0}(1+\sqrt{r\varepsilon_0})^3\frac{m}{m_s}r^{-2}\left\{1-\left(\frac{\delta_s}{\delta_g}\right)^{3/2}\exp[(\delta_s-\delta_g)\varepsilon_0]\right\}^{-1}$$
$$\times\int_0^{1-\varepsilon_0}\frac{\exp[(\delta_s-\delta_g)(\omega+\varepsilon_0)]-1}{\exp[\delta_s(\omega+\varepsilon_0)]-1}\frac{\sqrt{\omega}\,d\omega}{r\omega+(2+\sqrt{r\varepsilon_0})^2} \tag{5.4.3}$$

In this expression there are, in principle, two independent parameters, E_0 and γ. Gortel and Kreuzer argue, however, that $E_0 \lesssim Q$ and γ^{-1} equals approximately 2–3 Å. The explicit definitions of the quantities in terms of which $t_{d(f)}^{-1}$ above is expressed are

$$r = (2m\omega_D/\hbar\gamma^2), \qquad \varepsilon_0 = |E_0|/\hbar\omega_D, \qquad \delta_{s,g} = \hbar\omega_D/k_B T_{s,g} \tag{5.4.4}$$

In Figure 5.6, the best fit obtained by Gortel and Kreuzer to the data of Cohen and King[229] is reproduced. They emphasize the following points:

1. The bound-state energy $|E_0|/k_B = 25$ K is about 25% less than the slope $Q/k_b = 31 \pm 1$ K, the slope generally increasing with increasing E_0.
2. Changing γ^{-1} from 1 to 3 Å increases $t_{d(f)}$ by a factor of 25. The best value $\gamma^{-1} = 2.5$ Å agrees with the related calculations of Bendow and Ying.[231]
3. Flash-desorption times in the one- and three-dimensional versions of the theory (with all parameters unchanged) are comparable.
4. Frenkel's equation $t_d = t_d^0 \exp(-Q/k_B T)$ can only be a valid parametrization over a limited range of flash temperatures for the system under study, from $T_s = T_g = 2$ K up to about $T_s = 15$ K. The desorption times are shorter for higher values of T_s.

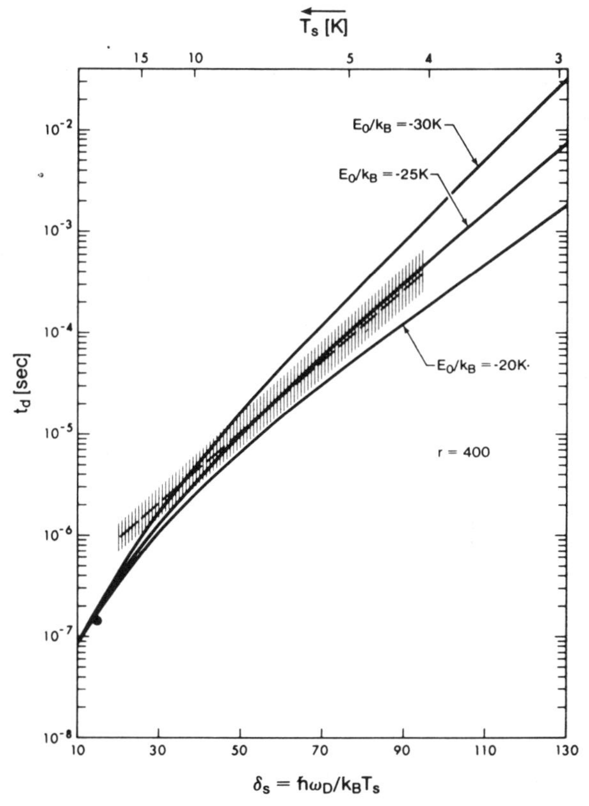

FIGURE 5.6. Best fit obtained by Gortel and Kreuzer to the data of Cohen and King, for flash desorption time vs. flash temperature T_s for He desorbing from constantan, with mass ratio $m/m_s = 0.065$; initial temperature $T_g \sim 2$ K.

5.4.2. Isothermal Desorption

The initial-value problem for isothermal desorption will now be formulated. At some time taken as $t = 0$ there is a sudden reduction in gas pressure, ideally to zero. The initial state for $t < 0$ is now one in which the gas and solid are in thermal equilibrium at temperature T. The final state is one in which all gas particles have been pumped out of the system. In equation (5.4.2) one must therefore take $n_0(t = \infty) = 0$, in which case the resulting isothermal desorption time in the one-dimensional model of Gortel and Kreuzer is[227]

$$t_{d(i)}^{-1} = 48\omega_D \sqrt{\varepsilon_0}(1 + \sqrt{r\varepsilon_0})^3 \frac{m}{m_s} r^{-2} \int_0^{1-\varepsilon_0} \frac{\sqrt{\omega}\, d\omega}{r + (2 + \sqrt{r\varepsilon_0})^2} \times \frac{1}{\exp[\delta_s(\varepsilon_0 + \omega)] - 1} \qquad (5.4.5)$$

It should be noted that the term $\exp[(\delta_s - \delta_g)(\varepsilon_0 + \omega)]/\{\exp[\delta_s(\omega + \varepsilon_0)] - 1\}$ in equation (5.4.3), that corresponds to readsorption of gas particles accompanied by spontaneous and stimulated emission of phonons, is absent in equation (5.4.5), the remaining term simply representing the absorption of phonons by gas particles desorbing from the solid surface. Gortel and Kreuzer have shown from their microscopic treatment that at very low temperatures, i.e., for $k_B T \ll |E_0|$, the isothermal desorption time $t_{d(i)}$ coincides with the adsorption time t_a.

To compare isothermal-desorption and flash-desorption times, one can first study the latter for an infinitesimal temperature step, i.e.,

$$t_{d(f)}^{-1}(T_s, T_s) = \lim_{T_s \to T_g} t_{d(f)}^{-1}(T_g, T_s)$$

$$= 48\omega_D \sqrt{\varepsilon_0}(1 + \sqrt{r\varepsilon_0})^3 \frac{m}{m_s} r^{-2}(\varepsilon_0 + \tfrac{3}{2}\delta_g)^{-1} \times \int_0^{1-\varepsilon_0} \frac{\varepsilon_0 + \omega}{\exp[\delta_g(\varepsilon_0 + \omega)] - 1} \frac{\sqrt{\omega}}{r\omega + (2 + \sqrt{r\varepsilon_0})^2}\, d\omega \qquad (5.4.6)$$

It should be noted first that, in this limit, no real time evolution takes place as nothing is done to the system. Nevertheless, this limit can be useful for comparison with isothermal-desorption times. For He desorbing from constantan, it turns out that $t_{d(i)}(T_s)$ is less than $t_{d(f)}(T_s, T_s)$ by an amount smaller than 1% for $T_s \lesssim 4$ K, increasing to some 30% at $T_s = 40$ K.

Another limit to which Gortel and Kreuzer[227] draw attention is

$$t_{d(f)}(T_g, T_s) \xrightarrow[T_g \ll T_s]{} t_{d(i)}(T_s)$$

if $\delta_g \varepsilon_0 \gg 1$.

5.4.3. Limitations of the Relaxation-Time Approach to Desorption

To find out the range of validity of the one-phonon second-order perturbation theory of desorption as a function of surface-potential depth, coupling of the adsorbate to the phonons of the solid, and temperature of the solid, Gortel and Kreuzer have performed a complete fourth-order calculation, including all one- and two-phonon processes.

The treatment is complex in detail, so the discussion below will be restricted to numerical results applicable to a special kind of gas–solid system, in particular, a system which develops just one bound state.

As regards the first example, Gortel and Kreuzer chose the range of the surface potential $\gamma^{-1} = 2.5$ Å and a ratio of gas-particle to solid-particle masses, $m/m_s = 0.0655$, appropriate for the He/constantan system for which the bound state is at a (renormalized) energy $\tilde{E}_0/\hbar\omega_D = -0.065$ with the Debye temperature for constantan being $\hbar\omega_D/k_B = 384$ K.

Their results demonstrate that for $\tilde{\varepsilon}_0 = |\tilde{E}_0|/\hbar\omega_D \lesssim 0.9$ and for $\delta = \hbar\omega_D/k_B T = 1$, the rate t_d^{-1} is dominated by the second-order contribution. For $\tilde{\varepsilon}_0 \rightarrow 1$, the second-order contribution goes to zero because a gas particle trapped in a bound state with $\tilde{\varepsilon}_0 > 1$ cannot be freed by absorbing a single phonon. For desorption from bound states with $\tilde{\varepsilon}_0 > 1$, two-phonon processes are necessary.

In the above example, the rate is dominated by the second-order contribution for $0 \lesssim \tilde{\varepsilon}_0 \lesssim 0.9$ and for $k_B T \lesssim \hbar\omega_D$. One can therefore expect the rate to drop appreciably as one-phonon processes become inoperative for $\tilde{\varepsilon}_0 > 1$.

Gortel and Kreuzer also consider another gas–solid system, which differs from the above case by γ^{-1} being taken as 0.5 Å. Their paper should be consulted for the details.

What should be stressed in their work is that the relaxation-time approach to desorption is valid for gas–solid systems for which the inequality $t_d \gg h/|\tilde{E}_0|$ is satisfied. These may be referred to as weakly coupled systems. One can trust a second-order calculation of the desorption time for such weakly coupled systems, provided $|\tilde{E}_0| \leq k_B T \lesssim \hbar\omega_D$ with fourth-order contributions important when $|\tilde{E}_0| < \hbar\omega_D$ and $k_B T \gtrsim \hbar\omega_D$.

The breakdown of the relaxation-time approach to desorption phenomena for strongly coupled gas–solid systems occurs when $h/\tilde{E}_0 \ll t_d$. It implies that transient phenomena and kinetic relaxation merge, resulting in a nonexponential time evolution. This in turn invalidates taking long time limits in the context of using Fermi's Golden Rule. Much work therefore remains to be done on the theory of desorption phenomena in strongly coupled systems.

5.5. COVERAGE-DEPENDENT REGIME

Most theories of physisorption kinetics and, it would seem, all microscopic kinetic theories of adsorption and desorption to date are restricted to situations of very low coverage. However, since Langmuir's pioneering work phenomenological theories have been developed that acount for saturation effects in the kinetics as an adsorbate approaches monolayer coverage. These theories are surveyed, for instance, in the books by Clark,[232] Tompkins,[233] and Wedler.[234]

If the coverage θ in the submonolayer range is defined as the fraction of the monolayer adsorbed, its rate of change with time is given by

$$\frac{d\theta}{dt} = r_a - r_d \tag{5.5.1}$$

where

$$r_a = P(2\pi m k_B T)^{-1/2} n_{max}^{-1} S(\theta, T) \tag{5.5.2}$$

is the rate of adsorption, with $S(\theta, T)$ the sticking coefficient and n_{max} the number density of adsorbed gas particles in a monolayer. Saturation is incorporated via an ansatz $S(\theta, T) \sim (1 - \theta)$. Also, for first-order desorption,

$$r_d = \theta / t_d \tag{5.5.3}$$

where t_d is the desorption time, which is usually given in the Frenkel–Arrhenius parametrization as

$$\begin{aligned} t_d &= t_d(T, \theta) \\ &= \nu^{-1}(T, \theta) \exp[Q(T, \theta)/k_B T] \end{aligned} \tag{5.5.4}$$

where $\nu = \nu(T, \theta)$ is called the prefactor and $Q(T, \theta)$, the isosteric heat of adsorption, is given by

$$Q(T, \theta) = k_B T^2 \left. \frac{\partial \ln P}{\partial T} \right|_{\theta} \tag{5.5.5}$$

Ultimately, the above coverage dependence arises from the two-body interaction between gas particles in the adsorbate, resulting in several different mechanisms to be discussed below following the account of Sommer and Kreuzer.[235] For strongly localized adsorption as it occurs in many chemisorption systems, only one particle can be adsorbed per adsorption site due to the finite size of the adsorbing particle and due to bond saturation restricting chemisorption typically to monolayer coverage. As the monolayer fills up, the long-range interaction between adsorbed particles can lead to a variety of ordered structures reflected, in particular, by characteristic changes in Q, ν, and t_d. Relevant experiments are reported by Bauer *et al.*,[236] Pfnür *et al.*,[237] and Bertel and Netzer.[238] A microscopic model has been analyzed by Zhdanov.[239]

Next, let us consider the adsorption of a gas onto a solid surface without pronounced adsorption sites. For physisorption systems, this implies that the surface potential is more or less uniform along the surface, i.e., it is a function only of the distance z above the surface. At high temperatures, gas particles in the adsorbate will then be highly mobile. With monolayer densities typically of the order of liquid densities, collisions of adparticles will probe predominantly the short-range repulsion between them, leading to a decrease in heat of adsorption as coverage builds up. A dramatic decrease occurs typically at monolayer completion, because the adparticles in the second layer are found to be more weakly bound to the adparticles in the first layer than the latter are to the metal. For multilayer adsorbates, Q will eventually approach the heat of vaporization of the corresponding liquid. The theory of Sommer and Kreuzer, to be treated below, demonstrates these features explicitly. At lower temperature, small lateral vibrations in the surface potential due to the lattice structure of the solid can lead to a two-dimensional commensurate crystallization in the adsorbate. In addition, the attractive part of the two-body interaction between adparticles can lead to a crystalline phase typically as an adsorbate superstructure, accompanied most likely by a rise in the heat of adsorption.

The mean-field theory of the kinetics of mobile multilayer adsorption, proposed by Sommer and Kreuzer, will now be summarized.

5.5.1. Mean-Field Theory of Physisorption

A gas particle adsorbing onto a clean surface finds itself in the bare surface potential $V_s(\mathbf{r})$. As the coverage builds up further, gas particles approaching the surface from the gas phase also experience the mutual forces due to the particles already adsorbed. These forces can be taken into account in the mean-field approximation by deriving, from a quantum-mechanical many-body Hamiltonian, spin-averaged temperature-dependent Hartree–Fock equations. Using this approach, they conclude that a good intuitive tool to understand multilayer physisorption in the mean-field approximation is given by the effective coverage-dependent surface potential

$$V_s(z, \theta) = V_s(z) + V_{mf}(z, \theta) \tag{5.5.6}$$

where, as usual, $V_s(z)$ is the bare surface potential seen by a single gas particle while $V_{mf}(z, \theta)$ is the potential representing the mean field generated by all other particles already adsorbed and calculated as a Slater average.[240]

5.5.1.1. Rate Equations

To describe the kinetics of adsorption and desorption, the static Hamiltonian of the gas–solid system must be supplemented by a term accounting for energy dissipation and supply for adsorption and desorption, respectively. For weakly coupled physisorbed gas–solid systems, this arises from coupling of the gas particles to the phonon bath of the solid, as first proposed by Lennard-Jones and co-workers.[241,242] The total Hamiltonian is expressed as

$$H = H_s + T + V_2 + V_s \tag{5.5.7}$$

where H_s is the phonon Hamiltonian of the solid, T is the kinetic energy of the gas particles, V_2 contains their mutual two-body interactions, and $V_s(\mathbf{r}, t)$ is the dynamic surface potential. For a molecular solid the latter quantity is given by

$$V_s(\mathbf{r}) = \sum_i V(\mathbf{r} - \mathbf{r}_i),$$

where $\mathbf{r}_i = \mathbf{r}_i(t)$ represent the positions of the thermally agitated constituent particles of the solid. For mobile adsorbates with surface potential a function of z only, one then argues that, with $\mathbf{u}(t)$ the displacement vector,

$$\begin{aligned} V_s(\mathbf{r}, t) &= V_s(\mathbf{r} - \mathbf{u}(t)) \\ &= V_s(z) - \mathbf{u}(t) \cdot \nabla V_s(z) + \cdots \end{aligned} \tag{5.5.8}$$

provided one-phonon processes dominate. Sommer and Kreuzer[235] develop this (see Appendix 6.6) to yield nonlinear rate equations.

5.5.1.2. Applications of Mean-Field Theory

At the time of writing, there do not seem to have been applications of mean-field theory to molecular desorption.

Sommer and Kreuzer[235] have applied their theory, for reasons that need not be elaborated here, to the case of a Fermionic gas particle ^{3}He desorbing from a Cu–Ni alloy, constantan. They predict transition from first-order desorption to zero-order evaporation at coverages slightly larger than a monolayer. For this system they have also predicted a compensation effect (see the theory of Peacock-Lopez and Suhl in Section 6.3.1 below). It is noteworthy that ^{4}He desorbing from constantan has been measured by Cohen and King[229] for θ equal to approximately 1.5.

A changeover from first- to zero-order desorption has been observed in the Xe–W(110) system by Opila and Gomer,[243] albeit at quantities of adsorbed gas such that uniform adsorption would be in the submonolayer regime. The results of Sommer and Kreuzer would then suggest that the zero-order desorption of Xe from W(110) takes place from the top of multilayer patches of adsorbate, one reason being that in this system the heat of adsorption also decreases only slightly on completion of the first three adlayers.

5.5.1.3. Summary and Range of Validity of Treatment

The main findings of Sommer and Kreuzer are:

1. Desorption rates increase with coverage in some systems, including ^{3}He on constantan, changing from first-order desorption at submonolayer coverage to zero-order evaporation above a monolayer.
2. A pronounced compensation effect is found when the desorption time has the Arrhenius form.

As regards the range of validity of the treatment, the following assessment is concerned with the assumptions made at the kinetics level.

For weakly coupled systems, the rate equations used are quite generally valid for Markov processes, as discussed, for instance, in the book by van Kampen.[175] The calculation of the transition probabilities R_{ij} as one-phonon processes has frequently been questioned. Gortel *et al.*[226] have shown corrections to be small for weakly coupled physisorbed systems.

Work on physisorption kinetics around monolayer coverage ought to include the possibility that the desorbing particle, in addition to absorbing phonons from the metal, can acquire energy from collective excitations in the adsorbate, presumably effectively below the ordering transition in the adsorbate. These could perhaps be usefully incorporated by means of the random-phase approximation. At higher coverages of a few layers, one expects intuitively that such collective excitations will become dominant.

As for generalizations, Sommer and Kreuzer suggest that a compensation effect (see Section 6.3 below) should be expected for phonon-mediated desorption around monolayer coverage in gas–solid systems where the phonon density of states in the surface region is not appreciably changed by the presence of the adsorbate. Local changes in surface properties can be very pronounced in chemisorbed gas–solid systems, in which case the above treatment is inappropriate.

5.5.2. Kramers' Theory

Kramers' theory, described in Section 4.10, has been applied by Redondo *et al.*[244] to obtain the coverage dependence of the desorption rate for Xe on W(111).

Most experimental studies of desorption processes involve the effects of adsorbate–adsorbate interactions leading, of course, to such coverage-dependent desorption rates. One experiment in which adsorbate–adsorbate interaction effects are often observed is the so-called temperature-programmed desorption (TPD). In a typical experiment, the surface is prepared with a given adsorbate coverage and the gas pressure monitored as the temperature is increased (usually linearly with time). From this type of experiment the strength of the adsorbate–surface interaction can be extracted, as well as some information about the distribution of adsorbate sites on the surface. Redondo *et al.* have used Kramers' theory to predict TPD spectra.

Experimental rates of desorption are usually expressed in the Arrhenius form (5.1.1). In general, both the preexponential factor A and desorption enthalpy E_d are functions of coverage θ, thereby reflecting adsorbate–adsorbate interactions. An objective of first-principles theory must naturally be to obtain an explicit expression for the rate that will allow A and E_d to be related to the specific adsorbate–surface and adsorbate–adsorbate interactions, or, more specifically, to parameters characterizing them. This would then allow such parameters to be extracted from, say, a fit to experimentally determined TPD rates.

Redondo *et al.* have extended the results of Kramers' theory given in Section 5.1 to include surface-coverage dependence of the desorption rate. They have applied this extension to Xe atoms using model adsorbate–adsorbate interactions.

5.5.2.1. Coverage-Dependent Kramers' Treatment

Adsorbate–adsorbate interactions influence the desorption rate via modification of the interaction potential $V_{\text{eff}}(z)$, and hence of the desorption enthalpy.

In what follows, the simplifying, plausible assumption will be made that the direct adsorbate–surface interaction is not changed by adsorbate–adsorbate coupling, so that the net potential seen by the desorbing particle is the superposition of its interaction with the surface and with the other adparticles. Although this assumption would obviously be vitiated by three-body or charge-transfer effects, it is likely to be appropriate quantitatively for adsorbates such as Xe.

The minimum free-energy arrangement of the adparticles on the surface must be determined in order to include coverage effects into D_e and $V_{\text{eff}}(z)$. To do so, Redondo *et al.* assume that the particles adsorb at well-defined periodic sites on the infinite surface. Next, a relatively small region of the surface is chosen, referred to as the desorption zone (in their detailed application to Xe on W, Redondo *et al.* took this area to contain nineteen adsorption sites), which is used to generate all possible arrangements of the adparticles for each coverage. In order to determine the configuration(s) exhibiting the minimum free energy, one must calculate the total energy associated with each configuration. Since the adparticle–surface interactions are the same for all arrangements, their sum over all the adparticles in the system can be omitted from the calculation of the total energy. This is written as a pairwise sum of adsorbate–adsorbate interaction energies. Redondo *et al.* divide this pairwise sum into two parts such that

$$E_{\text{total}} = \sum_{\substack{\text{desorption}\\ \text{region}}} (V_{\text{in-in}} + V_{\text{in-out}}) \tag{5.5.9}$$

where E_{total} is the total adsorbate–adsorbate interaction energy of the desorption region, $V_{\text{in-in}}$ represents the adsorbate interactions within the desorption region, and $V_{\text{in-out}}$ corresponds to the interaction energy of an adparticle inside the desorption region and all the other adparticles outside this zone. To evaluate the second term in equation (5.5.9), a mean-field approximation is employed in which each site outside the desorption region is assumed to contain θ adparticles, θ being the coverage. The value of E_{total} for the different configurations is then calculated by combining an appropriate adsorbate–adsorbate interaction potential with the arrangement of the adparticles in the desorption region.

Once the total energies for all possible arrangements at a given θ have been evaluated, the configurations are subdivided into groups according to their E_{total} values. The free energy associated with each group of configurations is then obtained as a function of temperature with the aid of

$$F = E_{\text{total}} - TS \tag{5.5.10}$$

and

$$S = k_B \ln \Gamma \tag{5.5.11}$$

where Γ is the number of configurations with energy E_{total}. The arrangement of adparticles (inside the desorption region) used to evaluate the desorption rate is chosen from the group of configurations corresponding to the lowest free energy. The validity of using only the minimum-free-energy configuration in calculating the rate of desorption is confirmed by Redondo *et al.* using detailed analysis.

After identifying the minimum-free-energy configurations (for given coverage and temperature range), the net potential felt by the desorbing particle is calculated as

$$V_{net}(z) = V_{eff}(z) + \sum_{j \neq 0}^{\text{adparticle}} V_{0j}(z) \tag{5.5.12}$$

where V_{eff} is the effective adsorbate–surface interaction potential for $\theta = 0$ and V_{0j} is the pairwise interaction between the desorbing particle and the jth adparticle on the surface. Finally, the potential energy in equation (5.1.23) is replaced by $V_{net}(z)$ and D_e by the corresponding well depth of the net potential to obtain the temperature dependence of the rate of desorption for a given coverage.

5.5.2.2. Temperature-Programmed Desorption Spectra

From the coverage dependence of the desorption rate $R(\theta)$ one can extract the corresponding θ-dependence of the preexponential factor A and the effective dissociation energy $D_e(\theta)$. Redondo *et al.* employed these relations to calculate the temperature-programmed desorption spectra by solving the Redhead equation discussed in Chapter 1, namely

$$\frac{d\theta}{dt} = -\theta A(\theta) \exp[-D_e(\theta)/k_B T] \tag{5.5.13}$$

This equation has been solved numerically for a linear time dependence of the temperature.

5.5.2.3. Results and Discussion

Redondo *et al.* studied the effect of coverage on desorption rate within the above framework by examining, as a model system, the desorption of Xe atoms from W(111) surfaces. As a result of the weak bonding ($D_e = 5$

kcal) one expects only a small overlap between the adsorbate and surface electronic wave functions, and hence the three-body and charge-transfer effects should be small. Consequently, the adsorbate–surface interaction can be obtained from low-coverage experimental results. However, the nature of adsorbate–adsorbate interaction needs further discussion.

While modification of the free-space interaction along the lines of the treatment in Section 2.1 would seem very appropriate here, the limiting cases discussed to date are:

1. An attractive Xe–Xe interaction obtained from gas-phase scattering data.
2. A purely repulsive Xe–Xe interaction based also on experimental gas-phase studies.

Redondo *et al.* note that, in line with the theory presented in Section 2.1, the presence of the surface will make the effective Xe–Xe interaction less attractive than in the gas phase, so that the actual potential should lie in between the two limits.

In calculating the desorption rate, Redondo *et al.* always focus on a site (cf. Figure 5.7; site 1, say) that happens to be occupied, and then calculate the rate of desorption from this site. They assume that the other possible adsorption sites are arranged in a periodic lattice forming a hexagonal pattern commensurate with the substrate (cf. Figure 5.7). The specific arrangement of occupied sites used in calculating the thermal-desorption spectrum is the most probable one (the lowest free energy) for a

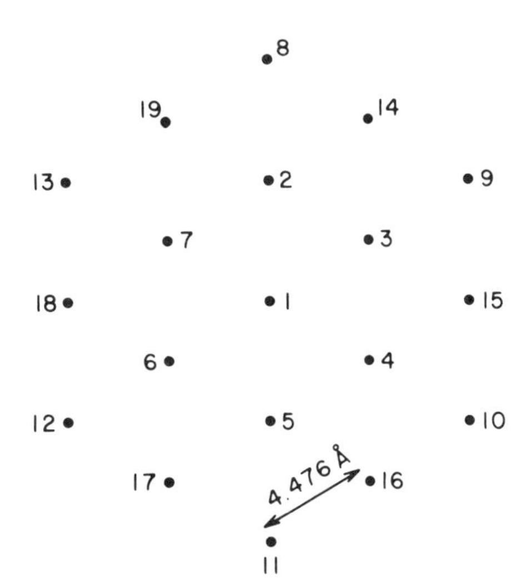

FIGURE 5.7. Depicts site (labeled 1) from which rate of desorption is calculated (after Redondo *et al.*[244]). Actual adsorption sites shown are for Xe on W(111) surface.

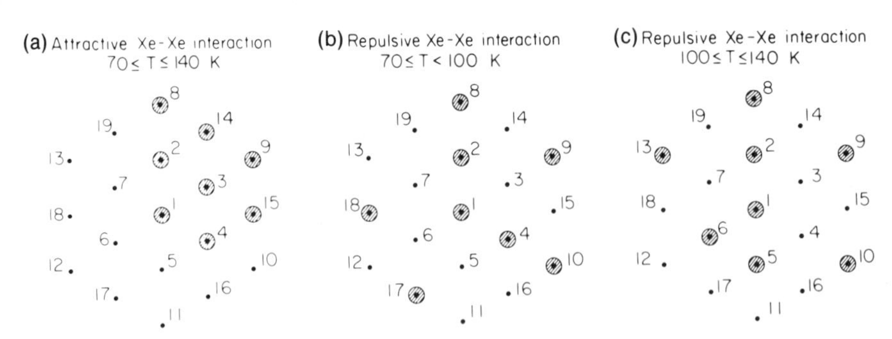

FIGURE 5.8. (a) Optimum arrangement for attractive Xe–Xe interatction. (b), (c) Minimum-free-energy configuration corresponding to repulsive Xe–Xe interaction. θ is taken as 0.42.

given coverage θ and temperature T. In order to calculate this most probable configuration, Redondo *et al.* consider the desorption region of Figure 5.7 containing eighteen additional sites. This includes all sites within 10 Å of site 1. Such an approach should suffice since the range of the Xe–Xe potential is only 8 Å (interaction with adsorbates outside the desorption region, shown in Figure 5.7, is included in the calculations via a mean-field description). The desorption rate is then calculated assuming that the configuration of the adsorbates does not change during the process of desorbing the adsorbate from site 1 [this assumption should be valid for Xe on W(111), because the diffusion barrier is much larger than $k_B T$].

5.5.2.4. Attractive Xe–Xe Interaction

For the attractive Xe–Xe potential, Redondo *et al.* used the form determined by Barker *et al.*[245] from gas-phase molecular-beam scattering studies. This potential has a well depth of 281 K (0.024 eV) and an equilibrium distance of 4.35 Å, slightly shorter than the nearest-neighbor site distance (4.47 Å) on W(111). The optimum arrangement for this attractive Xe–Xe interaction is shown in Figure 5.8(a) and is independent of temperature over the range between 70 and 140 K. Essentially, the system has condensed to the lowest enthalpy configuration. At higher temperatures the entropy would lead to a more dispersed configuration.

Figure 5.9 presents the results of Redondo *et al.* as a function of temperature for different coverages. At a given temperature, the desorption rate increases with decreasing coverage. This behavior is a consequence of the attractive Xe–Xe interactions that lead to a higher effective bond energy as the coverage is increased. This trend is apparent from an examination of Figure 5.10, where the variation in bond energy is given as a function of coverage. The dependence of the preexponential factor on the coverage is also shown in Figure 5.10, where Redondo *et al.* find that $A(\theta)$ varies but little with coverage. It is therefore reasonable to assume a constant preexponential factor throughout the range of coverage studied. The conclusion of this work is that the most important factor governing the coverage dependence of the overall rate of desorption (for an attractive Xe–Xe interaction) is the variation of the well depth.

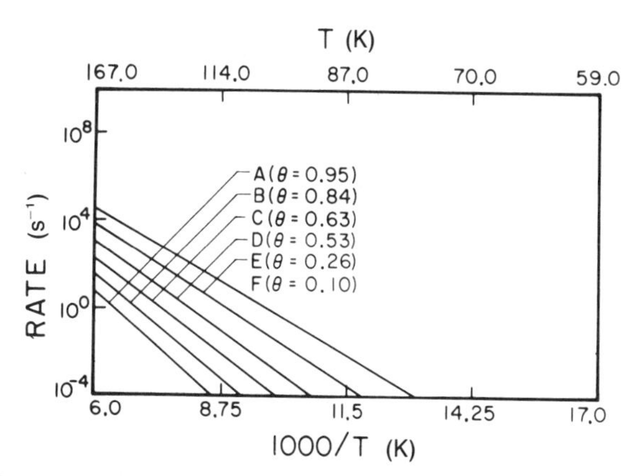

FIGURE 5.9. Results of Redondo *et al.*[244] for rates of desorption of Xe from W(11$\bar{1}$) as a function of temperature for different coverages (using attractive Xe–Xe interaction).

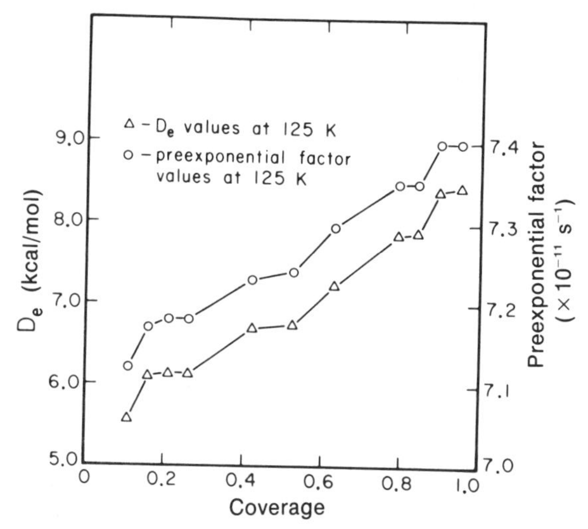

FIGURE 5.10. Variation of effective dissociation energy and preexponential factor as a function of coverage for desorption of Xe from W(111) using attractive Xe–Xe interaction (after Redondo *et al.*[244]).

5.5.2.5. Repulsive Xe–Xe Interaction

The behavior of the desorption rate as a function of coverage for a repulsive adsorbate–adsorbate interaction was studied using the Xe–Xe interaction potential reported by Leonas, as obtained from gas-phase molecular-beam experiments. This potential includes only repulsive interactions between the electron clouds of two Xe atoms as they are brought close together. As in the attractive-interaction case, this potential is short range (at 8 Å, the interaction energy is 10^{-6} eV) and hence Redondo *et al.* used the same desorption region as in Figure 5.7.

Their minimum-free-energy configurations corresponding to a repulsive Xe–Xe interaction are summarized in Figures 5.8(b) and (c). In contrast to the results obtained for the attractive potential, it is found that for some coverages the optimum configuration depends strongly on temperature. In particular, for certain coverages it is possible to find two different optimal arrangements of adparticles in the desorption region, depending on the temperature. This is due to the increased importance of the entropy term in the free energy.

The variation of desorption rate as a function of temperature is shown in Figure 4 of Ref. 244. For a given temperature, the rate increases with increasing coverage. This behavior is opposite to that found for the attractive potential. As expected, the difference between the attractive and repulsive potentials is most marked for high coverage. For instance, when $\theta = 0.95$ the rates of desorption differ by at least six orders of magnitude while for $\theta = 0.1$ they differ by 0.5 orders of magnitude. Figure 5 of Ref. 244 shows the effective dissociation energy and preexponential factor as a function of coverage. The dissociation energy again depends strongly on coverage while the preexponential factor remains practically constant. The dominant factor in the behavior of the total rate of desorption is the decrease in effective dissociation energy with increasing coverage.

Comparison with TPD curves of Xe on W(110) [not W(111), as in the calculations] shows that the effective Xe–Xe potential should be intermediate between the attractive and repulsive functions. A rough estimate indicates that the Xe–Xe well depth must be reduced by a factor of about 5.

Subsequently Joyce, Grout and March (1986; to be published) have performed related calculations, but now using the Xe–Xe interaction according to the theory of Mahanty and March, discussed in Chapter 2, Section 2. As anticipated above, the results thereby obtained do indeed lie intermediate between the attractive and repulsive potential functions described above, and give confidence in their first-principles interaction potential as a useful starting point for such desorption studies. Related calculations have also been carried out by Joyce *et al.* for rare-gas atoms other than Xe.

5.6. DESORPTION INDUCED BY ELECTRONIC TRANSITIONS

To conclude this chapter, electronically induced desorption will be reviewed briefly. Major emphasis in this section will be focused on identifying the mechanisms responsible for desorption of molecules due to photon, electron, and ion bombardment of surfaces.

A convenient starting point is a discussion of models of electron-induced desorption, reviewed by Gomer.[246]

5.6.1. Menzel–Gomer–Redhead Model

The basic argument is as follows. Elementary considerations indicate that direct momentum transfer from approximately 100 eV electrons to even the lightest adsorbates is wholly insufficient to lead to anything but slight vibrational excitation, so that desorption or rearrangement must result from electronic excitation of the adsorbate or the adsorbate–substrate bond. In this respect, Gomer likens electron-stimulated desorption to analogous processes in isolated molecules. In the latter, the primary excitation event is necessarily irreversible. If the molecule is ionized, it will remain so; if it is excited to an antibonding state, it will dissociate, or remain excited until deactivated radiatively.

For a molecule adsorbed on a metal surface, however, ionization, or excitation of the substrate–adsorbate bond, can be followed by neutralization or by a "bond-healing" transition due to the availability of conduction electrons in the metal. Thus the overall ionization and desorption cross-sections would be expected to be less than for comparable processes in isolated molecules, even though the primary excitation cross-sections should be comparable if not identical. Thus the electron-stimulated desorption cross-section should have the form

$$\sigma = \sigma_e P \tag{5.6.1}$$

where σ_e is a primary excitation cross-section and P an escape probability.

5.6.1.1. Desorption of Neutral Species

The situation is depicted by a potential-energy diagram, as in Figure 5.11, which shows excitation to an antibonding curve. The initial excitation is Franck–Condon-like, namely vertical, and is followed by descent of the adsorbate A along the repulsive curve labeled $(M + A)^*$. Transition to the bonding curve $M + A$ can best be represented by noting that the latter is merely the lowest number of an infinite manifold $\{M^* + A\}$ of curves which differ from $M + A$ by some excitation in the metal M, which does not affect

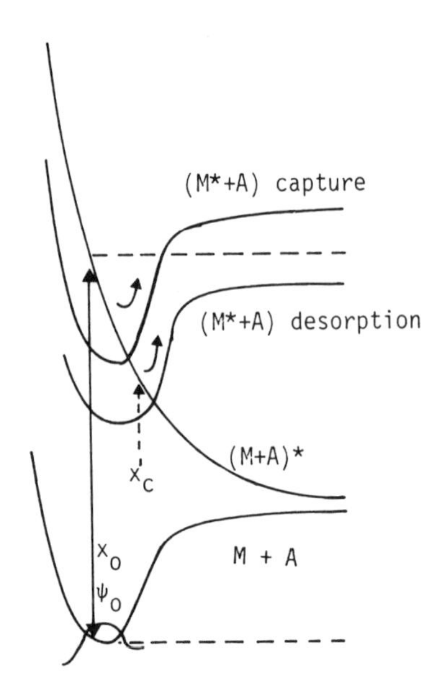

FIGURE 5.11. Critical distance x_c beyond with a transition will still lead to desorption.

the bonding. In other words, a bond-healing transition from the antibonding to the bonding state must be energy-conserving and thus transfers the appropriate amount of energy to an electronic excitation in the metal. Since there is an infinite manifold of curves M* + A, the transition can occur in principle anywhere along the trajectory but not all transitions will result in recapture. Figure 5.11 shows that there is a critical distance x_c beyond which a transition will still lead to desorption (along a copy of the ground-state curve) because the adsorbate A has sufficient kinetic energy for escape. The quantity P can thus be expressed in the form

$$P = \exp\left(-\int_0^{t_c} dt/\tau\right) = \exp\left(-\int_{x_0}^{x_c} \frac{dx}{v(x)\tau(x)}\right) \tag{5.6.2}$$

where $\tau(x)$ is the lifetime with respect to the transition, assumed to be distance-dependent, and $v(x)$ the velocity of A along the repulsive curve. A detailed analysis of equation (5.6.2), assuming an exponential dependence of τ on distance from the surface and an exponential form of the curve $(M+A)^*$ in the region of interest, has been given by Menzel and Gomer.[248]

Following Gomer, it will be assumed that τ is constant over the range

x_0 to x_c and that the repulsive curve can be taken to be linear over this small range. With these approximations P reduces to

$$P = \exp(-\Delta t/\tau) \tag{5.6.3}$$

where Δt is the time required for A to move from x_0 to x_c along the repulsive curve $(M+A)^*$, and is given by

$$\Delta t = (2m\Delta x/S_r)^{1/2} \tag{5.6.4}$$

Here $\Delta x = x_c - x_0$ and S_r is the slope of the potential curve $(M+A)^*$, i.e.,

$$S_r = \Delta E/\Delta x \tag{5.6.5}$$

Equations (5.6.3) and (5.6.4) can now be combined to yield

$$P = \exp[-(2m\Delta x/S_r)^{1/2}/\tau] = \exp(-cM^{1/2}) \tag{5.6.6}$$

where M is the molecular weight and c a constant. This mass-dependence, predicted by Menzel and Gomer, has been found in a number of cases.

It is worthwhile estimating Δx, even crudely. Clearly, it could be determined quantitatively if the shapes of relevant potential-energy curves were known. If the repulsive curve is taken to be linear and the ground-state curves $M^* + A$ are approximated by triangular wells truncated on the right side, it is easy to show that

$$\Delta x = a/(1 + S_r/S_c) \tag{5.6.7}$$

where a is the half-width of the triangular well and S_c the slope of the outgoing branch of this well, namely

$$S_c = H_a/a \tag{5.6.8}$$

where H_a is the binding energy in the ground state. Since $S_r \gg S_c$ in general,

$$\Delta x \sim a(S_c/S_r) \tag{5.6.9}$$

If one takes $a = 2$ Å and $S_c/S_r = 1/4$, then $\Delta x = 0.5$ Å. While this value changes from one example to another, Δx is of the order of 1 Å or less.

5.6.1.2. Ionic Desorption

The above treatment has dealt with excitation to a repulsive, but neutral state. Figure 5.11 shows excitation to an ionic curve $M^- + A^+$, assumed to be located so that vertical excitations from the ground-state curve put the system on the repulsive part of the ionic curve. The situation depicted there differs from the previous case in one important respect. Recapture will occur only for transitions at $x < x_c$, as before, but desorption as an ion requires that no transitions occur along the entire trajectory. Evidently this increases the effective width of the zone in which neutralization can occur; it is limited only by the increase in τ with distance from the surface, and the increasing velocity of the departing ion. The matrix elements for neutralization are probably much greater than those for transition from an antibonding to the bonding curve. Thus, at a given distance, $\tau_{\text{neutralization}} \ll \tau_{\text{antibonding–bonding}}$ and one would expect cross-sections for ion formation to be much smaller than for neutral desorption. Equation (5.6.6) then also predicts a much greater isotope effect in ionic desorption. Both these expectations are borne out by experiment.

The mass dependence of P can be used to separate σ into σ_e and P, provided cross-sections can be determined for different isotopes. Hence equation (5.6.6) gives

$$c = \frac{\ln(\sigma_1/\sigma_2)}{M_2^{1/2} - M_1^{1/2}} \tag{5.6.10}$$

where the subscripts refer to the isotopic species 1 and 2. Several points are noteworthy here. Excitation cross-sections are generally quite high, with the exception of that for O^+ formation for oxygen adsorbed on W(110). Quite generally, values of P for ionic desorption are much less than for neutral products, as discussed above, and σ_e for ions and neutrals in the same system generally differ, indicating that ion neutralization is not the main source of neutrals. The mechanism for Kr desorption is somewhat different from those treated so far and is discussed by Knotek.[253]

It is clear from Figure 5.11 that the energy distribution of ions is a measure of the ground-state vibrational wave function, although skewed by variations in neutralization probability, and it is possible to obtain a rough idea of the slope S_r from this distribution.

The actual nature of the antibonding curves, details of the antibonding–bonding transition, or of neutralization in the case of ions, have not so far been specified. There are many possibilities for the antibonding–bonding transition. For instance, if the ground state corresponds to a localized electron-pair bond, the antibonding state could be a triplet state, and bond-healing a triplet–singlet transition. In the case of isolated molecules this would be very slow, but there are obviously numerous ways of achieving such transitions via conduction electrons. In the case of reneutralization, the transition could be resonance tunneling, possibly to an excited state of A, or an Auger transition to the ground state.

In all these cases, the designation of the final state after transition as M + A (or possibly M* + A) is justified if the energy of Auger electrons is included as an excitation of the metal M. It is easy to show that, regardless of which particular Auger or other transition occurs, the final energy of M is the same. Obviously, the states M + A of given total energy are highly degenerate due to the degeneracy of M.

Gomer stresses that it is easy to understand from the model why electron-stimulated desorption cross-sections should be sensitive to details of binding. First x_0, and thus x, will depend on binding, and second, the matrix elements for the various transitions will also depend on geometry, number of neighbors, etc. Similar factors enter, at least in part, the anisotropies of the angular distribution of desorption products.

A slight generalization of the curves in Figure 5.11 also allows one to understand why electron-induced rearrangements are possible. Excitation from one stable binding mode can occur to an intermediate excited state, followed by deexcitation into another binding state. To date, the rearrangements that have been observed consist mainly of the dissociation of diatomic molecules by electron impact. Rubio *et al.** find the state of adsorbed CO on W(110) to be characterized by desorption of CO and CO^+. (See Gomer[246] for detailed discussion of this and also of related behavior observed with adsorbed N_2 on tungsten.)

5.6.1.3. Transcending the Menzel–Gomer–Redhead Model

Refinements of the Menzel–Gomer–Redhead model can, of course, be contemplated, even though it provides a most useful perspective from which to view electron-stimulated desorption from metals. A number of points, which might eventually require transcending their model, will be briefly referred to below.

The first point is that it is obviously a simplification to consider a single antibonding or even a single ionic curve. There will be many such curves, even barring counting metal excitations irrelevant to the adsorbate–metal surface bond. In the repulsive region, curves corresponding to different states may lie very close to each other and it is possible that well-defined transitions to a single excited state do not occur.

Brenig[(247)] has reconsidered the recapture mechanisms proposed by Menzel, Gomer, and Redhead.[(248,249)] His arguments do not seem at the time of writing to be decisive and have been partly answered by Gomer.[(250)] The work of Freed[(251)] should also be consulted in this connection.

* J. Rubio, J. M. López-Sancho and M. P. López-Sancho, *J. Vac. Sci. Technol.* **20**, 217 (1982).

5.6.2. Knotek–Feibelman Mechanism of Initial Excitation

The original Menzel–Gomer–Redhead model[248,249] did not consider excitation processes in detail but contented itself with assuming that they occur.

It has been found experimentally that core-ionization processes, despite their intrinsically smaller cross-sections when compared to primary-valence processes, can contribute significantly to the desorption of ions. This fact was emphasized in the work of Knotek and Feibelman.[252,253]

Their so-called mechanism, unlike the Menzel–Gomer–Redhead model designed for chemisorption on metals, stemmed from their observation that when the surface of an ionic oxide TiO_2 was subjected to electron bombardment, O^+ ions were generated, despite the fact that oxygen is present in TiO_2 as O^{2-}. They argued that the ejection of three electrons by conventional impact ionization was unlikely and proposed instead the interatomic Auger process shown in Figure 5.12. Since the conduction band of Ti (present as Ti^{+4} in TiO_2) is empty, ejection of a core electron can only be followed by Auger filling via oxygen electrons. If two Auger

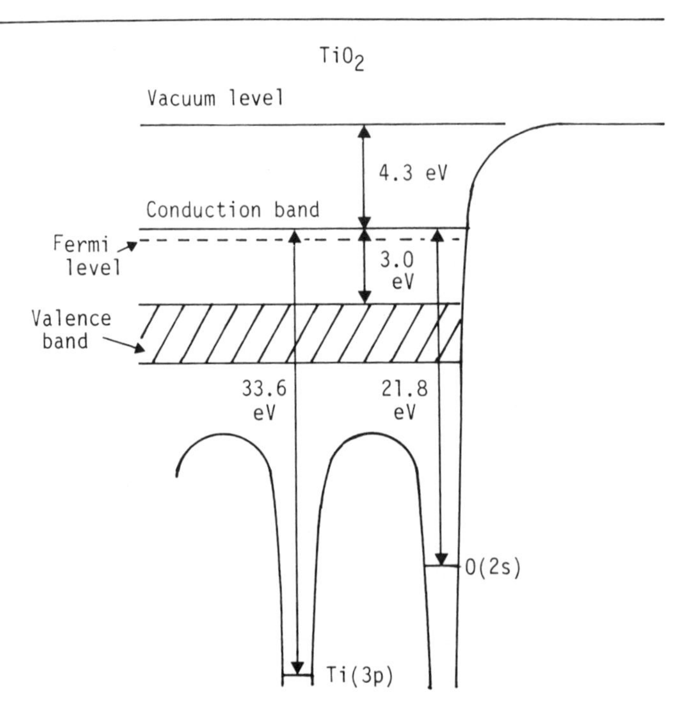

FIGURE 5.12. Proposed interatomic Auger process.

electrons are ejected, O^{2-} is converted to O^+. Since the O^+ ion has a Ti^{+4} ion as its nearest neighbor, there is strong Coulomb repulsion and O^+ will be ejected. Furthermore, in substances like TiO_2, and in contrast to metals, reneutralization is inherently unlikely since there are no available electrons. Examination of O^+ yields as a function of electron energy showed that the onset of O^+ generation corresponds to the energy required for Ti $3p$-ionization. It was argued by Knotek and Feibelman that electron energies just sufficient for O $2s$-ionization did not contribute appreciably because the process O $2s^12p^6 \rightarrow$ O $2s^22p^3 + 2e^-$ was just barely exoenergetic, thus leaving only a very small amount of phase space to the Auger electrons. These workers anticipated similar processes in ion generation from chemisorbed layers on metals and also in the production of neutral O from metal surfaces.

There is convincing evidence for the Knotek–Feibelman excitation mechanism for ion generation in the case of oxides. In the case of oxygen and other substances adsorbed on metals, there is now evidence that interatomic Auger processes also contribute to ionization but are not solely responsible. According to Knotek and Feibelman, ionization of O $2s$ following by intra-atomic Auger processes may also make a contribution to O^+ (and O) desorption on metals. The work of Franchy and Menzel further suggests that intramolecular Auger processes are important in ion generation from adsorbed molecules. For adsorbed CO, production of CO^+ is enhanced by ejection of an electron from C $1s$ while O^+ production is strongly enhanced by O $1s$-excitation.

Gomer[246] has considered these and other matters in detail and has examined energy thresholds in chemisorption. The interested reader is referred to his review and to the numerous other relevant articles in the same volume, devoted to desorption induced by electronic transitions.

5.6.3. Structural Information from Photon-Stimulated Desorption

Two examples of angle-resolved photon-stimulated desorption which have been studied are:

O^+ from oxygen adsorbed on W(111).
O^+ from CO adsorbed on Ru(001).

The ion-desorption patterns were in excellent agreement with previous ESDIAD data. Moreover, measurement of ion yield vs. photon energy clearly demonstrated the role of the metal-core hole excitations in photon-stimulated desorption of O^+ from O/W, while intramolecular excitations are responsible for photon-stimulated desorption of O^+ from chemisorbed CO.

Madey *et al.*[254] have compared electron- and photon-stimulated desorption for H_2O adsorbed on Ti. This illustrated the power of photon-stimulated desorption in its sensitivity to a specific bonding ligand (OH) in different confiugrations, namely OH adsorbed on Ti(001) as Ti–OH, in comparison to condensed water layers (HOH).

5.6.4. Theory of Angle-Resolved Electron- and Photon-Stimulated Desorption

Reviews of the theory have been given by Madey[255] and by Niehus.[256]

The work of Gersten *et al.*[257] should be referred to as the first in which model calculations using dynamic arguments allowed the ion angular distribution to be related to the positions of adatoms in different surface-binding sites. They report angular distributions for O^+ ions desorbing from O atoms adsorbed at different sites on W(100) and W(111). Preuss performed what appear to be improved calculations using screened atomic Coulomb potentials for the repulsive interactions between ion and surface and generated model ion-desorption patterns; the data agree closely with experimental results.

Clinton[258] presented a quantum-scattering theory of electron-stimulated desorption in which he assumes that the repulsive final-state potential energy acting on a desorbed atom is a sum of central potentials. He concludes that desorption will occur in a direction for which the vibrational wave function of the substrate–atom bond attains its maximum value, namely in the direction of the chemical bond "broken" by the excitation. He indicates that ESDIAD processes should be dominated by initial-state (ground-state) structures of molecules on surfaces. Subsequently, Clinton[259] used a frozen orbital approximation to show that the instantaneous force on an electron-stimulated desorbed ion will be repulsive, in the direction of the original bond. He concludes that ESDIAD images the initial-state vibrational wave function. These calculations, however, do not include reneutralization processes nor the influence of the image potential on ion trajectories (see Section 5.6.4.1 below).

Even in the initial-state picture in which the desorption impulse is along the bond direction, final-state effects will influence the trajectories and the resultant ion-desorption patterns, as discussed by Madey.[255] Possible final-state effects include anisotropy in the ion-reneutralization rate, "focusing" effects due to curvature in the final-state potential, and deflection of the escaping ions by the electrostatic image potential. Since no detailed knowledge of final-state potentials is available at present, only model calculations can be employed by which to consider their influence.

There have been several studies in which evidence for anisotropies in ion-neutralization rates has been reported. In these cases, high-energy ion beams scattered from single-crystal surfaces were preferentially neutralized along scattering azimuths where the scattered ion interacts most strongly with substrate atoms. It appears that such effects are most important near grazing scattering angles and should not play a major role in producing contrast in ESDIAD patterns.

5.6.4.1. Influence of Image Potential on Ion Trajectories

Several workers have considered the influence of the image potential on ion trajectories. Their conclusion is that the image potential causes an increase in polar angle for ion desorption. Clinton's calculations indicate that an ion desorbing at an initial angle α_i with respect to the surface normal will arrive at the detector with an apparent desorption angle α_0 given by

$$\cos \alpha_0 = \cos \alpha_i \left\{ \frac{1 + V_I/[(E_K - V_I)\cos^2 \alpha_i]}{1 + V_I/(E_K - V_I)} \right\}^{1/2} \tag{5.6.11}$$

Here, V_I is the screened image potential at the initial ion–surface separation z_0 and E_K is the final, measured kinetic energy of the ion. It is noteworthy that V_I is always a negative quantity so that

$$|V_I|/(E_K - V_I)| \leq 1 \qquad \text{and} \qquad \alpha_0 \geq \alpha_i \tag{5.6.12}$$

A discussion of the location of the image plane and the assumptions leading to equation (5.6.11) can be found in the papers of Clinton and Gadzuk.

Equation (5.6.11) has a meaningful solution if the following inequality is satisfied:

$$\left| \frac{V_I}{(E_K - V_I)\cos^2 \alpha_i} \right| \leq 1 \tag{5.6.13}$$

This means that there will be a cutoff angle for ion desorption defined by the value of α_i for which $\alpha_0 = 90°$. For values of α_i greater than the cutoff angle, the ions will follow low trajectories and be recaptured by the surface. The possibility of beam damage induced by bombardment of surface species by low-energy electron-stimulated desorbed ions is high.

As discussed by Madey,[255] it appears that the major final-state perturbation of ion trajectories is in the polar direction. In general, for a perfectly planar surface or for desorption along an azimuth of symmetry, the perturbation of the azimuthal angle should be minimal.

Madey *et al.* note that photodissociation studies of nonlinear gas-phase triatomic molecules (NO_2, NOCl) demonstrate that dissociation occurs rapidly, before statistical equilibrium is attained. The angular energy of fragments indicates that dissociation occurs from "bent" final-state structures. A modified impulsive model for NOCl dissociation, in which fragment recoil lies approximately along the direction of the breaking NO–Cl bonds, is consistent with the data.

5.6.5. Electron-Stimulated Molecular Desorption from Metals

The electron- and photon-stimulated desorption (ESD/PSD) of ions and neutrals has been shown to be a sensitive probe for the study of surface–adsorbate interactions and bonding, and even for the determination of such physical properties as the chemisorption bond angle (see Chapter 2 for H_2O on transition metals) and bond site.

However, full understanding of the ESD/PSD mechanism has not been achieved at the time of writing. Nevertheless, comparative investigations of Ramaker *et al.*[260] between molecular dissociation in the gas phase and chemisorption on the surface have been very helpful in understanding the desorption mechanisms in molecularly chemisorbed systems (such as CO and H_2O on metals).

5.6.5.1. ESD for Molecules from Metals

ESD N_2^+ and CO^+ ion yields have been reported from N_2 and CO/Ru and also neutral N_2 and CO desorption yields have been published.[254] A detailed investigation of the CO and CO^+ thresholds, and comparison with satellite peaks in the UPS spectrum for CO/Ru, has again pointed to the importance of the $1h$ $1e$ excited states.

It was explained in Chapter 2 that bonding of CO with the metal surface involves an interaction and charge donation to the metal via the 5σ orbital, and back-bonding with the 2π orbital. The 2π interaction with the metal leads to $2\pi_b$ and $2\pi_a$ (bonding and antibonding) orbital combinations with the Fermi level lying between these two orbitals. It has been suggested that an electron in the $2\pi_a$ orbital or a hole in the π orbital is involved.

Chapter 6

Catalysis

This chapter first reviews some definitions and concepts of heterogeneous catalysis and catalytic science, following the review by Spencer and Somorjai.[13] These authors point out that the word "catalyst" was invented by Berzelius[261] to describe the property of certain substances which facilitate chemical change without being consumed during the reaction. Catalysts aid the attainment of chemical equilibrium by reducing the potential-energy barriers in the reaction path.

Following the introduction to definitions and concepts in Section 6.1, a number of topics have been selected on which to focus attention. Future progress in the areas surveyed above in Chapters 1 through 5 will throw light on the problems of catalysis. However, at the present stage of development of the subject, most attention will be focused on three topics:

1. First, an example of the way X-ray photoemission spectroscopy (XPS) can be used for *in-situ* species-resolving studies of surface reactions, in particular those inaccessible to thermal-desorption studies, and then a further example in which photoelectron spectroscopy can be employed in relation to heterogeneous catalysis.
2. The theory of the compensation effect, presented in Section 6.3 below. Though the example chosen may not be one of quantitative relevance, the basic approach is well worth focusing on as an example of the principles that appear to underlie this effect.
3. The way in which Woodward–Hoffman rules may be relevant to single-crystal catalysts, as set out in Section 6.4.

6.1. SOME DEFINITIONS AND CONCEPTS

Catalysis is clearly a kinetic phenomenon. The speed of a catalyzed reaction is often described in terms of a "turnover rate," which determines how many reactant molecules are converted on an "active site" or on a unit catalytic surface, per second, at a given temperature, pressure, and concentration of reactant and products.

Since the catalytic action occurs at specific sites on solid surfaces, referred to already as "active sites," the rate can be significantly increased by using high-surface-area catalysts.

6.1.1. Specific Rate

The specific rate, namely the number of reactant or product molecules converted or produced per unit catalyst area per second (or the number of reactant or product molecules per surface site per second) at a given temperature, pressure, and at a given set of reactant and product concentrations, provides a figure of merit to compare the "activity" of different catalysts for the same reaction under similar experimental conditions. Although only a fraction of the catalyst surface is active during the reaction, the fraction is seldom known and thus the specific rate provides a lower limit to the activity of the catalyst.

To be of interest in chemical technological processes, specific rates must be in the range of 10^{-4}–10 site^{-1} s^{-1}. As a consequence, most catalytic reactions that involve the breaking or rearrangement of C–H, C–C, H–H, C=O, and N≡N bonds are carried out in the temperature range of 400–900 K, since activation energies range from minimum values of 10–15 kcal mol^{-1} (for hydrogenation of unsaturated hydrocarbons) to maxima as high as 40–50 kcal mol^{-1} (for dehydrocyclization or hydrogenolysis of hydrocarbons). There are exceptions, of course, to the above rather general statements.

Multiplication of the specific rate by the reaction time used in a given study gives the turnover number, N_t say (the number of product molecules per catalyst surface site). The value of N_t must be greater than unity to ensure that the reaction is catalytic. If $N_t < 1$, we may be dealing with a stoichiometric reaction and the solid surface may be acting as a reactant rather than a catalyst. Therefore, to demonstrate catalytic behavior, the reaction must be carried out for a sufficiently long time to obtain a turnover number greater than unity.

It should be stressed here that many of the chemical reactions of interest lead to the formation of several different, but all thermodynamically feasible, products. It is frequently more important to produce one of the many different molecules selectively, rather than increasing the overall rate. A selective catalyst will facilitate the formation of one product molecule while inhibiting the production of other, different molecules, even though the formation of all the species is thermodynamically feasible. Thus the blocking of reaction pathways, which would lead to the formation of unwanted molecules, is as important an attribute of a good catalyst as is lowering of the potential-energy barrier along the reaction pathway to the product molecule being sought.

6.1.2. Reaction Probability

The reaction probability, P_r say, is defined as the number of product molecules formed per number of reactant molecules incident on the catalyst surface. It is obtained by dividing the specific rate of catalysis by the incident reactant flux. The value of P_r is generally low, in the range 10^{-12} to 10^{-6}, for most catalytic reactions in the pressure and temperature ranges generally employed in the chemical industry. The reason for this is the long surface residence time for the adsorbed molecules and reaction intermediates (usually in the range of 10^{-1} to 10^{3} s), that leads to a surface which is covered and inaccessible to the impinging molecules. The desorption of the product molecules from the catalyst surface frees surface sites onto which the reactant molecules may adsorb. The desorption is always endothermic and appears to be the slow step in the catalytic process for many reactions at high pressures.

The long surface residence times of adsorbed intermediates have another consequence. The molecules may diffuse over long distances (10–10^4 Å) within their surface lifetime, visiting many catalytic sites where they can undergo consecutive reactions and molecular rearrangements. A catalyst surface may thus possess many active sites, all of which are likely to be accessible to the adsorbed reactants.

The coverage of adsorbed molecules, σ (molecules cm^{-2}), is usually determined by multiplying the incident flux F (molecules $cm^{-2}\ s^{-1}$) by the surface residence time t (s), i.e.,

$$\sigma = Ft \tag{6.1.1}$$

The flux can be estimated from the reactant pressure, $F = P(2\pi MRT)^{-1/2}$, and the residence time $t = t_0 \exp[\Delta H_{ads}(\theta)/RT]$ depends on the heat of adsorption ΔH_{ads}, the temperature T, and the preexponential factor t_0 (which varies in the range 10^{-17} to 10^{-13},[262] depending on the degree of mobility of the adsorbed molecules). The heat of adsorption is a measure of the strength of the chemical bond between the adsorbed molecule and the surface,* it depends on both the coverage and the structure of the catalyst surface. As the coverage is increased, the adsorbed molecules are packed

* Compare the discussion in Chapter 1, Section 1.1.

closer together in the surface monolayer. If the interaction among the adsorbates is repulsive, the heat of adsorption per molecule decreases as the coverage in the monolayer is increased. This is the case for molecular CO adsorption on most metal surfaces.[263] If the adsorbate–adsorbate interaction is attractive, the heat of adsorption per molecule may increase with coverage. This occurs, for example, during the growth of oxide monolayers on metal surfaces.[264]

The surface structure sensitivity of bonding of adsorbed molecules and, as a result, the structure sensitivity of catalytic reactions is well documented.[265,266] The surface structure is heterogeneous on the atomic scale, as depicted in Figure 6.1 following Spencer and Somorjai,[13] and there are many surface sites that are distinguishable by their number of nearest neighbors; there are surface atoms that can form terraces, steps, and kinks, and these structures may in turn contain point defects, including vacancies and adatoms. The binding of adsorbed atoms or molecules at each site may be different, giving rise to heats of adsorption that vary with the surface structure. This effect is particularly noticeable in adsorption and desorption studies carried out on single-crystal surfaces. These surfaces can be prepared with different relative concentrations of terrace, step, and kink sites. The structure sensitivity of bonding can lead to structure sensitivity in many catalytic reactions.[267]

It is often thought that the most active catalysts consist of surfaces that can form bonds of intermediate strength with adsorbates. If the surface chemical bonds are too strong, the reaction intermediates either have surface lifetimes which are much too long, or they may simply never come off the surface, forming instead stable surface compounds such as oxides, carbides, or nitrides. On the other hand, if the surface bonds to adsorbates are too weak, the needed chemical bond-breaking or rearrangements might not occur before desorption. Chemical experience is in

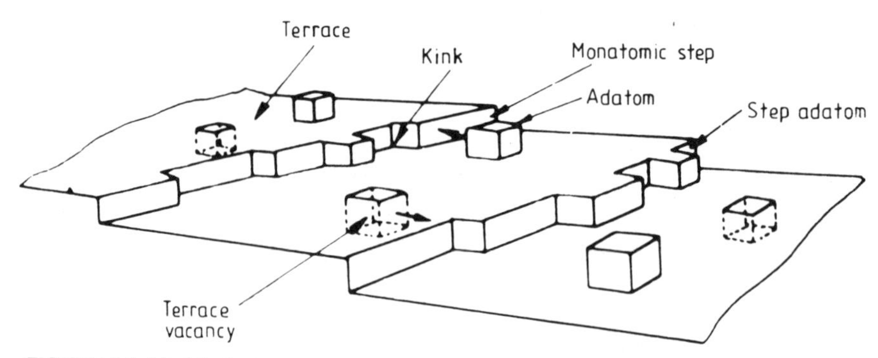

FIGURE 6.1. Model of a heterogeneous solid surface, depicting different surface sites. These sites are distinguishable by their number of nearest neighbors (after Spencer and Somorjai[13]).

agreement with the contention that intermediate bond strength leads to superior catalyst activity. The structure of the catalyst surface, however, is one of the key factors that control chemical selectivity.[13]

6.1.3. Elementary Reaction Steps

The catalytic process may be subdivided into several elementary reaction steps: (a) adsorption, (b) surface diffusion,* (c) molecular rearrangements at active surface sites, and (d) desorption.* There are several techniques capable of studying the elementary reaction step. Predominant among them are molecular-beam surface scattering and thermal desorption.* These methods of investigation are discussed in reviews by King,[268] and by Ceyer and Somorjai.[269]

Unfortunately, the catalyst does not have an indefinite life but becomes poisoned or deactivated after a certain time. Spencer and Somorjai comment that it is thought to reflect good catalyst performance if seven pounds of products are produced per pound of catalyst before the catalyst must be regenerated. Catalyst deactivation and regeneration are important and major fields of catalysis science. There are many possible reasons for deactivation. For instance, the catalyst may restructure as a consequence of selective adsorption of impurities from the reactant stream, such as sulfur or nitrogen. Alternatively, carbon may be deposited, blocking the active surface, due to side reactions leading to complete dehydrogenation and decomposition of organic molecules during hydrocarbon conversions. The use of additives (introduced either during catalyst preparation or with the reactants) to prevent restructuring of the catalyst or blocking of active sites is being explored intensively by many research groups.

The field of catalysis has been transformed as a result of developments in surface science, which have permitted molecular-level scrutiny of the catalyst surface. The structure, composition, and oxidation of atoms on the catalyst surface can now be correlated with the macroscopic reaction parameters, namely the specific rate, activation energy, and selectivity. As a result, there have been rapid advances in understanding many technologically important catalysts, and design of improved catalysts becomes feasible as knowledge gained from the study of surface science is increasingly utilized.

Spencer and Somorjai present "case histories" of a number of important catalytic processes which illustrate the above comments. One of these was presented in some detail in Chapter 1; five more are given in their review.[13]

* A brief discussion of surface diffusion is given in Section 4.8 while desorption studies are treated at some length in Chapter 5.

6.2. XPS STUDIES OF DESORPTION AND DISSOCIATION PROCESSES

The work of Umbach and Menzel[270] will now be summarized. They emphasize that the static characterization of adsorbate species has improved greatly over the past decade or so due to the large number of surface spectroscopic techniques that have become available (see, for example, Appendix 1.1). However, the investigation of surface kinetics has not seen a corresponding development, even though kinetics lie at the heart of the problem of catalysis. To date, most kinetic investigations of adsorbate layers are conducted by indirect methods, such as thermal desorption (discussed in Chapter 5) combined with monitoring the species appearing in the gas phase. While this may be adequate for the purposes of the study of desorption itself, as shown earlier, conclusions are difficult or impossible to draw about the kinetics of processes confined to the adsorbate layer, such as dissociation. In fact, even the mere existence of dissociation can remain open if only indirect methods are available, as evidenced by the system CO/W(110).

Umbach and Menzel[270] discuss the applicability of XPS as a fully quantitative technique for studying desorption and dissociation processes. They give a potential-energy diagram of CO on a clean W(110) surface, which is reproduced in Figure 6.2. However, they point out that the reaction coordinate in this figure is an extremely complicated cut through

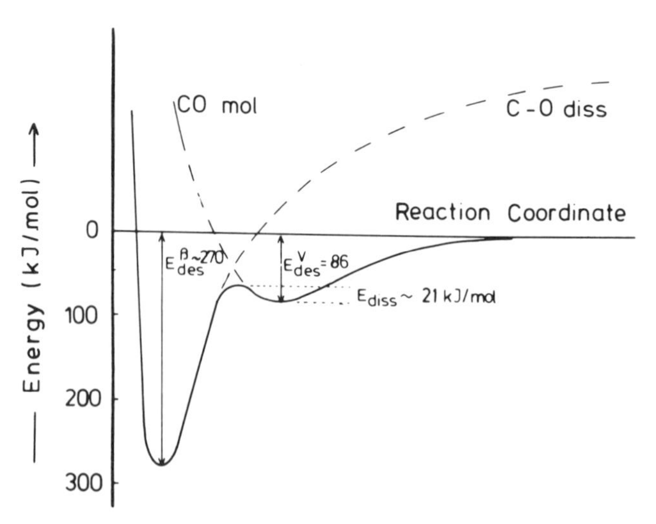

FIGURE 6.2. Potential energy diagram for CO on a clean W(110) surface (after Umbach and Menzel[270]).

the potential hypersurface of the system, involving participation of many degrees of freedom.

To summarize, Umbach and Menzel show that XPS is a useful quantitative technique for *in-situ* species-resolving studies of surface reactions, in particular those inaccessible to thermal-desorption experiments. However, their paper should be consulted for a discussion of limitations, precautions, and experimental necessities when this technique is used in such kinetic studies.

It is relevant here to note that many chemical reactions can be viewed as the crossing of a free energy barrier separating two regions of relative stability. In many cases the relevant motion from the standpoint of the reaction is of a diffusional nature along a single reaction coordinate.

The dynamics of barrier-crossing reactions have been characterized by a number of theoretical methods based on the solution of stochastic equations of motion, as in the work of Skinner and Wolynes[(271)] and Brenig *et al.*[(272)] A related method is summarized in Appendix 4.3.

6.2.1. Photoelectron Spectroscopy in Relation to Heterogeneous Catalysis

Tysoe *et al.*[(273)] discuss photoelectron spectroscopic results in relation to heterogeneous catalysis. They point out that the surface chemistry of some hydrocarbon molecules on transition-metal surfaces is of interest because of potential relevance to certain important classes of catalytic reactions. Most single-crystal studies on the group VIII metals have been carried out on Ni and Pt, while Pd, though a superior selective hydrogenation catalyst, has received less attention to date.

It is clear from the earlier chapters of this book that much interest has been focused on chemical identity and bonding geometry of the chemisorbed species. However, to take a particular example, for acetylene Demuth's low-temperature results for Pd(111) were interpreted in terms of an essentially acetylenic surface species.[(274,275)] On the other hand, an EELS investigation[(276)] (cf. Appendix 1.1) favored the view that extensive rehybridization occurs in the adsorbed state although calculations[(277,278)] suggest that this effect is likely to be weak at low temperatures.

The UPS results of Tysoe *et al.*[(273)] can be summarized as follows. At 175 K, acetylene adsorbs as a "flat-lying" and possibly somewhat distorted molecule with charge transfer from adsorbate to metal ($\pi \rightarrow d$ bonding). Between 220 K and 270 K an intramolecular rearrangement occurs resulting in a "perpendicular" vinylidene species, while at higher coverages there is evidence that trimerization occurs to form benzene. Proposed structures for the 300 K vinylidene phase and the low-temperature acetylene phase are shown in Figure 6.3 following Tysoe *et al.*[(273)]

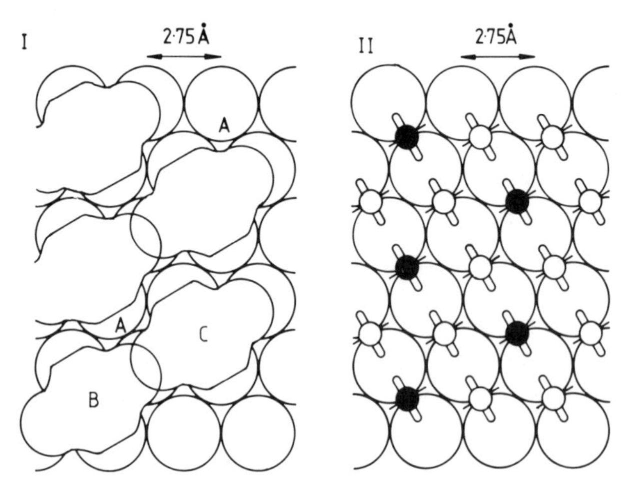

FIGURE 6.3. Proposed structures for the 300 K vinylidene phase (I) and the low-temperature acetylene phase (II) (after Tysoe *et al.*[273]).

The observations relating to benzene formation can be accounted for by the scheme in Figure 6.4, suggested by Tysoe *et al.*[273] Low-temperature benzene evolution appears to be associated with a tilted product molecule formed from acetylenic species on an initially crowded surface. As the surface becomes deplected of reactant and product molecules, the remaining benzene can adopt a flat configuration (more strongly chemisorbed) eventually desorbing at 500 K. This is consistent with the observation that benzene formation only occurs for $\theta(C_2H_2) > \frac{1}{3}$ at which coverage the surface cannot accommodate "flat-lying" benzene.

The homogeneously catalyzed reaction of acetylene to yield benzene at around 300 K is a well-known process for transition-metal cluster compounds, as discussed, for example, in the book by Kochi.[279] However,

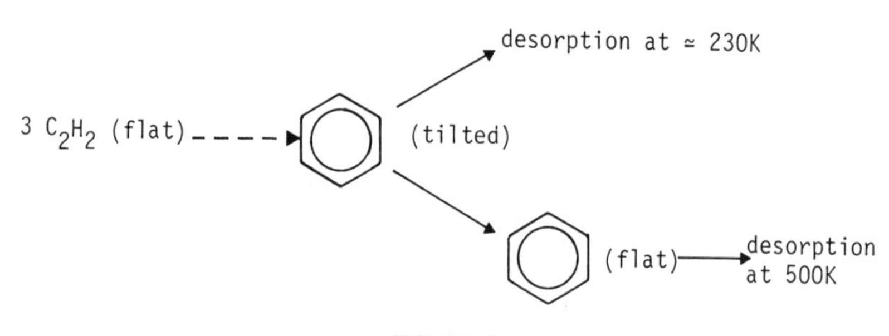

FIGURE 6.4

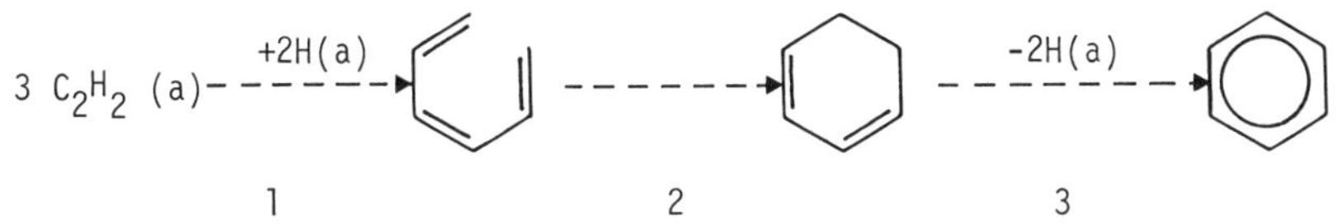

$$3\,C_2H_2\,(a) \xrightarrow[1]{+2H(a)} \text{(hexatriene)} \xrightarrow[2]{} \text{(cyclohexadiene)} \xrightarrow[3]{-2H(a)} \text{(benzene)}$$

FIGURE 6.5

heterogeneously catalyzed aromatizations have usually been known only at much higher temperatures on bifunctional catalysts (e.g., at approximately 550 K on Pt–SiO_2).[280] The mechanism of the low-temperature aromatization found by Tysoe *et al.*[273] may involve initial formation of a 1,3,5-hexatriene, which then undergoes closure to cyclohexadiene (a reaction that proceeds easily at low temperatures even in the absence of catalysts[281]). This could be followed by palladium induced dehydrogenation and eventual desorption of benzene. Referring to Figure 6.5, reproduced from the work of Tysoe *et al.*,[273] step 1 of this mechanism need not proceed via the simultaneous interaction of three acetylenic species; indeed this seems inherently less likely than a series of bimolecular reactive encounters. Some support for this view derives from the observation of a C_4 product species and the fact that this species appears to act as a precursor of the C_6 product. Step 2 provides a basis for understanding why the initially formed benzene might be tilted with respect to the surface. It is also consistent with the known structure and reactive properties of C_6 alkyne-bridged clusters of, for example, molybdenum and cobalt, studied by Knox *et al.*[282] A further point of interest is that $Os_3H_2(CO)_9C_6H_4$ has been shown by Goudsmit *et al.*[283] to contain a benzene molecule in which the ring is tilted at 70° with respect to the plane containing the metal atoms.

6.3. COMPENSATION EFFECT[284]

Peacock-López and Suhl[285] have discussed the compensation effect in thermally activated processes by counting the ways in which the heat bath can furnish the energy necessary to surmount the barriers.

The background to their work, to be discussed below, can be summarized in the following manner. In most thermally activated processes, a plot of the logarithm of the reaction rate against the inverse of the absolute temperature is a straight line. This "Arrhenius plot" can be written in the form

$$\ln k = \ln A - \beta\varepsilon_b, \qquad \beta = (k_B T)^{-1} \tag{6.3.1}$$

where ε_b, the slope, is the so-called activation or barrier energy, while k is the reaction rate; A is a preexponential factor in the expression for the reaction rate $k = A\exp(-\beta\varepsilon_b)$.

It frequently happens that the Arrhenius plots of different members of a family of reactions are straight lines with different slopes, which (when extrapolated if necessary) intersect at a common point, at least approximately. The temperature coordinate of this point is often referred to either as the compensation temperature or the isokinetic temperature T_c. Such behavior is represented phenomenologically by setting

$$A = A_0 \exp(\beta_c \varepsilon_b) \tag{6.3.2}$$

which yields

$$k = A_0 \exp[-(\beta - \beta_c)\varepsilon_b] \tag{6.3.3}$$

namely a compensation effect with isokinetic temperature $1/k_B\beta_c$. When the effect is present, a plot of $\ln A$ vs. the different values of ε_b for the family members gives a straight line of slope β_c.

The compensation effect was first reported long ago by Constable[286] in relation to dehydrogenation of ethanol on a copper catalyst. At first the effect was thought to be limited to heterogeneous catalysis, but it was subsequently found in other thermally activated rate processes as well.

By varying the surface treatment of the copper catalyst, Constable found a variation of ε_b from 20 to 24 kcal/mol, accompanied by a tenfold change in A and a compensation temperature of 875 K. Cramer[287] has reviewed numerous examples and discussed a statistical distribution in the properties of catalytically active centers as a possible explanation. However, this model has a restricted range of validity. In particular, it is unable to deal with cases in which the substrate is fixed and the reactants are different, as is the situation, for example, in the decomposition of ethanol, n-propanol, and n-butanol on Nd_2O_3, discussed by Cramer,[287] or in the dissociative adsorption reactions of methane, ethane, and propane on nickel.[288]

A further example of the compensation effect is ethane hydrogenolysis on silica-supported metals, reviewed by Sinfelt.[289] In this case the reaction rate is measured at fixed partial pressure of ethane and hydrogen. The experimental results show activation energies between 21.4 and 58 kcal mol^{-1}, and prefactors between 4.5×10^{20} and 3.7×10^{33} mol s^{-1} cm^{-2}. The metals cobalt, nickel, and rhenium had a compensation temperature of 1255 K, while the corresponding value for ruthenium, rhodium, palladium, osmium, iridium, and platinum was 380 K. It is noteworthy that the measurements were carried out at temperatures between 400 and 600 K. Thus, measurements were taken at temperatures lower than the compensation temperature in one case, and at temperatures higher than this temperature in the other. The compensation temperatures (as in most cases) had to be found by extrapolation.

Some concepts and a model will now be examined. The total probability of a reaction is evidently the sum of the probabilities that the reactants absorb one, two, three, etc., excitations with a total energy exceeding the barrier energy. The probability for absorption of exactly n excitations with energy ε is the square of the T-matrix element* for this process, multiplied by $\rho_n(\varepsilon)$, the density of states of n excitations with total energy ε. Peacock-López and Suhl[285] show (see below) that the sum over all n tends to behave like $\exp[-(\beta-\beta_c)\varepsilon]$, where $\beta_c = 1/k_B T_c$ is (a) a function of a suitable coupling parameter (not necessarily the coupling strength itself), and (b) a function of the way the single-excitation density $\rho_1(\varepsilon)$ depends on ε. In addition β_c, or equivalently the compensation temperature, may be weakly dependent on ε. Peacock-López and Suhl therefore propose that the compensation effect may be regarded as resulting from "dynamic" entropy of the heat bath. It results from counting the number of ways in which the heat bath can furnish the energy needed to overcome the barrier.

6.3.1. Theoretical Basis†

The argument below closely parallels that of Peacock-López and Suhl.[285] The probability of escape over the barrier may be written as a sum of all possible processes leading to escape. This sum is arranged to take advantage of features possessed by the transition matrix elements. The general expression for the transition rate per unit time is

$$\frac{2\pi}{\hbar}\sum_{f,i}|\langle f\,|\,T\,|\,i\rangle|^2\frac{\exp(-\beta E_i)}{Z}\,\delta(E_f-E_i) \qquad (6.3.4)$$

(cf. the "Golden Rule," which replaces T by the perturbation V in lowest order in this expression), where $\{i\}$ denotes the complete set of quantum numbers of the system in the reactant well and $\{f\}$ is the set of quantum numbers of the reacted system; $\langle f\,|\,T\,|\,i\rangle$ is the T-matrix element while Z is the reactant-plus-bath partition function. Expression (6.3.4) assumes that the conditions for the validity of absolute rate theory are met, i.e., that quasi-equilibrium conditions prevail during the reaction. Thus one supposes that thermal equilibration among the reactant levels is very much faster than the rate of transition to the reaction products. When this is not

* See Appendix 6.2 for a simple discussion of this quantity.

† This section contains more advanced theory than appears in most of this book. The reader interested in the results should proceed to Section 6.3.2.

the case, the relation between prefactor and T-matrix elements is less direct.

In the present case, following Peacock-López and Suhl one can take $|i\rangle$ and $|f\rangle$ to be direct products of reactant states and bath-excitation states. The latter are either inherently Boson excitations (such as phonons) or behave very nearly as such (such as hole–electron pairs or magnons; for a review of such collective excitations in solids reference may be made to the book by March and Parrinello[290]). Hence one may set

$$|i\rangle = |E_i^0, \{n_k + m_k\}\rangle, \qquad |f\rangle = |E_f^0, \{m_k\}\rangle \tag{6.3.5}$$

$$E_i = E_i^0 + \sum_k (m_k + n_k)\varepsilon_k \tag{6.3.6}$$

$$E_f = E_f^0 + \sum_k m_k \varepsilon_k \tag{6.3.7}$$

where the bath in its initial and final states is characterized by the excitation number sets $\{m_k + n_k\}$ and $\{n_k\}$, respectively, E_i^0 is the reactant energy in a bound state of the reactant well, E_f^0 is the energy of the reaction products, k labels the excitations, and ε_k denotes the energy of the kth of these excitations. If tunneling is neglected, then $E_i^0 < \varepsilon_b$ and $E_f^0 > \varepsilon_b$, where ε_b is the barrier energy. Transition occurs by absorption of the excess n_k excitations.

The nature of the bath–reactant coupling normally possesses the following properties.

(a) $|\langle f|T|i\rangle|^2$ depends on E_i^0, E_f^0, and the sets $\{n_k\}$ and $\{m_k\}$.

(b) As a result of the manner in which the order of summation is arranged in expression (6.3.4), $|\langle f|T|i\rangle|^2$ will depend on $\{n_k\}$ mainly through the sum $n = \sum_k n_k$.

Related to this is the fact that it does not depend much on k. This is particularly clear when $\mathbf{k}$ is considered to be a wave number and the reaction–bath coupling is highly localized. The strongest dependence on n is through the appropriate coupling parameter raised to the nth power; this will become clear from an example given below. Any residual dependence of the T-matrix on the individual distribution of the values of m_k (rather than on just $\sum m_k$) can be expressed as a minor modification of the simple excitation-state density.

(c) Except in the Born approximation for T, $|\langle f|T|i\rangle|^2$ will also depend on $\sum \varepsilon_k n_k = E_f^0 - E_i^0 = \varepsilon$. Evidently, this quantity occurs in the energy denominator of all higher terms in the perturbation series for T. However, this dependence is not strong and in the case where the absorption of n quanta appears only beyond the Born term, integration over intermediate states will tend to reduce the ε-dependence to a simple logarithmic form.

Thus, Peacock-López and Suhl argue that the sum (6.3.4) is adequately represented by

$$\frac{2\pi}{\hbar}\sum_i \int_{\varepsilon_b - E_i^0}^{\varepsilon_c} d\varepsilon \exp(-\beta\varepsilon) R_f(E_i^0 + \varepsilon)$$
$$\times \sum_{\text{all}\{n_k\}} |T(E_i^0, n, \varepsilon)|^2 \frac{\exp(-\beta E_i^0)}{Z_r} \times \delta\left(\varepsilon - \sum_k n_k \varepsilon_k\right) \qquad (6.3.8)$$

where Z_r is the partition function of the reactants alone. In the above result, $R_f(E)$ is the final-state density of the reaction products at energy E, and ε_c is a cutoff to be discussed below.

Summation over the n_k is most usefully carried out in the order

$$\sum_{|n_k|} = \sum_n \sum_{\Sigma n_k = n} \qquad (6.3.9)$$

In this case

$$\sum_{n_1} \cdots \sum \delta\left(\varepsilon - \sum_k \varepsilon_k n_k\right) = \rho_n(\varepsilon) \qquad (6.3.10)$$

which is the density of states of n independent excitations of total energy ε or, in other words, the convolution of n single-excitation densities of state

$$\rho_n(\varepsilon) = \int_{\Sigma_i \varepsilon_i = \varepsilon} \rho(\varepsilon_1)\rho(\varepsilon_2)\cdots\rho(\varepsilon_n)d\varepsilon_1\cdots d\varepsilon_n \qquad (6.3.11)$$

For simplicity, consider only one bound state in the well and choose its energy to be zero. Then the result for the rate R is

$$R = \frac{2\pi}{\hbar}\int_{\varepsilon_b}^{\varepsilon_c} d\varepsilon \exp(-\beta\varepsilon) R_f(\varepsilon) \times \sum_n |T(n, \varepsilon)|^2 \rho_n(\varepsilon) \qquad (6.3.12)$$

Let

$$\tilde{\rho}_1(p) = \int_0^\infty \exp(-p\varepsilon)\rho_1(\varepsilon)d\varepsilon \qquad (6.3.13)$$

Then the Laplace transform of $\rho_n(\varepsilon)$ is $[\tilde{\rho}_1(p)]^n$. Therefore the rate takes the form

$$R = \frac{1}{i\hbar}\int_{\varepsilon_b}^{\varepsilon_c} d\varepsilon \exp(-\beta\varepsilon) R_f(\varepsilon) \int_{c-i\infty}^{c+i\infty} dp \exp(p\varepsilon) \sum_{n=1}^\infty |T(n, \varepsilon)|^2 [\tilde{\rho}_1(p)]^n \qquad (6.3.14)$$

where the integration contour runs to the right of all singularities of the integrand.

The compensation effect is now examined. Peacock-López *et al.* consider the sum

$$\sum_{n=1}^{\infty} |T(n, \varepsilon)|^2 \tilde{\rho}_1^n \equiv S \tag{6.3.15}$$

as a function of the complex variable $\tilde{\rho}_1$. Its radius of convergence in the $\tilde{\rho}_1$ plane is

$$\mathcal{R}(\varepsilon) = \lim_{n \to \infty} |T(n, \varepsilon)|^{-2/n} \tag{6.3.16}$$

which is a weakly varying function of ε. Peacock-López *et al.* assume $\mathcal{R}$ to be finite (true in the example treated below). Since the $|T(n, \varepsilon)|^2$ contributions are positive, the point $\tilde{\rho}_1 = \mathcal{R}(\varepsilon)$ in the $\tilde{\rho}_1$ plane is a singularity of the sum S, and will be assumed a simple pole. This pole dominates the contour integral (6.3.14) for sufficiently large ε. If it is indeed a simple pole at

$$p = \beta_c(\varepsilon) \tag{6.3.17}$$

it yields a residue proportional to $e^{\beta_c(\varepsilon)\cdot\varepsilon}$. The dependence of β_c on ε is expected to be unimportant, especially at large values of ε. If $R_f(E)$ in equation (6.3.14) is also slowly varying with E, the result of the integration for $\beta \gg \beta_c$ is very nearly

$$R = \mathcal{P} \exp[(\beta_c - \beta)\varepsilon_b] \tag{6.3.18}$$

where the preexponential factor is $\mathcal{P} = (2\pi/\hbar)\bar{R}_f$ times a nonexponential function of β_c. The upper cutoff in the integral (6.3.14) can be neglected for $\beta > \beta_c$.

The important conclusion to be drawn from the work of Peacock-López and Suhl summarized above is that there will be a compensation effect for any family of reactions in which ε_b varies from one member to another, while the effective coupling to the bath remains the same for all members.

6.3.2. Compensation Temperature T_c

A power-law form for $\rho_1(\varepsilon)$ frequently occurs, for example, in the cases of sound waves in fluids [$\rho_1(\varepsilon) \sim \varepsilon^2$], electron–hole pairs [$\rho_1(\varepsilon) \sim \varepsilon$], and magnons* in continuum theory ($\rho_1 \sim \varepsilon^{1/2}$). Under such circumstances

* That catalytic effects on ferromagnetic substances may be aided by interaction with spin waves is suggested by work discussed in Appendix 5.1, where strongly paramagnetic metals are also treated.

$$\tilde{\rho}_1(p) = \frac{\Gamma(\nu+1)}{\varepsilon_0^{\nu+1}} \frac{1}{p^{\nu+1}} \tag{6.3.19}$$

where Γ denotes the gamma function, $\rho_1(\varepsilon) = \varepsilon^{\nu}/\varepsilon_0^{\nu+1}$, and ε_0 determines the energy scale.

Consequently, the dominant pole is $[\Gamma(\nu+1)]^{1/\nu+1}\varepsilon_0^{-1}$ and so

$$T_c = \frac{\varepsilon_0}{k_B[\Gamma(\nu+1)]^{\nu+1}} \tag{6.3.20}$$

For the case of $\rho_1(\varepsilon)$ with an upper cutoff ε_u, such as phonons or magnons in crystalline solids, the inequality $\tilde{\rho}_1(0) > \mathcal{R}$ will not hold for too small a value of $\rho_1(\varepsilon)$, or too small a radius of convergence. For a Debye spectrum with cutoff energy ε_d one has $\rho_1(\varepsilon) = 3N\varepsilon^2/\varepsilon_d^3$, where in the simplest approximation the numerical factor N is 3. For large N one obtains $k_B T_c \simeq \varepsilon_d/(6N)^{1/3}$. As N decreases, $k_B T_c$ increases and tends to infinity at $N = 1$ as $k_B T_c \simeq \frac{3}{4}\varepsilon_d/(N-1)$.

Peacock-López *et al.* give a more general discussion of the reaction rate, as well as specific examples, but the reader is referred to their paper for further details. In the present context, though, one other point from their work is that they comment specifically on the case of reactants coupled to a metal catalyst. In this case, the coupling to the conduction electron density, say $n(\mathbf{r})$, has the form $\gamma = \int V(\mathbf{R} - \mathbf{r})n(\mathbf{r})d\mathbf{r}$, which seemingly involves only one electron–hole pair at a time. However, the residual shielded Coulomb interaction of the electron taken to the nth order, together with γ taken only to first order, effectively couples the reactants to $n+1$ electron–hole pairs. The dominant pole, according to Peacock-López and Suhl, is then modified roughly by the ratio (Coulomb energy to kinetic energy)$^{1/2}$, but remains independent of γ. They point out that an analogous result holds for a weak linear coupling W to phonons of a significantly anharmonic solid. Interaction between the lattice phonons taken to nth order, together with W taken to first order, gives an effective coupling to $3n+1$ phonons. Once again, the sum $\Sigma\,\rho_n(\varepsilon)$ is essentially exponential and yields a compensation effect independent of W.

6.4. RELEVANCE OF WOODWARD–HOFFMANN RULES TO SINGLE-CRYSTAL CATALYSTS

Park *et al.*[291] studied the adsorption of NO and CO on Pt(410) using X-ray photoemission spectroscopy (XPS). They found that the (410) surface exhibits unusual activity for NO and CO bond-breaking. CO is found to adsorb molecularly on Pt(410) at 300 K, but it partially dissociates on heating to 500 K. NO is found to adsorb dissociatively under all conditions studied. By comparison, the low-index faces of platinum give negligible dissociation of CO and also of NO up to 350 K.

Park *et al.*[291] stress that the (410) surface is also unusual from an orbital-symmetry standpoint: in fact it was predicted to be unusually active for NO and CO bond-breaking based on the Woodward–Hoffmann rules,[292] as outlined below. The results of Park *et al.*, at the very least, point to the possibility that orbital-symmetry conservation methods can yield valuable insight into the nature of the active sites on single-crystal catalysts.

The idea that Woodward–Hoffmann rules, namely orbital-symmetry conservation rules, might be applicable for reactions on solid surfaces will be developed below following the approach of Banholzer *et al.*[293] It will be seen that, according to their analysis, CO and NO bond-breaking is symmetry forbidden on Pt(110) and Pt(111). There is a partial symmetry constraint on Pt(100) (see, for instance, Pearson).[294] Pt(410), on the other hand, is one of the few faces where the reaction is symmetry allowed.

It is not known to date whether orbital-symmetry effects are important compared to differences in bond energy between different crystal faces. However, if symmetry effects are important one might anticipate that Pt(410) will be more active for NO and CO bond-breaking than any of the low-index faces.

6.4.1. Classification of Symmetry-Allowed or Forbidden Reactions

Woodward and Hoffmann,[292] and also Fukui,[295] found that in determining the influence of structural properties on the rates of reactions in organic and inorganic systems, even though bond energy can play a significant role, more often orbital-symmetry constraints are the dominating influence. In fact, Woodward and Hoffmann propounded their rules based on symmetry effects alone.

Reactions can be classified into those that are symmetry allowed and those that are partially or totally forbidden. A symmetry-allowed reaction occurs via a direct transfer of electrons from the occupied levels in the reactants to the occupied levels of the products, while symmetry-forbidden reactions involve an indirect transfer, where electrons are promoted into an excited state during the electron-transfer process. It costs energy to promote the electrons. Hence there is an activation barrier associated with a symmetry constraint.

It should be mentioned here that previous work concerned with applying symmetry-conservation ideas to catalytic reactions was conducted by Balazs and Johnson,[296] Mango,[297] and Messmer and Bennett.[298] The specific example considered by Banholzer *et al.*,[293] whose work is summarized below, is that of the plane-to-plane variation in reactivity of heterogeneous catalytic systems. A reaction on a catalytic surface is analogous to the unimolecular decomposition of a small inorganic molecule. Woodward–Hoffmann methods are very successful for such molecules. However, it should be mentioned that since energy-level spacings are much smaller on a metallic surface than on a small metal cluster, it is not known whether symmetry methods are as effective in catalytic systems.

The work of Banholzer *et al.*[293] employs symmetry-conservation methods to make predictions about the relative activity of various faces of platinum for nitric-oxide decomposition.

6.4.2. Symmetry-Conservation Rules

Symmetry rules in chemistry are often developed from perturbation theory. The electronic wave function for the system at each point along the reaction coordinate is expressed as a linear combination of reactant molecular orbitals (MOs). As the reaction proceeds, there is a net transfer of electrons from "donor" to "acceptor" MOs. If the donor and acceptor MOs have the same symmetry (i.e., irreducible representation), then a direct transfer of electrons is possible. In this case, the activation energy for the reaction will be low (for an exothermic case). However, if the donor and acceptor orbitals are of different symmetry, then a direct transfer of electrons is forbidden. In order for the reaction to proceed, electrons must either be pulled from an excited state, or the electronic structure of the reactants must be distorted. A significant activation barrier is often associated with this process.

6.4.3. Use of a Correlation Diagram in Understanding the Activation Barrier

One can gain insight into the magnitude of the activation barrier with the aid of a "correlation diagram." Banholzer *et al.*,[293] and the work presented below, use the correlation diagram for the reaction between two ethylene molecules to form a cyclobutane molecule, by way of illustration; this approach originates from the book by Woodward and Hoffmann.[292] The case in Figure 6.6 is that in which the two ethylene molecules start out parallel to each other and are aligned so that the line between their centers is perpendicular to the molecular axes. The reaction is presumed to occur by a simple broadside collision, where electrons are removed from the π-orbital on each ethylene to form carbon–carbon bonds in the cyclobutane.

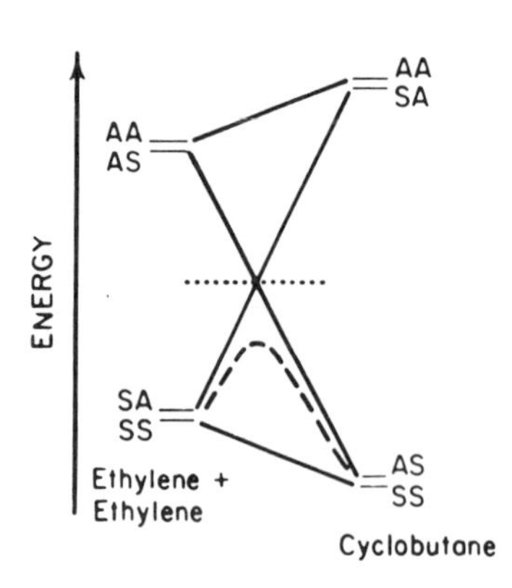

FIGURE 6.6. Correlation diagram for the reaction between two parallel ethylene molecules to form a cyclobutane molecule (after Woodward and Hoffmann[292]). Levels on left represent energy levels of reactants and on right those of products. Solid lines connecting various levels are symmetry-allowed electron-transfer pathways (after Banholzer *et al.*[293]).

There are three mirror planes in the ethylene–ethylene complex: the plane containing the molecular axes, the plane perpendicular to the axis that passes through the center of each carbon–carbon bond, and the plane perpendicular to the other two and situated midway between the two ethylene molecules. One can use group theory[299] to show that there are four possible configurations of the π-MOs in the ethylene–ethylene complex: an SA configuration (i.e., symmetric about the plane through the midpoint of the ethylene bond and antisymmetric about the plane between the ethylenes), an SS configuration, an AS configuration, and an AA configuration. The π-orbitals on the ethylene are symmetric about a plane perpendicular to the midpoint of the ethylene bond, but they can be symmetric or antisymmetric about the plane halfway between the two ethylenes. Thus if one expands the ground-state wave functions for the π-electrons on each ethylene in the MOs of the entire complex, the SA and SS states are found to be occupied while the AS and AA states are empty. Conversely, the carbon–carbon bonds formed during the reaction can be symmetric or antisymmetric about the plane through what was the ethylene bond, but they must be symmetric about what was the plane between the two ethylenes. Hence the AS and SS states are occupied in the ground state of cyclobutane, while the SA and AA states are empty. Therefore, if the reaction is to proceed, one must transfer electrons from an SA to an AS state. Here lies the problem. One can only get a direct transition between states of the same symmetry, according to group theory. Hence in order that the reaction can proceed, one must first form an AS state, which involves a considerable activation barrier.

The above argument has assumed that the two reactants come together in a perfectly symmetric configuration. In reality, they will do so with some asymmetry. It can be shown that the asymmetry provides a perturbation, which allows the system to follow a lower curve, such as the dashed curve in Figure 6.6. It should be noted that it is no longer necessary to form the AS state at the beginning of the reaction. However, the transition state still has considerable AS character. Hence there will be a substantial barrier to the reaction.

In surface studies, it is useful to consider the symmetry of the individual donor and acceptor orbitals. For this, one decomposes the wave function of the reactants and the products into reactant MOs. For the example in Figure 6.6, the SA state decomposes into σ- and π-orbitals while the AS state goes into σ- and π^*-orbitals. One must therefore transfer electrons from a π-orbital to a π^*-orbital for the reaction to proceed. Figure 6.7[293,300] shows the two orbitals schematically; the π-orbital is symmetric about a plane perpendicular to the molecular axis, while the π^*-orbital is antisymmetric about the same plane. Transfer of electrons from a symmetric to an antisymmetric orbital is symmetry forbidden.

A similar type of argument can be given for a surface reaction. As the reaction proceeds, electrons flow in and out of the various orbitals (bands) in the surface and into and out of orbitals in the adsorbate. Hence, one may expect to gain some insight into the variation of the rate of reaction with crystal face by comparing the symmetry of the possible donor and acceptor orbitals in the surface to those in the adsorbates. In particular, it seems clear that one should look for transitions between symmetric and antisymmetric states and note that such transitions are symmetry forbidden.

Of course, it must be borne in mind that a reaction involving a surface has the unique feature that there are many energy levels and the energy-level spacings are small. As a result, it takes much less energy to distort the electronic levels in the surface than in the customary situations where symmetry arguments are known to work so well. One can find examples where the curve crossing in the correlation diagram will be below the initial state, and hence there will not be a symmetry-induced barrier. Still, it seems that, in most cases, energy must be expended to distort the electronic structure of the surface. Thus Banholzer *et al.*[293] argue that the symmetry of the orbitals in the surface should affect the relative catalytic

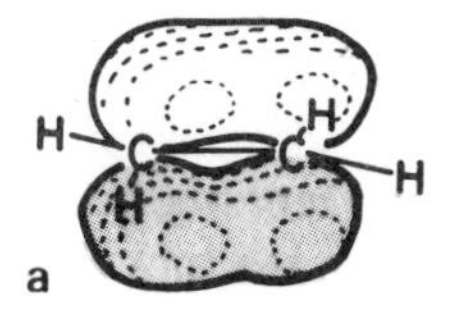

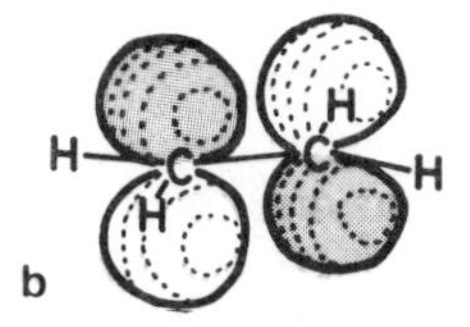

FIGURE 6.7. (a) π- and (b) π^*-orbitals of ethylene adapted from Jorgensen and Salem.[300] Negative lobes are depicted by clear areas while positive lobes are indicated by shaded regions (after Banholzer *et al.*[293]).

activity of the various crystal planes. At the time of writing, it does not appear to be known whether symmetry effects are important compared to the effects of differences in bond energy with crystal face.

6.4.4. Can Symmetry Effects Alone Explain Nitric-Oxide Decomposition on Platinum?

Banholzer *et al.*[293] next study to what extent symmetry effects alone can explain the plane-to-plane variations seen during nitric-oxide decomposition on platinum. Their basic assumption is that the crucial step in the decomposition process is a concerted reaction where adsorbed NO is converted to chemisorbed nitrogen and oxygen. This reaction is treated as a simple unimolecular decomposition, to which available methods[301,302] can be applied directly. The effects of variations in bond energy with crystal face are neglected. The NO dissociation step is modeled as one where electrons are transferred from the bonding orbitals of the NO to the metal, and from the metal to the antibonding orbitals of the NO. If it is assumed that the oxygen and nitrogen end up with closed shells, then the $M \rightarrow \sigma^*$ and $M \rightarrow \pi^*$ transfers will dominate, M denoting metal or surface.

Figure 6.8 is a diagram of the σ^*- and π^*-orbitals of NO. The σ^*-orbital is rotationally symmetric about the molecular axis and is approximately antisymmetric about the midpoint of the NO bond. When the reaction commences, the NO is bound with its axis perpendicular to the surfaces. Both π^*s are equivalent. However, the NO molecule bends to move the oxygen toward the surface as the reaction proceeds. The π^*-orbitals split into $\pi^*_\perp$- and $\pi^*_\parallel$-orbitals, as shown in Figure 6.9. The $\pi^*_\perp$-orbital is antisymmetric about the plane labeled I in Figure 6.9 (the plane perpendicular to the surface passing through the molecuar axis). It stays antisymmetric as the reaction proceeds. The $\pi^*_\parallel$- and σ^*-orbitals are symmetric about plane I and remain symmetric everywhere along the reaction coordinate. Hence, from the arguments set out above, if there are appropriately oriented symmetric and antisymmetric orbitals in the surface, with energies near the Fermi edge, then the dissociation process will be symmetry allowed. If not, it will be symmetry forbidden.

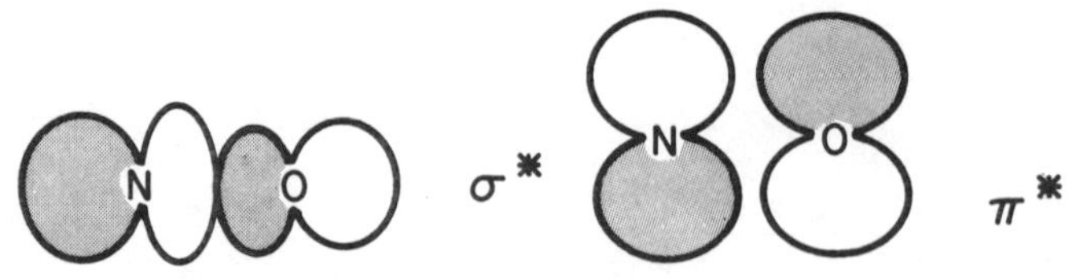

FIGURE 6.8. Schematic representation of the σ^*- and π^*-orbitals of NO. Shaded areas depict positive lobes while clear areas represent negative regions (after Banholzer *et al.*[293]).

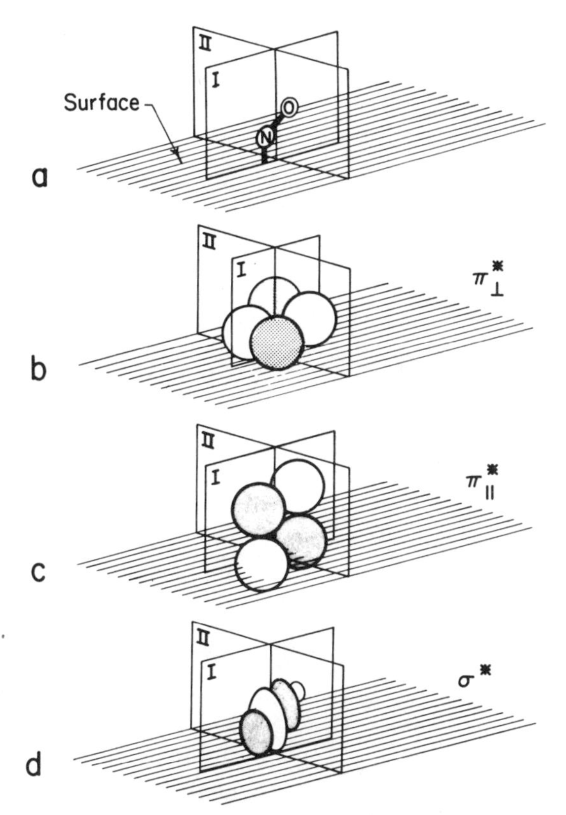

FIGURE 6.9. Orientation of (b) $\pi^*_\perp$, (c) $\pi^*_\parallel$, and (d) σ^* as the NO bends as shown in (a). Positive lobes are indicated by clear areas while negative lobes are depicted by shaded regions. Plane labeled I is perpendicular to the surface and goes through the molecular axis. Plane labeled II is perpendicular to both plane I and the surface, and it goes through the center of the NO bond (after Banholzer *et al.*[293]).

To proceed further, Banholzer *et al.*[293] input information about the electronic structure of the surface. There are several surface-band structures available for platinum surfaces[303–306]; some calculational details are different but all the band structures have the same symmetry properties. The interested reader can find the details in Appendix 6.1, where symmetry rules are developed based on Bond's picture of the surface electronic structure,[303] in which it is assumed that bands are formed as a symmetry-adapted linear combination of filled atomic orbitals.

This is how the conclusions drawn at the beginning of Section 6.4 were reached, namely that NO bond-breaking is symmetry forbidden on Pt(110) and Pt(111). While there is a partial symmetry constraint on Pt(100), using a Pt(410) surface the reaction is symmetry allowed. These conclusions are in accord with experiment and it therefore appears that symmetry arguments may well have a valuable role to play in catalytic systems. The way in which, by lowering the surface symmetry, terraces and steps can play a part in enhancing surface activity is also considered in Appendix 6.1, following Banholzer *et al.*[293]

6.5. ELECTRONIC STRUCTURE STUDY OF A "POISONED" CATALYST SURFACE*

Feibelman and Hamann[307] have reported self-consistent linearized-augmented plane-wave calculations of the electronic-structure perturbations induced by a catalytic "poison" S on a Rh(001) surface. The motivation for this work was that studies of adsorption and kinetics on transition-metal surfaces show that the addition of less than 0.1 monolayer of a "poison" species such as S can cause drastic reductions in reactant sticking probabilities, coverages, and reactivities.[308] While in some cases additive effects may result from blocking a small number of "defect sites" essential for surface chemistry, in the methanation of CO the evidence suggests that poisons act at low coverage by perturbing surface electronic structure over distances larger than those to their nearest neighbors.

The calculations reported by Feibelman and Hamann were carried out for relatively low coverages ($\frac{1}{4}$ and $\frac{1}{3}$ monolayer) of S on thin Rh(001) films. They identify features of the electronic structure of S-covered surfaces that seem relevant to their decreased chemical activity and distinguish effects of S atoms that can manifest themselves at low coverages from those that cannot.

A fairly widespread view has been that poisoning by an electronegative atom such as S is the result of its removing charge from the surface that would otherwise facilitate a rate-limiting reaction step. For example, in CO dissociation, the C–O bond is supposedly weakened by the transfer of a d-electron from the surface into the antibonding $2\pi^*$-orbital. One finds that coadsorption of low coverages of K, an electropositive atom, causes a lowering of the C–O stretch vibration frequency,[309] presumably because of increased $2\pi^*$-occupancy. Thus for an electronegative additive like S, one anticipates less charge transfer into the $2\pi^*$-orbital, a stronger CO bond, less CO disssociation, and the inhibition of reactions that depend on its occurrence.

However, Feibelman and Hamann[307] emphasize that this charge-transfer mechanism does not explain how S atoms can inhibit chemistry beyond a screening length. Since their electronic-structure calculations

* **Note added in proof.** Highly relevant work is that of E. Wimmer, C. L. Fu, and A. J. Freeman (*Phys. Rev. Lett.* **55**, 2618, 1985) following that of J. K. Norskov, S. Holloway, and N. D. Lang (*Surf. Sci.* **137**, 65, 1984).

show that the charge associated with a S atom is screened to zero at distances larger than those to the neighboring metal atoms, charge transfer cannot be the source of the poisoning effect of S at low coverages. Indeed, they argue that, generally, any "long-range" effect of electronegative surface additives must stem from an electronic perturbation that is not screened. They therefore focus in their calculation on the local density of states (LDOS) at the Fermi level. They argue that this "energy-resolved charge density" has obvious significance for chemistry. It governs the degree to which energy must be supplied to remove electrons from occupied states and transfer them into unoccupied states. The lower the Fermi-level LDOS, the less well the surface can respond to the presence of reactants. At the same time, a large LDOS at E_f indicates that considerable energy can be gained by a reaction which moves the states that lie at E_f to lower energies. In addition to its bearing on chemical activity, the LDOS at E_f is a quantity that is unscreened. Screening is the tendency of the electron gas to seek total local-charge neutrality. The charge corresponding to any narrow energy window is not screened. Thus the LDOS at E_f is a quantity whose perturbation by S atoms can be expected to extend considerably beyond a screening length.

Feibelman and Hamann present results which show that the Fermi-level local density of states is substantially reduced by the S, even at nonadjacent sites. It is stressed that this LDOS governs the ability of the surface to respond to the presence of other species. Therefore, because of the importance of LDOS, Appendix 6.2 outlines a method whereby its direct calculation is feasible.

6.6. CHEMICAL OSCILLATIONS

Chemical oscillations usually occur in rather complex reactions only. For that reason, the oxidation of carbon monoxide over catalysts like platinum, which is governed by relatively simple rate equations, is of special interest. An explanation for oscillations in the rate of this process was offered by Dauchot and Van Cakenberghe,[310] who interpreted their experiments in terms of variation in surface temperature in the course of the reaction.

Subsequently, Turner *et al.*[311] made a further study of the oxidation of CO over both Pt and Pd, and noticed a different type of oscillation, which is not explained by the Dauchot–Van Cakenberghe mechanism. The reason is that the period of the oscillation (up to 20 min) is so much longer than any conceivable thermal delay time.[314]

Taking into account that two oxidation states exist on the surface of Pt and Pd, Turner *et al.*[311] have proposed that the oscillation they observe comes about in the following way. Starting with a relatively small CO coverage and a relatively large oxygen coverage of type I, say, some of that oxygen combines with the adsorbed CO in a Langmuir–Hinshelwood process (cf. Chapter 1) and some of it converts to oxygen of type II. Were it not for the latter process, a steady CO_2 production rate could be maintained so long as the CO partial pressure is not excessive. However, conversion to type II reduces the sites available to type I oxygen. The effect of the conversion is therefore somewhat equivalent to increasing type II coverage at fixed partial pressures. After the switch,* there is little type I oxygen. The type II oxygen now gradually diminishes by oxidizing the CO, its replenishment rate from type I being very small. Eventually, type II oxygen is so sparse that this state is unstable, the system returns to dominant type I coverage, and the cycle resumes. Evidently the duration of the high type II coverage phase must be dominated by the CO oxidation rate from this phase, and if this state is assumed to be tightly bound the rate will have a high activation energy. The duration should therefore vary exponentially with inverse temperature, the slope of a logarithmic plot vs. T^{-1} giving the activation energy. The duration of the type I-rich phase of the cycle will show a more complicated dependence. First of all it must, of course, depend on the Langmuir–Hinshelwood rate of CO_2 formation from phase I. But it must also depend sensitively on the rate of conversion to type II, since oscillations would not occur if this rate were zero. Less obviously, it also depends on CO and O_2 adsorption rates (as well as on CO and O_2 partial pressures). Again the logarithmic plot vs. T^{-1} will be linear, but with slope equal to a linear combination of the several activation energies.

6.6.1. Nature of Rate Equations

The work of Suhl[312] will now be presented by outlining the main points; the detail is given in Appendix 6.3. Suhl, together with the earlier workers, assumes that the type II oxygen O_{II} occupies subsurface sites and that, in doing so, it interdites the use of surface sites directly above it for CO or O type I (O_I) adsorption, but that occupation of surface sites by CO and O type I does not interdite the use of the sites directly below it for $O_I \rightarrow O_{II}$ conversion. If θ_1, θ_2, and θ_3 are the fractional coverages of O_I, CO, and O_{II}, the exclusion factors in the rates of dissociative adsorption of O_2, simple adsorption of CO, and conversion to O_{II} are then, respectively,

$$(1 - \theta_1 - \theta_2 - \theta_3)^2, \qquad (1 - \theta_1 - \theta_2 - \theta_3), \qquad (1 - \theta_3)$$

In other words, O_{II} deactivates the site above it with respect to adsorption, while the presence of an adsorbate does not interfere appreciably with

* A bistability point is reached at which the coverage switches to CO dominance.[312]

O_I–O_{II} conversion. The rate equations are given by

$$\dot{\theta}_1 = r_O(1 - \theta_1 - \theta_2 - \theta_3)^2 - k_1\theta_1\theta_2 - a(1 - \theta_3)\theta_1 \tag{6.6.1}$$

$$\dot{\theta}_2 = r_{CO}(1 - \theta_1 - \theta_2 - \theta_3) - k_1\theta_1\theta_2 - c\theta_2 \tag{6.6.2}$$

and

$$\dot{\theta}_3 = a(1 - \theta_3)\theta_1 - k_2\theta_2\theta_3 \tag{6.6.3}$$

In these equations, r_O is the rate of dissociative adsorption of oxygen, r_{CO} the adsorption rate for CO, $k_{1,2}$ the Langmuir–Hinshelwood rates for CO_2 formation from CO and O_I and from CO and O_{II}, respectively; a is the $O_I \rightarrow O_{II}$ conversion rate, while c denotes the CO desorption rate. Direct desorption of oxygen is neglected. The equations expressing the Dauchot–Van Cakenberghe mechanism are similar, except that the O_{II} phase is neglected by them. The equation for θ_3 is replaced by one for the temperature coupled to θ_1 and θ_2 through the heat of reaction. The Dauchot–Van Cakenberghe oscillations are obtained by a theory that treats the rates of variation of θ_1, θ_2, and the temperature all on the same footing and the simultaneous dynamics of these three degrees of freedom must be fully taken into account.

In contrast, the mechanism of Turner *et al.*[311] invokes the additional oxidation state and dismisses the temperature variation as incidental. A very much simpler theory can be derived for that mechanism, because a and k_2 are much smaller than any other rate. For this reason, one may assume that θ_1 and θ_2 rapidly adjust themselves to the prevailing value of θ_3. In other words, the values of θ_1 and θ_2 may be obtained by solving equations (6.6.1) and (6.6.2) with $\dot{\theta}_1$ and $\dot{\theta}_2$ equated to zero. The solution gives θ_1 and θ_2 as algebraic functions of θ_3, and is substituted in (6.6.3), which is then a single first-order differential equation whose solution is not, in itself, oscillatory. Oscillations arise because in a certain region of parameter space, θ_1 and θ_2 as functions of θ_3 have three branches stretched between two branch points. Actually, there are six branches but only three are relevant. As θ_3 varies according to the differential equation it may, again for suitable parameter values, reach one of the branch points of (θ_1, θ_2). At this point (θ_1, θ_2) switches to the other stable branch. On that other branch θ_3 reverses its direction of change, i.e., $\dot{\theta}_3$ changes sign, and eventually reaches the other branch point. There (θ_1, θ_2) returns to the first branch and the cycle resumes.

Suhl points out that such an analysis completely disregards the detailed dynamics of the switch between branches and a few comments are needed to justify this. Because a and k_2 are very small, θ_3 must vary slowly. Consider now equations (6.6.1) and (6.6.2), treating θ_3 as constant, retaining $\dot{\theta}_1$ and $\dot{\theta}_2$, but neglecting $a\theta_1(1-\theta_3)$. These equations may be analyzed in almost the same manner as the corresponding equations of the work of Dagonnier and Nuyts[313] in the isothermal case, except that the presence of the desorption term $c\theta_2$ in (6.6.2) moves the stable CO-rich singular point away from $\theta_1 = 0$, $\theta_2 = 1$. The motion in the θ_1, θ_2 phase plane is sketched by Suhl, whose result is reproduced in Figure 6.10 for small c/k and $p_{CO}/p_O < 1$. The physical region is now the regime $\theta_1 > 0$, $\theta_2 > 0$, $\theta_1 + \theta_2 < 1 - \theta_3$, and a phase point starting out in this region never leaves it. As in the study of Dagonnier and Nuyts,[313] there are at most four singular points, two saddles X_{+1} and X_0^-, a stable node X_{-1}, and X_0^+, which can be an ordinary node or a spiral point depending on the parameters. (If a temperature lag is introduced, it is this latter point that can spawn a limit cycle around itself.) Suppose the phase point has reached X_0^+. This corresponds to the upper branch referred to above. On this branch, for suitable parameter values $\dot{\theta}_3$ turns out to be positive, so θ_3 slowly increases. This causes X_0^+ and X_0^- to slowly approach each other and finally to annihilate. The phase point now has no choice but to go to the only remaining node, X_{-1}. Because θ_1 is small there, $\dot{\theta}_3$ is negative, so θ_3 now diminishes. This causes X_0^+ and X_0^- to reappear, but the phase point stays at X_{-1}. As θ_3 diminishes X_0^- and X_{-1} approach each other and finally

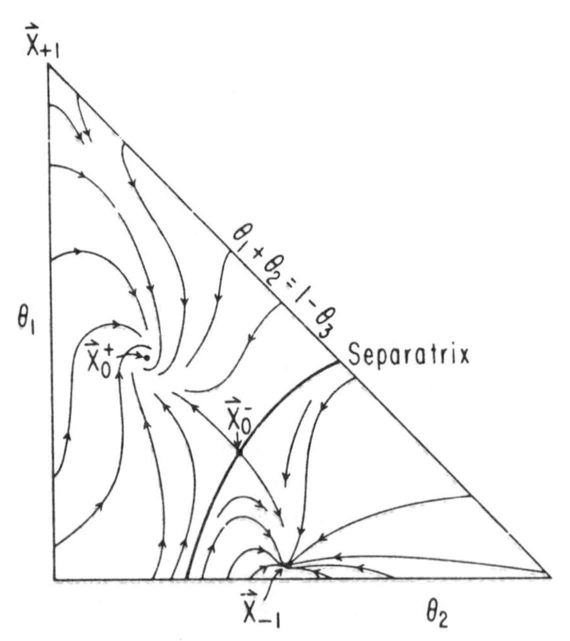

FIGURE 6.10. Motion of a phase point in the θ_1, θ_2 phase plane. Result sketched is for case of small c/k and $p_{CO}/p_O < 1$ (after Suhl[312]).

annihilate. The phase point is now obliged to return to X_0^+ where $\dot{\theta}_3$ is again positive, so that the cycle is resumed.

This oscillation is clearly distinct from the limit cycle that can develop around X_0^+ due to temperature lag, but it is quite possible that if such a lag is introduced both types of oscillation will occur. In that case, one would see a short-period oscillation superposed on the motion along the upper branch, and no oscillation along the lower branch (since the simple node X_{-1} cannot go unstable). Finally, it is noteworthy that if the merging of X_0^+ and X_0^- were to occur precisely, without fluctuation, the motion of the phase point toward X_{-1} would be quasi-chaotic[313] and would take a very long time, so that the hypothesis $\dot{\theta}_3 \ll \dot{\theta}_1, \theta_2$ would be violated. In practice, fluctuation effects will, of course, make the transition quite rapid.[314]

The main steps in the analysis of Suhl are presented in Appendix 6.3.

In connection with the two-state theory developed by Suhl, it is relevant that a further study of CO on Ni has been carried out by Craig[315] using electron- and ion-induced desorption. Thermal-desorption spectra indicate the presence of two binding states. The O species is observed to be released from state 1, the state of higher binding energy, while state 2, a more loosely bound state, is the source of CO.[312] O^- species appear to be released from both states.

After this rather specific example, the section on chemical oscillations is concluded by referring to the extensive review by Gurel and Gurel,[316] which brings experimental results on a number of systems into some contact with work on mathematical models.

Appendixes

APPENDIX 1.1
SURFACE CHARACTERIZATION TECHNIQUES USED TO DETERMINE STRUCTURE AND COMPOSITION OF SOLID SURFACES (Adapted from Spencer and Somorjai[13])

Method	Abbreviation	Physical process	Information gained
Low-energy electron diffraction	LEED	Elastic back-scattering of low-energy electrons	Atomic surface structure of surfaces and adsorbates
Auger electron spectroscopy	AES	Electron emission from surface atoms excited by electron, X-ray, or ion bombardment	Surface composition
High-resolution electron energy-loss spectroscopy	HREELS	Vibrational excitation of surface atoms by inclastic reflection of low-energy electrons	Structure and bonding of surface atoms and adsorbates
Infrared spectroscopy	IRS	Vibrational excitation of surface atoms by absorption of infrared radiation	Structure and bonding of adsorbates

APPENDIX 1.1 (Continued)

Method	Abbreviation	Physical process	Information gained
X-ray and ultraviolet photoelectron spectroscopy	XPS, UPS (ESCA)	Electron emission from atoms excited by X-rays or ultraviolet radiation	Electronic structure and oxidation state of surface atoms and adsorbates
Ion-scattering spectroscopy	ISS	Elastic reflection of inert-gas ions	Atomic structure and composition of solid surfaces
Secondary ion mass spectroscopy	SIMS	Ion-beam-induced ejection of surface atoms as positive and negative ions	Surface composition
Extended X-ray absorption fine-structure analysis	EXAFS	Interference effects in photoemitted electron wave function in X-ray absorption	Atomic structure of surfaces and adsorbates
Thermal desorption spectroscopy	TDS	Thermally induced desorption or decomposition of adsorbates	Adsorption energetics and composition of adsorbates

APPENDIX 2.1
DENSITY MATRIX AND LINEAR-RESPONSE FUNCTION OF THE INFINITE-BARRIER MODEL OF A METAL SURFACE

The density matrix and linear-response function of the infinite-barrier model of a metal surface are used in several applications in earlier chapters, so it is worthwhile summarizing the main steps in their derivation by Moore and March.[317]

The tool employed is the so-called canonical density matrix $C(\mathbf{r}, \mathbf{r}', \beta)$

defined by

$$C(\mathbf{r}, \mathbf{r}', \beta) = \sum_{\text{all } i} \psi_i^*(\mathbf{r})\psi_i(\mathbf{r}')\exp(-\beta E_i), \qquad \beta = (k_B T)^{-1} \tag{A2.1.1}$$

where ψ_i and E_i are the wave functions and corresponding energies of the barrier model. The appropriate Schrödinger equation $H\psi = E\psi$ yields, in a straightforward manner, that $C(\mathbf{r}, \mathbf{r}', \beta)$ satisfies the so-called Bloch equation

$$H_\mathbf{r} C(\mathbf{r}, \mathbf{r}', \beta) = -\frac{\partial C}{\partial \beta} \tag{A2.1.2}$$

It also follows from the definition (A2.1.1) that

$$C(\mathbf{r}, \mathbf{r}', 0) = \delta(\mathbf{r} - \mathbf{r}') \tag{A2.1.3}$$

which is simply an expression of the completeness relation for the eigenfunctions $\psi_i(\mathbf{r})$.

Moore and March[317] exploit the analogy between the classical equation of heat conduction[318] and the above Bloch equation to calculate the canonical density matrix exactly for noninteracting electrons in not only the infinite-barrier model, but also in a finite-step model of a metal surface. By restricting the considerations below to the infinite-barrier limit, these workers select the z-axis perpendicular to the planar metal surface and set

$$C(\mathbf{r}, \mathbf{r}', \beta) = (2\pi\beta)^{-1}\exp(-|\mathbf{x} - \mathbf{x}'|^2/2\beta) f(z, z', \beta) \tag{A2.1.4}$$

where $\mathbf{x}$ denotes a two-dimensional vector in the (x, y) plane taken to lie in the planar metal surface. It is easy to show by direct solution of the Bloch equation that for free electrons

$$f(z, z', \beta) = (2\pi\beta)^{-1/2}\exp(-|z - z'|^2/2\beta)$$

Substitution in the Bloch equation easily gives a differential equation for f in the case of a general potential energy $V(z)$, namely

$$\frac{1}{2}\frac{\partial^2 f(z, z', \beta)}{\partial z^2} = \frac{\partial f(z, z', \beta)}{\partial \beta} + V(z) f(z, z', \beta) \tag{A2.1.5}$$

For the infinite-barrier limit, $z' < 0$ and $z \le 0$, the solution is readily verified to be

$$f(z, z', \beta) = \frac{\exp(-|z - z'|^2/2\beta)}{(2\pi\beta)^{1/2}} - \frac{\exp[-(z + z')^2/2\beta]}{(2\pi\beta)^{1/2}} \tag{A2.1.6}$$

in agreement with Brown *et al.*[319] In the other regions, $f=0$ in the infinite-barrier limit.

The diagonal form of the Bloch density matrix $C(\mathbf{r},\mathbf{r},\beta)$ for the infinite-barrier limit follows:

$$C(\mathbf{r},\mathbf{r},\beta)=(2\pi\beta)^{-3/2}[1-\exp(-2z^2/\beta)], \qquad z\leq 0 \tag{A2.1.7}$$

Laplace inversion of $C(\mathbf{r},\mathbf{r},\beta)/\beta$ leads to the electron density $\rho(z)$ of the infinite-barrier model:

$$\rho(z)=\frac{k_f^3}{3\pi^2}\left[1-\frac{3j_1(2k_f z)}{2k_f z}\right], \qquad z<0 \tag{A2.1.8}$$

This agrees with the earlier result of Bardeen[320] by direct summation of the squares of the normalized wave functions up to the Fermi level $E_f=\hbar k_f^2/2m$. In equation (A2.1.8), $j_1(x)$ is the first-order spherical Bessel function $(\sin x - x\cos x)/x^2$.

Moore and March also record the static random-phase approximation for the linear-response function $F(\mathbf{r},\mathbf{r}',E_f)$ given by[321]

$$\frac{\partial F}{\partial E}=2\,\mathrm{Re}\left[G_0(\mathbf{r},\mathbf{r}',E)\frac{\partial\rho(\mathbf{r}',\mathbf{r},E)}{\partial E}\right] \tag{A2.1.9}$$

where G_0 is the Green function for the infinite-barrier model while ρ is the Dirac density matrix constituting the off-diagonal generalization of equation (A2.1.8). Their infinite-barrier result is

$$F(\mathbf{r},\mathbf{r}',E_f)= \tag{A2.1.10}$$

$$\begin{cases} -\dfrac{k_f^2}{(2\pi)^3}\left[\dfrac{j_1(2k_f s)}{s^2}+\dfrac{j_1(2k_f s')}{s'^2}-\dfrac{2j_1(2k_f\{s+s'\})}{ss'}\right] & \text{for } z,z'<0 \\ 0 & \text{otherwise} \end{cases}$$

where $s=|\mathbf{r}-\mathbf{r}'|$ and $s'=[|\mathbf{x}-\mathbf{x}'|^2+(z+z')^2]^{1/2}$. Here $\mathbf{x}$ and $\mathbf{x}'$ are vectors in the plane parallel to the surface.

This result (A2.1.10) was employed in Section 2.2.2 of the main text.

APPENDIX 2.2
INFLUENCE OF FERMI-SURFACE TOPOLOGY ON ASYMPTOTIC DISPLACED-CHARGE ROUND ADATOMS

Flores *et al.*[322] have generalized the free-electron model discussed in Section 2.2 of the main text and based on a spherical Fermi surface separating occupied and unoccupied electron states in **k**-space. Their generalization enables one to take account of the influence of the topology of this constant-energy surface on the anisotropy of the displaced charge $\Delta\rho(\mathbf{r})$ in the asymptotic region far from the defect. Provided the perturbing potential $V(\mathbf{r})$ satisfies some reasonable requirements pertaining to its analytic properties, Flores *et al.* show that

$$\Delta\rho(\mathbf{r}) \sim r^{-n} \times \text{oscillatory function} \tag{A2.2.1}$$

For impurities in bulk metals, n assumes values between 1 and 5 in different directions for Fermi surfaces with particular topologies. However, for a planar metal surface the displaced charge round an adatom has a shorter range for V embedded in a surface than for the bulk metal, in most but not all cases. For a closed Fermi surface with nonzero curvature, $n = 5$ for the parallel configuration, as derived in the body of the text for the special case of a spherical Fermi surface.

APPENDIX 2.3
CORRELATION EFFECTS FOR HYDROGEN CHEMISORBED ON TRANSITION METALS

Due to widespread interest in the chemisorption of H on normal and transition metals,[323–325] the theory of correlation effects in this system will be briefly outlined below. By utilizing pseudopotentials, the local-density formulation has been used to obtain the density of states and the chemisorption energy as a function of distance of the atom from the surface,[326–329] together with a good picture of the properties associated with the adatom. However, computer-time limitations presently preclude the application of such methods to complex systems.

Therefore, attention here will be paid to the analysis within the tight-binding framework, which is evidently closely related to the chemical point of view. The conclusion that emerges is that, while correlation effects are rather small for the case of H chemisorbed on most transition metals, in the specific case of W and to a lesser extent for Mo, the appropriate values of the bandwidth and the coupling strength between H and the surface atoms produce many-body effects in the density of states that may be observable in photoemission spectra.[330]

A summary of the basic approach used and the main findings will be given below. The starting point is an Anderson–Hubbard-type Hamiltonian.[331,332] Second quantization is not strictly necessary to understand the gist of what follows; nonetheless its use will be convenient in writing down this Hamiltonian. For the reader who wishes to use the technique, it may be understood by referring to introductory accounts such as those in the books by Avery[333] and March *et al.*[334] The Hamiltonian then reads

$$H = \sum_{k,\sigma} E_k C^\dagger_{k\sigma} C_{k\sigma} + \varepsilon_H \sum_\sigma A^\dagger_\sigma A_\sigma$$

$$+ \sum_{k,\sigma} V_k (A^\dagger_\sigma C_{k\sigma} + \text{h.c.}) + U A^\dagger_\sigma A_\sigma A^\dagger_{-\sigma} A_{-\sigma} \tag{A2.3.1}$$

The important quantities entering H are: E_k, the substrate energy levels; ε_H, the adatom level; V_k, measuring interaction between the adatom and substrate; U, the intra-atomic interaction for the adatom; and σ, the spin variable.

The zero-order starting point of the method consists in assuming degenerate substrate levels. In this limit, the above Hamiltonian has been solved analytically by Brenig and Schönhammer.[335] Baldo *et al.*[330] introduce the following operators:

$$I = \sum_\sigma (A^\dagger_\sigma A_\sigma + B^\dagger_\sigma B_\sigma) \tag{A2.3.2}$$

$$\begin{aligned} \kappa^\sigma_1 &= \tfrac{1}{2} i (A^\dagger_\sigma B_\sigma - B^\dagger_\sigma A_\sigma) \\ \kappa^\sigma_2 &= \tfrac{1}{2} (A^\dagger_\sigma B_\sigma + B^\dagger_\sigma A_\sigma) \\ \kappa^\sigma_3 &= \tfrac{1}{2} (A^\dagger_\sigma A_\sigma - B^\dagger_\sigma B_\sigma) \end{aligned} \tag{A2.3.3}$$

which form a closed SO(4) Lie algebra (see, for example, Cornwell).[336] Here

$$B_\sigma = \frac{1}{\bar{V}} \sum_k V_k B^\dagger_{k\sigma}, \qquad \bar{V} = \left(\sum_k V_k^2 \right)^{1/2} \tag{A2.3.4}$$

The starting Hamiltonian can then be expressed only in terms of these operators and the single-particle Green function $\hat{G}(\omega)$ can be evaluated. It satisfies

$$\begin{pmatrix} \hat{\Sigma} & 0 \\ 0 & -1\omega \end{pmatrix} \hat{G}(\omega) = 1, \qquad \hat{\Sigma} = \begin{pmatrix} M_{AA} - \omega & \bar{V} \\ \bar{V} & -\omega \end{pmatrix} \tag{A2.3.5}$$

$\hat{G}$ being an $N \times N$ matrix where N denotes the number of single-particle levels. The explicit expression for M_{AA} in the simplest case considered by Baldo *et al.*[330] is, with $\langle n_{-\sigma} \rangle \equiv \langle A^\dagger_{-\sigma} A_{-\sigma} \rangle$,

$$M_{AA}(\omega) = \frac{U^2 \omega}{4(\omega^2 - 9\bar{V}^2)} + U \langle n_{-\sigma} \rangle \tag{A2.3.6}$$

Baldo *et al.*[330] discuss how to proceed in order to calculate higher-order contributions to $M_{AA}(\omega)$. These involve higher-order correlation functions, which are calculable using algebraic techniques. For further results and discussion the reader should consult Baldo *et al.*[330]

APPENDIX 2.4
LIFETIMES OF EXCITED STATES OF MOLECULES WELL OUTSIDE METAL SURFACES: A MODEL CALCULATION

It is still hardly a simple matter to determine accurately the bond length of a diatomic molecule outside a metal surface. It seems possible that spectroscopic observations of lifetimes of excited states of molecules may eventually provide one route to such a bond-length determination. We dicussed in Chapter 2 how the equilibrium bond length would alter for a diatomic molecule well outside a planar metal surface. The experimental techniques developed by Kuhn *et al.*[337] for studying molecules at well-defined distances from metal surfaces seem to afford the possibility of studying whether appreciable changes can occur in diatomic bond lengths.

For the above reasons, Babiker and March[338] have conducted a model calculation of the hydrogen-molecule ion, even though their prime interest was in neutral molecules. For atomic excited states, Babiker[339] has discussed their lifetimes outside a perfectly conducting planar metal surface. Babiker and March[338] studied the configuration in which the hydrogen-molecule ion was parallel to, but far from the conduction electron spill-out region of, the planar metal surface.

In their work, it was further assumed that the internuclear distance is the same in the ground and excited states of the molecule. In fully quantitative work it may be necessary to relax this assumption. The question of the applicability of the Born–Oppenheimer approximation to molecular excited states seems to lie at the heart of this matter and may require further clarification.

The model calculation presented below will appeal to analytical results near the united atom limit. This will serve to establish the method; no doubt computer calculations can be employed to extend the results to realistic internuclear distances.

Lifetime of H_2^+ United Atom Limit

Change in Transition Frequency with Internuclear Distance

When the internuclear distance R is small, the H_2^+ ion approaches the electronic structure of He^+. This ion evidently has a nonrelativistic ground-state energy of $-Z^2e^2/2n^2a_0$ with $Z=2$ and principal quantum number $n=1$. Throughout we use units in which $e^2/a_0=27$ eV is the unit of energy (atomic units) while the unit of length is the Bohr radius a_0. In these units the velocity of light $c=137$. It is then possible to express the electronic energy of the ground state of H_2 as

$$E_{1s}(R) = -2+\tfrac{8}{3}R^2+\text{higher order terms} \tag{A2.4.1}$$

This result is easily verified by noting that, at small separations, the spherical average of the two protons can be taken in the form

$$V(r) = \begin{cases} -2/R, & r < R/2 \\ -2/r, & r > R/2 \end{cases} \tag{A2.4.2}$$

First-order perturbation theory yields

$$\Delta E_1 = \int_0^{R/2} \frac{8}{\pi}\left[\frac{2}{r}-\frac{2}{R}\right]\exp(-4r)4\pi r^2\,dr \tag{A2.4.3}$$

which leads directly to the term in R^2 on the right-hand side of equation (A2.4.1).

The first excited-state energy (say $2p_z$) in the united He^+ ion is obtained by using strictly degenerate perturbation theory. However, since $\psi_{2p} \sim N_{2p}r$ when r is small, for all three degenerate functions ψ_{2p_x}, ψ_{2p_y}, and ψ_{2p_z}, the change in the first excited-state energy can be expressed as

$$\Delta E_2 = -N_{2p}\int_0^{R/2}\left(\frac{2}{R}-\frac{2}{r}\right)r^2\frac{4\pi}{3}r^2\,dr \tag{A2.4.4}$$

which yields

$$\Delta E_2 \simeq \left(\frac{N_{2p}^2\pi}{120}\right)R^4 \tag{A2.4.5}$$

Thus the lowest-order correction to the energy of the first excited state ($2p_z$, say) is $O(R^4)$ and hence the analog of equation (A2.4.1) is

$$E_{2p_z}(R) = -\tfrac{1}{2}+O(R^4) \tag{A2.4.6}$$

The transition frequency is therefore given by

$$\hbar\omega_0(R) = \tfrac{3}{2}-\tfrac{8}{3}R^2+\text{higher order terms} \tag{A2.4.7}$$

Change in Dipole Matrix Element with Internuclear Distance

The change in the dipole matrix element arises due to changes in wave functions ψ_{2p_z} and ψ_{1s}. We write

$$\langle z\rangle_R^2 = \langle z\rangle_{R=0}^2 + \Delta(R) \tag{A2.4.8}$$

and, for consistency, we evaluate Δ to $O(R^2)$. Let us consider first the change in the $2p_z$ wave function, which can be expressed as

$$\Delta\psi_{2p_z}(r, R) = \int G(\mathbf{r}, \mathbf{r}')\left(\frac{2}{r'} - \frac{2}{R}\right) H(r', R/2)\psi_{2p_z}(\mathbf{r}')d\mathbf{r}' \tag{A2.4.9}$$

Here $G(\mathbf{r}, \mathbf{r}')$ is the Green function for the He^+ ion, and if we expand in spherical harmonics, only the term in $P_1(\cos\theta)$ in G contributes to equation (A2.4.9). The quantity H is the usual Heaviside function. Using the fact that $\psi_{2p_z}(r')$ is proportional to r' at small r' and the p-component of $G(\mathbf{r}, \mathbf{r}')$ to rr' at small r and r', we readily obtain that, as $R \to 0$, the change in the $2p_z$ wave function is $O(R^4)$. Thus to calculate $\langle z\rangle_R^2$ to $O(R^2)$ we only require the change in the ground-state wave function ψ_{1s} to $O(R^2)$.

Hameka[340] has given the explicit form of the s-wave contribution to G and his result can be inserted into the analog of equation (A2.4.8) for the $1s$ wave function to obtain the corresponding change Δ_{1s}. Alternatively, one can use first-order perturbation theory and proceed as follows. If H' is the perturbation due to separating the protons of He, then

$$\Delta\psi_{1s}(r, R) = \sum_{j\neq 0} \frac{H'_{j0}(R)}{E_j^0 - E_0^0}\psi_j^0(\mathbf{r}) \tag{A2.4.10}$$

The quantity $\langle z\rangle_R$ is given by

$$\langle z\rangle_R = \int d\mathbf{r}\psi_{1s}(\mathbf{r}, R)z\psi_{2p_z}(\mathbf{r}) = \langle z\rangle_{R,0} + \Delta(R) \tag{A2.4.11}$$

where

$$\Delta R = \sum_{j\neq 0}\int d\mathbf{r}\frac{H'_{j0}}{E_j^0 - E_0^0}\psi_j^0(\mathbf{r})z\psi_{2p_z}^0(\mathbf{r}) \tag{A2.4.12}$$

We now consider

$$H'_{j0}(R) \simeq \int_0^{R/2}\psi_j^0(0)\left(\frac{2}{r} - \frac{2}{R}\right)\psi_0^0(0)4\pi r^2 dr \tag{A2.4.13}$$

which yields

$$H'_{j0} \simeq \frac{\pi}{3}R^2\psi_j^0(0)\psi_0^0(0) \tag{A2.4.14}$$

Substitution in equation (A2.4.12) enables the expression to be put in the following form:

$$\Delta(R) = \frac{\pi R^2}{3} \int d\mathbf{r} G_1(0, r) \psi_0^0(0) z \psi_{2p_z}^0(r) \tag{A2.4.15}$$

in which $G_1(0, r)$ is Hameka's Green function for the ground state of the He^+ ion and is given by

$$G_1(0, r) = 2\exp(-2r)[\gamma - \tfrac{5}{2} + 3r + \ln(4r)] \tag{A2.4.16}$$

where γ is Euler's constant. If the explicit standard forms of $\psi_0^0(0)$ and $\psi_{2p_z}^0$ are inserted into equation (A2.4.15), one obtains

$$\Delta(R) = R^2 \frac{2^3}{9\sqrt{2}} \int_0^\infty dr \exp(-3r) r^4 [\gamma - \tfrac{5}{2} + 3r + \ln(4r)] \tag{A2.4.17}$$

Using standard integrals, it is found that

$$\Delta = R^2 \frac{\sqrt{2}}{36} \frac{2^4}{3^5} [110 + 24 \ln(\tfrac{4}{3})] \simeq 0.31 R^2 \tag{A2.4.18}$$

Thus, to the required order, we finally have

$$\langle z \rangle_R = \langle z \rangle_{R,0} + 0.31 R^2 + O(R^4) \tag{A2.4.19}$$

Change in Decay Rate with Internuclear Distance

According to Babiker,[339] the decay rate* is given by

$$\Gamma = \frac{4}{3} \frac{e^2}{\hbar c^3} [\omega_0(k)]^3 \langle z \rangle_R^2 \{1 + F[2\omega_0(R) z / c]\} \tag{A2.4.20}$$

where F is known in terms of Bessel functions and is given by

$$F(x) = \tfrac{1}{3}[J_2(x) + J_0(x)] \equiv \frac{\sin x}{x^3} - \frac{\cos x}{x^2} \tag{A2.4.21}$$

* It is of interest to note here the study by A. P. Alivisatos, D. H. Waldeck, and C. B. Harris (*J. Chem. Phys.* **82**, 541, 1985) of the effect of a metal reflecting surface on the lifetime of an oscillating dipole.

The transition frequency given by equation (A2.4.7) and the dipole matrix element given by equation (A2.4.19) enable the small-R limit of the decay rate to be obtained in the form

$$\Gamma(R, z) = \Gamma(R, z=\infty)\left[1 - \frac{8R^2}{\hbar\omega_0(0)} + \frac{0.62R^2}{\langle \mathfrak{z}\rangle_0}\right]\{1 + F[2z\omega_0(R)/c]\} \tag{A2.4.22}$$

We must stress here that Stark-effect terms arising from the ionic charge of H_2^+ and its image in the conductor are not considered, since the present model calculation is to be regarded as preliminary to obtaining Γ for neutral molecules.

If we consider first the limit $z \to \infty$ in equation (A2.4.22), then we clearly have a calculation of the lifetime of a free H_2^+ ion, namely when the metal surface has no influence. The decay rate is expected to decrease as we separate the protons by a small distance from the united atom limit. This is consistent with the requirement that, as R tends to infinity, Γ must become the lifetime of the hydrogen atom, which is in fact $\frac{1}{16}$ of that for the united ion He^+. Presently, we see no reason for other than monotonic behavior as a function of R. It should be possible in the future, following the above outline, to use available numerical information on H_2^+ for arbitrary R to calculate the quantity $\Gamma(R, z=\infty)$. The effect of the metal surface may then be represented by rewriting equation (A2.4.20) as

$$\frac{\Gamma(R, z)}{\Gamma(R, z=\infty)} = 1 + F\left[\frac{2z\omega_0(R)}{c}\right] \tag{A2.4.23}$$

which relates the decay rate of the diatomic system in the presence of the metal surface to the decay rate (at the same internuclear distance) in free space. The influence of separating the protons from the united ion limit is seen to be essentially described as a phase shift in the function $F(x)$.

APPENDIX 3.1
BONDING OF WATER MOLECULES TO Pt: A MODEL OF THE ICE–*d*-ELECTRON METAL INTERFACE

Painter *et al.*[341] have contrasted the important role the directionality of the d-orbitals in Pt plays in the bonding of water molecules to this metal, with that for the noble metals, with full d-shells and therefore in which the directionality of the bonding is less important. This is done in the context of a model they put forward for the ice–d-electron metal interface. A qualitative explanation of the different orientations of the growth of ice on Pt(111) and Ag(111) surfaces is thereby proposed.

The motivation for the work of Painter *et al.* was the experimental finding that bulk (multilayer) ice grows on an Ag(111) surface with two orientations, namely parallel to or rotated by 30° from the Ag(111) hexagonal net, according to measurements of Firment and Somorjai.[342] In contrast, on Pt(111) only the 30° orientation has been observed by the same workers[343] and also by Fisher and Gland.[344] The size of the hexagonal net of the ice surface is 4.52 Å, which, within the experimental uncertainty, is equal to the separation of water molecules in the (0001) plane of bulk (I_h) ice.

Lattice Matching and Domain Sizes

Earlier investigations of the metal–metal interface by Frank and Van der Merwe[345] and by Ecob and Ralph[346] prompted Painter *et al.* to examine the degree of matching between the hexagonal nets of ice and the metals. Figures A3.1.1 and A3.1.2 show the Pt and Ag lattices superimposed on the ice lattice for the two possible orientations dictated by symmetry, namely 0° and 30°. To be definite, these figures have been drawn with the preferred bonding site for a water molecule situated directly above a metal atom, but because of lattice periodicity the following arguments hold for any preferred site. For the preferred bonding site of the water molecule on the metal surface, it can be seen from Figures A3.1.1 and A3.1.2 that favorable sites for the water molecules occur more frequently in the 30° orientation than in the 0° case. In addition, study of these figures shows that the degree of matching in the 30° orientation is significantly better for Pt than for Ag.

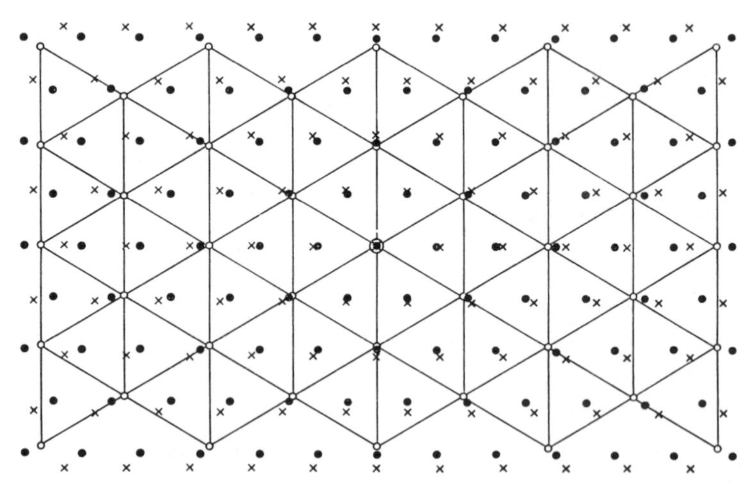

FIGURE A3.1.1. First layer of molecules of hexagonal ice (○) together with the Pt(●) and Ag(×) atoms in the 30° orientation.

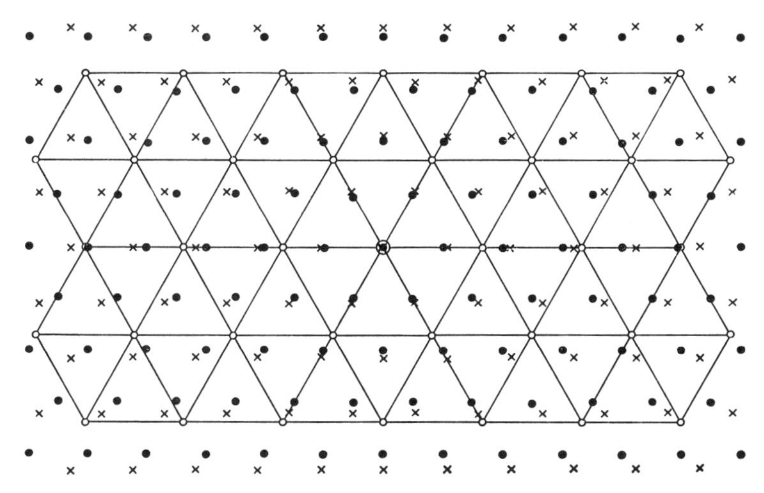

FIGURE A3.1.2. As for Figure A3.1.1 but in the 0° orientation. Experimentally, this orientation is only observed for ice on Ag.

The point already made about the greater importance of directionality of the *d*-orbitals in Pt than in Ag is borne out by the experimental results shown schematically in the above figures.

For Pt, Figure A3.1.1 shows a region corresponding roughly to one Frank–Van der Merwe domain.[343–345] Within this domain there is little deviation of the water molecules from the preferred adsorption sites. There does not seem to be any experimental evidence to date on the domain size for Ag. However, since the degree of matching in the 30° orientation is not as good as that for Pt, it is to be expected that the domain size in this geometry will be smaller for Ag than for Pt.

In this connection, Painter *et al.*[341] draw attention to the relevance of the fact that for rare gases on metals the domain size is large, presumably due to the absence of chemical bonding.

Thermodynamic Considerations

In the model of Painter *et al.*,[341] it seems clear that the internal energy will be lower in the 30° than in the 0° orientation. Moreover, the stronger the interaction of the water molecules with the metal surface, the greater will be the tendency of the molecules to be adsorbed in a preferred direction. It was apparent from Figure A3.1.1 that chemical bonding is more significant for Pt than for Ag, so it is therefore natural in this metal that the LEED data for Pt reveal only the 30° orientation.

However, for Ag, one can anticipate that entropy will also be an important factor. The LEED experiments suggest that the entropy term for Ag will tend to favor the 0° orientation at elevated temperatures.

Absence of Macroscopic Electric Polarization

The above arguments, involving chemical bonding of the water molecules to the Pt(111) surface in the 30° orientation, will evidently lead to some degree of ordering of the orientation of the water molecules with respect to the metal surface. As a result of the rules of Pauling,[347] such a model must result in a net electric polarization of the first few layers of the ice lattice. Since bulk ice is nonpolar, it is clear that the dipole moments generated in subsequent layers must result in a cancellation of this polarization. The model proposed by Painter *et al.* is best described in terms of lattice defects. This seems natural enough when one recalls that in bulk ice there is disorder, which is usefully accounted for in terms of Bjerrum defects.[348] The presence of the metal surface, when chemical bonding is important as in the case of Pt, can be viewed as reducing the orientational disorder in the ice lattice in the surface layers.

One should add here that Ibach and Lehwald[349] have considered the growth of ice on a Pt(100) reconstructed surface. They concluded that, starting from the infinitely dilute case, as growth continues there is a change in orientation of the water molecule, from clusters of H– or O–pointing toward the metal surface to all the bonding being through the H–atoms.

APPENDIX 3.2
PARAMETRIZATION OF THE TIGHT-BINDING MODEL OF ETHENE

The tight-binding Hamiltonian introduced to analyze the energy levels of the orbitals of ethene as a function of the molecular geometry is discussed in this Appendix.

By exploiting the symmetry of the system (C_{2V}) the Hamiltonian can be written in the following block-diagonal form:

$$\left[\begin{array}{cccc|cccc|cc|cc}
E_H & V_1 & \sqrt{2}V_2\mu & \sqrt{2}V_2\delta & \cdot & \cdot & \cdot & \cdot & \cdot & \cdot & \cdot & \cdot \\
 & E_s+V_4 & V_5 & \cdot & \cdot & \cdot & \cdot & \cdot & \cdot & \cdot & \cdot & \cdot \\
 & & E_p+V_3 & \cdot & \cdot & \cdot & \cdot & \cdot & \cdot & \cdot & \cdot & \cdot \\
 & (B_2) & & E_p+V_6 & \cdot & \cdot & \cdot & \cdot & \cdot & \cdot & \cdot & \cdot \\
\hline
 & & & & E_H & V_1 & \sqrt{2}V_2\mu & \sqrt{2}V_2\delta & \cdot & \cdot & \cdot & \cdot \\
 & & & & & E_s-V_4 & -V_5 & \cdot & \cdot & \cdot & \cdot & \cdot \\
 & & & & & & E_p-V_3 & \cdot & \cdot & \cdot & \cdot & \cdot \\
 & & & & & (A_1) & & E_p-V_6 & \cdot & \cdot & \cdot & \cdot \\
\hline
 & & & & & & & & E_H & \sqrt{2}V_2\nu & \cdot & \cdot \\
 & & & & & & & (A_2) & & E_p+V_6 & \cdot & \cdot \\
\hline
 & & & & & & & & & & E_H & \sqrt{2}V_2\nu \\
 & & & & & & & & & & (B_1) & E_p-V_6
\end{array}\right]$$

where the transformation property of each of the block matrices is given in parentheses.

The eigenvalues of the matrix transforming like A_1 include the energies of the σ_{ss}, σ_{cc}, and π_{cc} orbitals, and those transforming like B_2, B_1, and A_2 include the energies of the σ_{ss}, σ_{CH}, and σ^*_{CH} levels, respectively. In each case the remaining eigenvalues correspond to energies of unoccupied orbitals.

In the above matrix ν, δ, and μ are the cosines of the angles formed between the C–H bond and the x, y, and z axes; the remaining symbols and their associated values are given in Table A3.2.1.

The parameters were fitted to the energy levels of the π_{cc}, σ^*_{CH}, σ_{cc}, and σ_{CH} orbitals calculated by Demuth.[350]

TABLE A3.2.1. The Various Parameters Involved in the Hamiltonian Matrix[a]

Interaction	Parameter (eV)
E_H $(=\langle 1s \mid H \mid 1s\rangle)$	−8.0
E_S $(=\langle 2s(A) \mid H \mid 2s(A)\rangle)$	−21.4
E_P $(=\langle 2p(A) \mid H \mid 2p(A)\rangle)$	−5.7
V_1 $(=\langle 1s \mid H \mid 2s\rangle)$	−6.0
V_2 $(=\langle 1s \mid H \mid 2p\rangle)$	−7.0
V_3 $(=\langle 2p_z(A) \mid H \mid 2p_z(B)\rangle)$	−7.0
V_4 $(=\langle 2s(A) \mid H \mid 2s(B)\rangle)$	−5.0
V_5 $(=\langle 2s(A) \mid H \mid 2p_z(B)\rangle)$	+6.0
V_6 $(= 2p(A) \mid H \mid 2p(B))$	−4.3

[a] The letters (A) and (B) in parentheses merely distinguish the two carbon atoms of the ethene molecule, $2p$ is directed along the C–H bond, while $2p$(A) and $2p$(B) are directed along the C–C bond.

APPENDIX 3.3
PARAMETRIZATION OF THE ANDERSON HAMILTONIAN DESCRIBING THE π-ENERGY LEVELS OF THE FREE ETHENE MOLECULE

Equation (3.4.3) presents the Hamiltonian H_m with which we shall describe the energy levels of the π-electrons in the free ethene molecule. Here we record how the four parameters ε_0, t, U, and J were determined.

First, Hiett *et al.*[106] determined the Hubbard U for the $2p$-orbitals occupied by electrons with opposite spin by using Slater atomic orbitals, and the result $U = 16.6$ eV was thereby obtained. The other parameters in H_m were determined using spectroscopic data. The singlet excitation energy was calculated as follows: the singlet wave function $|\phi\rangle$ was constructed in the form

$$|\phi\rangle = \alpha\,|1100\rangle + \beta\,|0011\rangle + \gamma\,|0110\rangle + \delta\,|1001\rangle \qquad \text{(A3.3.1)}$$

where a general wave function is written as $|\psi_{A\sigma}\psi_{A\sigma\_}\psi_{B\sigma}\psi_{B\sigma\_}\rangle$.

Parameters α, β, γ, and δ were determined by solving the equation

$$H_{\text{m}}|\phi\rangle = E\,|\phi\rangle \qquad \text{(A3.3.2)}$$

This gives a 4×4 matrix equation with the following solutions for the energy:

$$E = \varepsilon_0, \qquad \varepsilon_0 + U, \qquad \varepsilon_0 + \frac{U \pm (U^2 + 16t^2)^{1/2}}{2} \qquad \text{(A3.3.3)}$$

The experimental singlet excitation is[105] 7.6 eV, which gives $t = -6.8$ eV.

The triplet excitation energy was calculated by using the wave function

$$|\phi_t\rangle = \eta\,|1010\rangle + \lambda\,|0101\rangle \qquad \text{(A3.3.4)}$$

The energy of the triplet state is $\varepsilon_0 + J$. Since the triplet excitation is 3.9 eV,[105] we obtain $J = -3.7$ eV.

Using the fact that the ionization potential is 10.5 eV[105] we obtained the following equation:

$$10.5 = (\varepsilon_0 + t) - 2\left(\varepsilon_0 + \frac{U - \sqrt{U^2 + 16t^2}}{2}\right) \qquad \text{(A3.3.5)}$$

which yields

$$\varepsilon_0 = -9.7 \text{ eV}$$

When a molecule approaches a metal surface, its ionization potential and Hubbard interaction are modified by the screening effect of the metal. These effects have been taken into account by reducing U from 16.6 eV to 10 eV while at the same time assuming that the value of the ionization and affinity levels of an atom, $\varepsilon_0 + U/2$, remains constant in both the free and chemisorbed molecule. This leads to $\varepsilon_0 = -6.5$ eV. Moreover, a reduced value of $J = -2.5$ eV was also used.

Finally, the effect of the increase in length of the C–C bond was taken into account by changing the hopping integral t. Thus the free-space value of t was modified by a factor $(d_0/d)^2$, where d_0 is the free-space bond length[350] (1.34 Å) and d is the bond length of the chemisorbed molecule.

APPENDIX 3.4
NONMAGNETIC HARTREE–FOCK SOLUTION TO THE HAMILTONIAN REPRESENTING CHEMISORPTION OF THE π-ELECTRONS ON A d-METAL

The Hamiltonian describing the π orbitals and their bonding to a metal was referred to in Section 3.4.4. Here, we show how to obtain the Hartree–Fock solution to that Hamiltonian.

Within this approximation the Hamiltonian can be written as

$$H = \sum_{k;\sigma} \varepsilon_k n_{k\sigma} + \sum_{\sigma} (\varepsilon_{A\sigma} n_{A\sigma} + \varepsilon_{B\sigma} n_{B\sigma}) + \sum_{\sigma} t(a^{+}_{A\sigma} a_{B\sigma} + a^{+}_{B\sigma} a_{A\sigma})$$
$$+ U[\langle u_{A\sigma}\rangle\langle u_{A\sigma^-}\rangle + \langle u_{B\sigma}\rangle\langle u_{B\sigma^-}\rangle] - J[\langle u_{A\sigma}\rangle\langle u_{B\sigma}\rangle + \langle u_{A\sigma^-}\rangle\langle u_{B\sigma^-}\rangle]$$
$$+ \sum_{k,\sigma} [V_{Ak}(a^{+}_{k\sigma} a_{A\sigma} + a^{+}_{A\sigma} a_{k\sigma}) + V_{Bk}(a^{+}_{k\sigma} a_{B\sigma} + a^{+}_{B\sigma} a_{k\sigma})] \qquad \text{(A3.4.1)}$$

where

$$\varepsilon_{k\sigma} = \varepsilon_0 + U\langle u_{A\sigma^-}\rangle + J\langle u_{B\sigma}\rangle$$

and

$$\varepsilon_{B\sigma} = \varepsilon_0 + U\langle u_{B\sigma^-}\rangle + J\langle u_{A\sigma}\rangle$$

The solution to Hamiltonian (A.3.4.1) was obtained using Green-function techniques. It can be shown[351] from the equation-of-motion method that the Green function $G_{AA}(\omega)$ [and similarly $G_{BB}(\omega)$] is given by

$$G^{\sigma}_{AA}(\omega) = \frac{\omega - \varepsilon_{B\sigma} - \Sigma^{\sigma}_{BB}}{(\omega - \varepsilon_{B\sigma} - \Sigma^{\sigma}_{BB})(\omega - \varepsilon_{A\sigma} - \Sigma^{\sigma}_{AA}) - (t + \Sigma^{\sigma}_{AB})^2} \qquad \text{(A.3.4.2)}$$

where

$$\Sigma^{\sigma}_{\alpha\beta} = \sum_k \frac{V_{\alpha k} V_{k\beta}}{\omega - \varepsilon_k + i\eta} \qquad \text{(A3.4.3)}$$

For a restricted Hartree–Fock solution, all the Green functions and self-energies Σ were taken to be independent of σ, and

$$\langle u_{A\sigma}\rangle = \langle u_{A\sigma^-}\rangle = \langle u_{B\sigma}\rangle = \langle u_{B\sigma^-}\rangle = \tfrac{1}{2}$$

Since Σ_{AB} is a small fraction of Σ_{AA}[352] it was neglected for simplicity. Further, it was assumed that

$$\Sigma_{AA}(\omega) = \Sigma_{BB}(\omega) = \sum_k \frac{|V_{kA}|^2}{\omega - \varepsilon_k + i\eta} \qquad \text{(A3.4.4)}$$

and to the same level of approximation used before[117]

$$V_{Ak} = \beta'\langle 0 | k \rangle \tag{A3.4.5}$$

where $|0\rangle$ is the metal atom to which the orbital A is bonded.

Knowing $\Sigma_{AA}(\omega)$, equation (A3.4.2) can be used to calculate the density of states from which the localized states can be readily determined.

The chemisorption energy of the molecule can be obtained from $\Sigma_{AA}(\omega)$ and $G_{AA}(\omega)$. This was achieved using the method of Einstein and Schrieffer,[48] as discussed by Takahashi.[352]

APPENDIX 4.1
THREE-DIMENSIONAL TREATMENT OF THE ESCAPE OF A BROWNIAN PARTICLE FROM A WELL

Following Banavar *et al.*,[353]* Figure A4.1.1 shows schematically the variation of the free energy $F(\mathbf{r})$ along the jump direction. The free energy at the equilibrium sites A and B is defined to be zero. At the saddle point, denoted by C, the free energy has the value F_c. If F is expanded about its values at the equilibrium position A and the saddle point C, one can write

$$F(\text{near } A) \simeq \tfrac{1}{2}M\omega_a^2 x^2 + \tfrac{1}{2}Ms_a^2 y^2 + \tfrac{1}{2}Mt_a^2 z^2 \tag{A4.1.1}$$

and

$$F(\text{near } C) \simeq F_c - \tfrac{1}{2}M\omega_c^2 \xi^2 + \tfrac{1}{2}Ms_c^2 \phi^2 + \tfrac{1}{2}Mt_c^2 \psi^2 \tag{A4.1.2}$$

The coordinates (ξ, ϕ, ψ) are displacements from the saddle point along the (x, y, z) directions. The quantities ω_a, s_a, t_a and ω_c, s_c, t_c in the above expansions are vibrational frequencies related to the curvatures of the free-energy surface near A and C, respectively.

The Fokker–Planck equation (see also Appendix 4.2) becomes

$$u\frac{\partial W}{\partial \xi} + v\frac{\partial W}{\partial \phi} + w\frac{\partial W}{\partial \psi} + \omega_c^2 \xi \frac{\partial W}{\partial u} - s_c^2 \phi \frac{\partial W}{\partial v} - t_c^2 \psi \frac{\partial W}{\partial w}$$
$$= \eta\left(u\frac{\partial W}{\partial u} + v\frac{\partial W}{\partial v} + w\frac{\partial W}{\partial w}\right) + 3\eta W + \eta^2 \Delta\left(\frac{\partial^2}{\partial u^2} + \frac{\partial^2}{\partial v^2} + \frac{\partial^2}{\partial w^2}\right) W \tag{A4.1.3}$$

* See also references to important early work given there.

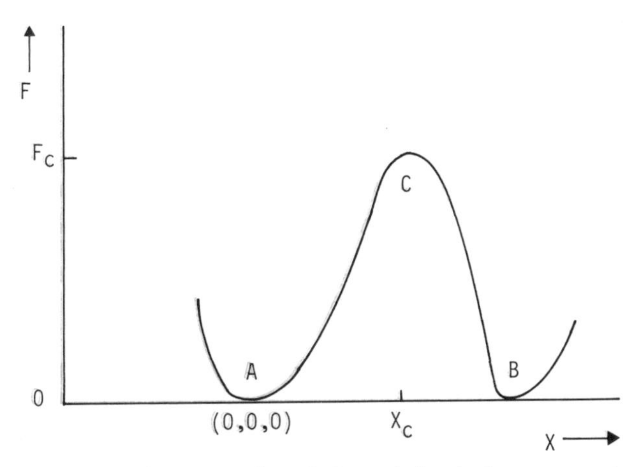

FIGURE A4.1.1. Schematic presentation of the variation in free energy along the jump direction (after Banavar *et al.*[353]).

in its steady-state form in the neighborhood of the saddle point C with quantity Δ defined by

$$\Delta = k_B T / M\eta \tag{A4.1.4}$$

$W(\xi, \phi, \psi : u, v, w)$ is the full distribution function over both space coordinates (ξ, ϕ, ψ) and the corresponding velocities (u, v, w).

Separation of variables in the form

$$W = X(\xi, u) Y(\phi, v) Z(\psi, w) \tag{A4.1.5}$$

in equation (A4.1.3) yields as usual, with ηn and ηm denoting separation constants,

$$\eta^2 \Delta \frac{\partial^2 X}{\partial u^2} = u \frac{\partial X}{\partial \xi} - (\eta u - \omega_c^2 \xi) \frac{\partial X}{\partial u} - \eta(1 + n + m) X \tag{A4.1.6}$$

$$\eta^2 \Delta \frac{\partial^2 Y}{\partial v^2} = v \frac{\partial Y}{\partial \phi} - (\delta v + s_c^2 \phi) \frac{\partial Y}{\partial v} - \eta(1 - n) Y \tag{A4.1.7}$$

and

$$\eta^2 \Delta \frac{\partial^2 Z}{\partial w^2} = w \frac{\partial Z}{\partial \psi} - (\eta w + t_c^2 \psi) \frac{\partial Z}{\partial w} - \eta(1 - m) Z \tag{A4.1.8}$$

Since Y and Z give rise to no flux in the ϕ and ψ directions, it is appropriate to set the separation constants equal to zero. This yields the

Maxwell–Boltzmann solutions for Y and Z in the form

$$Y = \exp[-\tfrac{1}{2}g^2(v^2 + s_c^2\phi^2)], \qquad Z = \exp[-\tfrac{1}{2}g(w^2 + t_c^2\omega^2)] \quad \text{(A4.1.9)}$$

with $g^2 = (\eta\Delta)^{-1} = M\beta$.

An expression for X is then obtained following the one-dimensional treatment of Chandrasekhar,[163]

$$X = \{\exp[-\tfrac{1}{2}g(u^2 - \omega_c^2\xi^2)]\} \left(\frac{q_1 - \eta}{2\pi\eta}\right)^{1/2} g \int_{-\infty}^{\infty} dy \exp\left[-\frac{g^2}{2\eta}(q_1 - \eta)y^2\right] \quad \text{(A4.1.10)}$$

with $q_1 = (\omega_c^2 + \eta^2/4)^{1/2} + \eta/2$.

The rate P at which a particle initially in the potential well at A escapes over the barrier at C is then obtained by writing the current density j in the x-direction at C in the vicinity of A, per particle, namely

$$P = \frac{\omega_a}{2\pi\omega_c}([\omega_c^2 + \eta^2/4)^{1/2} - \eta/2]\frac{s_a}{s_c}\frac{t_a}{t_c}\exp(-F_c/k_B T) \quad \text{(A4.1.11)}$$

This differs from the two-dimensional result of Donnelly and Roberts[354] by the factor t_a/t_c and from the one-dimensional result by the factor $(s_a t_a/s_c t_c)$. In the limit when the friction is large ($\eta \gg \omega_c$), equation (A4.1.11) reduces to

$$P = \frac{\omega_a\omega_c}{2\pi\eta}\frac{s_a}{s_c}\frac{t_a}{t_c}\exp(-F_c/k_B T) \quad \text{(A4.1.12)}$$

This is the limit in which the Fokker–Planck equation reduces to the Smoluchowski equation.[355] On the other hand, as $\eta \to 0$

$$P = \frac{\omega_a}{2\pi}\frac{s_a}{s_c}\frac{t_a}{t_c}\exp(-F_c/k_B T) \quad \text{(A4.1.13)}$$

an upper limit to the rate P. The diffusion constant D is then given by

$$D = GPa^2 \quad \text{(A4.1.14)}$$

where a is the mean jump length and, for the case illustrated in Figure A4.1.1, is equal to $2x_c$; G is a numerical factor of order unity and depends on the geometry of the substrate atoms. The diffusion coefficient can be fitted to the form

$$D = D_0\exp(-E/k_B T) \quad \text{(A4.1.15)}$$

The main results are in keeping with the work of Kramers,[162] even though he used what appeared to be a one-dimensional treatment.

APPENDIX 4.2
DYNAMICS OF A BROWNIAN PARTICLE IN A FLUID: DERIVATION OF STOCHASTIC ELEMENTS

Kramers' theory[162] plays an important role in the study of the motion of adparticles. It is therefore of interest to summarize here an investigation into the motion of a Brownian particle in a fluid without the *a priori* introduction of purely stochastic elements. The argument below follows the work of Lebowitz and Rubin.[356]

Kinetic arguments were used in the main text when discussing Brownian motion and all these descriptions were of a stochastic nature *ab initio.*

Despite this indeterminacy in the motion of an individual Brownian particle, it is still possible when there are a very large number of such particles to give a deterministic equation for the time evolution of their spatial and velocity distribution $f(\mathbf{R}, \mathbf{V}, t)$. This function is defined here such that the fraction of Brownian particles in the macroscopically very small volume element of phase space $d\mathbf{R}d\mathbf{V}$, which is sufficiently large nonetheless to contain many particles, is given by $f(\mathbf{R}, \mathbf{V}, t)d\mathbf{R}d\mathbf{V}$. It is well known to be convenient mathematically to work in terms of an ensemble of systems each containing a single Brownian particle, since one is assuming that in the real system the Brownian particles move independently, with the distribution function defined in terms of probabilities, and that they are assumed to satisfy the same equation as the "coarse-grained" f defined above.[357,358]

When use is made of the natural assumption that a Brownian particle has mass M, which is much greater than the mass of the fluid particles m and which in turn makes the fractional change in their velocities small, on a molecular time scale, and furthermore results also in their moving on the average at speeds far below those of the fluid molecules, we obtain the Fokker–Planck equation [cf. equation (4.10.2)] for $f(\mathbf{R}, \mathbf{V}, t)$,[163,359] namely

$$\frac{\partial f(\mathbf{R}, \mathbf{V}, t)}{\partial t} + \mathbf{V} \cdot \frac{\partial f}{\partial \mathbf{R}} + \frac{1}{M} \mathbf{X} \cdot \frac{\partial f}{\partial \mathbf{V}} = \left[\zeta \frac{\partial}{\partial \mathbf{V}} \cdot (\mathbf{V} f) + D \nabla_{\mathbf{V}}^2 f \right]$$

$$= D \frac{\partial}{\partial \mathbf{V}} \cdot \left[f \frac{\partial}{\partial \mathbf{V}} \ln(f/f_0) \right] \qquad \text{(A4.2.1)}$$

Here $\mathbf{X}$ denotes the external force acting on the Brownian particle; ζ is the friction and D the diffusion constant in velocity space, of the Brownian particles in the fluid. The friction constant ζ has been eliminated in the last line of this equation by making use of the Einstein relation

$$D = (k_B T/M)\zeta \tag{A4.2.2}$$

f_0 is the equilibrium distribution function of the Brownian particles, the velocity dependence of which is $\exp(-\frac{1}{2}\beta MV^2)$; $\beta = (k_B T)^{-1}$ with T the temperature of the host fluid. The right-hand side of equation (A4.2.1) represents the effect of the fluid on the Brownian particles. It is a special case of a collision term and will sometimes be denoted below by δf.

The approach of Lebowitz and Rubin aims at dispensing with the *a priori* introduction of a stochastic interaction. They therefore start from the Liouville equation for the distribution function μ of the assembly of host fluid plus the Brownian particle. A transport equation for f will then result after integration over the variables of the fluid particles within certain limits involving the size of the fluid system and the time scale. Such limits are necessary to derive an irreversible transport equation from a reversible Liouville equation and have been discussed extensively in the literature on irreversible processes.[360,361]

The equation obtained by Lebowitz and Rubin for f then proves to be of the same form as equation (A4.2.1) to lowest order in the mass ratio of fluid to Brownian particle with an explicit, though unevaluated, molecular expression for D.

It turns out that there are higher-order corrections to equation (A4.2.1). These higher-order terms have relevance to the Kirkwood theory of liquids,[362–364] which uses the Fokker–Planck equation to describe the time evolution of the low-order distribution functions of a liquid. Lebowitz and Rubin also treat the case where the Brownian particle has internal structure.

They consider first the case when the Brownian particle is subject to some external force, such as a constant electric field **E**, which does not act on the fluid particles. This enables one to examine the equation satisfied by the stationary nonequilibrium distribution f, to terms linear in **E**. This distribution represents a balance between the effect of the acceleration by **E** and scattering by the fluid particles. It will thus contain the collision term representing the effect of the fluid but will avoid some of the difficulties enountered in deriving the general time-dependent equation for f.[365,366] The method employed by Lebowitz and Rubin is similar to that developed by Kohn and Luttinger[365] for deriving the quantum transport equation of an electron moving in the field of stationary impurities. Discussion of the collision term for the time-dependent case is given elsewhere.[356]

General Formulation

The Hamiltonian for the system of host fluid and Brownian particle, excluding the external field, has the form

$$H = \tfrac{1}{2}MV^2 + \left[\sum_{i=1}^{N} \tfrac{1}{2}v_i^2 + \sum_{i<j} \phi(r_{ij})\right] + \sum_{i=1}^{N} u(\mathbf{R}_i)$$
$$= H_1 + H_l + U, \qquad \mathbf{R}_i = \mathbf{r}_i - \mathbf{R} \tag{A4.2.3}$$

where $\mathbf{r}_i$ and $\mathbf{v}_i$ are the position and velocity of the ith fluid particle. The three terms comprising H are, respectively, the kinetic energy of the Brownian particle, the Hamiltonian of the N fluid particles (of unit mass), and the interaction between them. The whole system is enclosed in a periodic box of volume Ω. The joint distribution function of the whole system obeys the Liouville equation

$$\partial\mu(x, y, t)/\partial t + (\mu, H) + M^{-1}\mathbf{E}\cdot\frac{\partial\mu}{\partial\mathbf{V}} = 0 \tag{A4.2.4}$$

where (μ, H) is the Poisson bracket between μ and H (expressed in terms of the velocity variables), and the abbreviated notation $x = (\mathbf{R}, \mathbf{V})$, $y = (\mathbf{r}_i \cdots \mathbf{r}_N, \mathbf{v}_i \cdots \mathbf{v}_N)$ has been used.

Since only terms linear in E will be considered, it is useful to write

$$\mu = \mu_0 + \mu' \tag{A4.2.5}$$

where

$$\mu_0 = Z^{-1}\exp(-\beta H), \qquad Z = \int \exp(-\beta H)dxdy \tag{A4.2.6}$$

μ' is linear in $\mathbf{E}$ and satisfies the equation

$$\partial\mu'/\partial t + (\mu', H) - \beta\mathbf{E}\cdot\mathbf{V}\mu_0 = 0 \tag{A4.2.7}$$

The distribution function of the Brownian particle f, normalized to unity, is given by

$$f = \int \mu dy = f_0 + f' \tag{A4.2.8}$$

On the assumption that one is dealing with a uniform system, f will be independent of $\mathbf{R}$ and contain a normalization factor Ω^{-1}, namely

$$f = \Omega^{-1}[\rho_0(\mathbf{V}) + \psi(\mathbf{V})] \tag{A4.2.9}$$

where $(\rho_0 + \psi)$ is the velocity distribution of the Brownian particle:

$$\rho_0 = (2\pi k_B T/M)^{-3/2} \exp(-\tfrac{1}{2}\beta M V^2) \tag{A4.2.10}$$

$$\begin{aligned} \psi &= \Omega \int \mu' dy \\ &= \int \mu' dy d\mathbf{R} \end{aligned} \tag{A4.2.11}$$

Integration of equation (A4.2.7) over y and $\mathbf{R}$ yields

$$\frac{\partial \psi}{\partial t} - \beta \mathbf{E} \cdot \mathbf{V} \rho_0 = -\frac{1}{M} \frac{\partial}{\partial \mathbf{V}} \cdot \left(\int \mu' \mathbf{F} dy d\mathbf{R} \right) \tag{A4.2.12}$$

where

$$\mathbf{F} = -\frac{\partial U}{\partial \mathbf{R}} = \sum_{i=1}^{N} \frac{\partial u(\mathbf{R}_i)}{\partial \mathbf{R}_i} \tag{A4.2.13}$$

Clearly, it is the right-hand side of equation (A4.2.12) that will become the collision term $\delta\psi$.

At this stage the conditional distribution function $P(x, y, t)$ will be considered. This function gives the probability density of finding the fluid at y given that the Brownian particle is at x,

$$P(x, y, t) = \frac{\mu(x, y, t)}{f(x, t)} = P_0(1 + \xi) \tag{A4.2.14}$$

where

$$P_0 = \frac{\mu_0(x, y)}{f_0} = \frac{\exp(-\beta H)}{\int \exp(-\beta H) dy} = C \exp[-\beta(H_l + U)] \tag{A4.2.15}$$

C being a constant for the uniform system. One also defines

$$P_0(1 + \eta) = \frac{\mu'}{\int \mu' dy} = \frac{\mu'}{\psi/\Omega} \tag{A4.2.16}$$

Normalization yields

$$\int P_0 dy = 1 \tag{A4.2.17}$$

$$\int P_0 \eta dy = 0 \tag{A4.2.18}$$

If equation (A4.2.16) is substituted into (A4.2.7), then

$$\eta\frac{\partial\psi}{\partial t}+\psi\frac{\partial\eta}{\partial t}+\psi(\eta,H)=-(\psi,H)-\eta(\psi,H)-\psi(\eta+1)(\ln P_0,H)$$
$$+\frac{1}{M}\frac{\partial}{\partial\mathbf{V}}\left(\int\mu'\mathbf{F}\,dy\,d\mathbf{R}\right) \qquad \text{(A4.2.19)}$$

where equation (A4.2.12) has been used to eliminate the field term.

Now from equation (A4.2.15)

$$(\ln P_0,H)=\beta\mathbf{F}\cdot\mathbf{V}=-\frac{1}{M}\mathbf{F}\cdot\frac{\partial}{\partial\mathbf{V}}\ln\rho_0 \qquad \text{(A4.2.20)}$$

and

$$\int P_0\mathbf{F}\,dy=0 \qquad \text{(A4.2.21)}$$

Dividing equation (A4.2.19) by ψ and writing out explicitly some of the Poisson-bracket expressions, one obtains

$$\eta\frac{\partial}{\partial t}\ln\psi+\frac{\partial\eta}{\partial t}+(\eta,H_2)=-\frac{1}{M}\frac{\partial}{\partial\mathbf{V}}\ln(\psi/\rho_0)\cdot\mathbf{F}$$
$$-\frac{1}{M}\left[\frac{\partial}{\partial\mathbf{V}}\ln(\psi/\rho_0)+\frac{\partial}{\partial\mathbf{V}}\right]\cdot[\mathbf{F}\eta-\langle\mathbf{F}\eta\rangle]$$
$$-\left[\mathbf{V}\cdot\frac{\partial\eta}{\partial\mathbf{R}}-\left\langle\mathbf{V}\cdot\frac{\partial\eta}{\partial\mathbf{R}}\right\rangle\right] \qquad \text{(A4.2.22)}$$

where

$$\langle\cdots\rangle=\int P_0\cdots dy \qquad \text{(A4.2.23)}$$

and H_2 is the Hamiltonian of the fluid in the presence of a fixed "Brownian" particle, which thus serves as a source of external potential:

$$H_2=H_l(y)+\sum u(\mathbf{r}_i-\mathbf{R}) \qquad \text{(A4.2.24)}$$

with $\mathbf{R}$ treated as a constant number, not a canonical coordinate. The Poisson bracket (η,H_2) in equation (A4.2.22) thus contains derivatives only with respect to the fluid variables $(\mathbf{r}_i,\mathbf{v}_i)$. The relation

$$\mathbf{V}\cdot\int P_0\frac{\partial\eta}{\partial\mathbf{R}}\,dy=-\beta\mathbf{V}\cdot\int P_0\mathbf{F}\eta y \qquad \text{(A4.2.25)}$$

has also been employed here.

The set of coupled equations (A4.2.22) and (A4.2.12) are quite general. Equation (A4.2.12) can be written in the form

$$\frac{\partial\psi}{\partial t} - \beta\mathbf{E}\cdot\mathbf{V}\rho_0 = -\frac{1}{M}\frac{\partial}{\partial\mathbf{V}}\cdot(\psi\langle\mathbf{F}\eta\rangle) \qquad \text{(A4.2.26)}$$

for ψ and η, which describe the linear deviation of μ from μ_0 caused by the electric field $\mathbf{E}$.

Following Lebowitz and Rubin, we shall now specialize these equations to determine the final steady-state value of ψ in the presence of a constant field $\mathbf{E}$. Effects originating from the finiteness of the system will evidently need to be eliminated to obtain a physically sensible result.

In order to get a transport equation for ψ, it turns out to be necessary to solve (formally) for η before going to the limit of an infinite system, substitute that result in equation (A4.2.26), and then proceed to the desired limit. This technique is discussed fully by Lebowitz and Rubin.

Subsequently, the next essential step is to display the mass M dependence by the scaling

$$\mathbf{V} = \mathbf{v}/M^{1/2} = \gamma\mathbf{v}, \qquad \gamma^2 = M^{-1} \qquad \text{(A4.2.27)}$$

If η is expanded in the power series

$$\eta = \gamma\eta_1 + \gamma^2\eta_2 + \cdots \qquad \text{(A4.2.28)}$$

Lebowitz and Rubin show how to calculate η_1 and η_2. It then transpires that the first correction to the Fokker–Planck equation comes from η_3 and, furthermore, that part of the terms coming from the average $\langle F\eta_3\rangle$ may be combined with those coming from $\langle F\eta_1\rangle$ to obtain again a Fokker–Planck equation with new values for the frictional force and for the diffusion tensor. The remaining terms lead to higher derivatives.

In summary, the main results of the argument of Lebowitz and Rubin demonstrate that the effect of the fluid on the distribution of heavy particles of mass M and small velocities $V \sim M^{-1/2}$ may be described to lowest order in M^{-1} by a Fokker–Planck-type collision term with friction constant ζ defined by

$$\zeta = \frac{M^{-1}\beta}{3}\int_0^\infty \langle\mathbf{F}\cdot\mathbf{F}(-t)\rangle dt \qquad \text{(A4.2.29)}$$

This force–force correlation function is the analog for the classical theory of the resistivity equation discussed in Appendix 6.2 and Chapter 4. In order for the result (A4.2.29) to exist, one takes the limit $N \to \infty$, $\Omega \to \infty$,

N/Ω constant, before extending the integration to infinity. When applicable, this expression is to be compared with the hydrodynamic expression for the friction constant of a sphere of radius σ, namely Stokes' Law:

$$\zeta = M^{-1} 6\pi\eta\sigma \tag{A4.2.30}$$

where η is the viscosity of the fluid. For the hydrodynamic expression to be valid, σ must be much greater than the mean free path of the fluid particles. The comparison is made in some detail by Lebowitz and Rubin,[356] and the reader is referred to their account for the details.

APPENDIX 4.3
VARIATIONAL PRINCIPLE FOR THE OPTIMUM REACTION COORDINATE IN DIFFUSION-CONTROLLED REACTIONS

There is interest in Kramers' approach[162] to activated processes, so a variational treatment in diffusion-controlled problems will be examined. The argument below follows that of Berkowitz *et al.*,[367] who consider an assembly of solute particles subject to the Smoluchowski equation[163,175]

$$\frac{\partial P}{\partial t} = -\nabla \cdot \mathbf{j} \tag{A4.3.1}$$

$$\mathbf{j} = -(\beta \mathbf{D} \cdot \nabla U)P - \mathbf{D} \cdot \nabla P \tag{A4.3.2}$$

where $P(\{\mathbf{r}_i\}, t)$ is the configuration-space distribution function of the particles, $\mathbf{j}$ is the diffusive flux, $\beta = (k_B T)^{-1}$, $\mathbf{D}$ is the diffusion tensor of the particles, and U is the potential of the mean force. The diffusion tensor is symmetric and related to the friction tensor $\mathbf{f}$ by

$$\mathbf{f} \cdot \mathbf{D} = \mathbf{D} \cdot \mathbf{f} = k_B T \mathbf{1} \tag{A4.3.3}$$

The flux may be written more compactly as

$$\mathbf{j} = -\exp(-\beta U)\mathbf{D} \cdot \nabla[P \exp(\beta U)] \tag{A4.3.4}$$

If this equation is multiplied on the left by $\mathbf{f}$ and equation (A4.3.3) then used, one readily finds

$$\mathbf{f} \cdot \mathbf{j} \exp(\beta U) = -k_B T \nabla[P \exp(\beta U)] \tag{A4.3.5}$$

Berkowitz *et al.*[367] now consider a possible trajectory connecting two points A (e.g., the reactant configuration) and B (e.g., the product configuration) and assume that all the particles are moving along this trajectory. Then the flux and differential line element can be rewritten as

$$\left.\begin{aligned} \mathbf{j} &= \mathbf{t}j \\ d\mathbf{s} &= \mathbf{t}ds \end{aligned}\right\} \tag{A4.3.6}$$

where $\mathbf{t}$ is a unit vector tangential to the path. Integration of equation (A4.3.5) over the path $\mathscr{P}$ leads to

$$\begin{aligned} \int_{\mathscr{P}} d\mathbf{s} \cdot \mathbf{f} \cdot \mathbf{j} \exp(\beta U) &= -k_{\mathrm{B}} T \int_{\mathscr{P}} d\mathbf{s} \cdot \nabla[P \exp(\beta U)] \\ &= -k_{\mathrm{B}} T [P \exp(\beta U)]_{\mathrm{A}}^{B} \end{aligned} \tag{A4.3.7}$$

For stationary reaction conditions, j is a constant, and equations (A4.3.6) and (A4.3.7) yield

$$j \int_{\mathscr{P}} \mathbf{t} \cdot \mathbf{f} \cdot \mathbf{t} \exp(\beta U) ds = -k_{\mathrm{B}} T [P \exp(\beta U)]_{\mathrm{A}}^{B} = \text{const} \tag{A4.3.8}$$

Therefore

$$j_{\mathscr{P}} = \frac{\text{const}}{\int_{\mathscr{P}} f_{tt} \exp(\beta U) ds} \tag{A4.3.9}$$

where $j_{\mathscr{P}}$ denotes the flux along trajectory $\mathscr{P}$ and $f_{tt} = \mathbf{t} \cdot \mathbf{f} \cdot \mathbf{t}$.

Now one defines the optimum reaction path, $\mathscr{P}^*$ say, to be that path over which the flux $j_{\mathscr{P}^*}$ attains its maximum value. From equation (A4.3.9), $\mathscr{P}^*$ is the path for which

$$\int_{\mathscr{P}} f_{tt} \exp(\beta U) ds = \text{minimum} \tag{A4.3.10}$$

Berkowitz *et al.*[355] term this integral the "generalized resistance,"[367] and the minimum condition of equation (A4.3.10) the minimum resistance principle.

Euler–Lagrange Equations

The principle can be written as

$$\delta I = \delta \int_0^{s_f} \exp(\beta U) f_{tt} ds = 0 \tag{A4.3.11}$$

which can now be used to find Euler equations for the minimum resistance path. Here, s_f is the total arc length of the path and this length is, in general, different for all paths over which variations are taken. In order to eliminate the variation of s_f, one considers s to be a function of a new parameter t, which varies between the fixed limits of 0 and 1. One can then write, with dots denoting derivatives with respect to t,

$$\delta I = \delta \int_0^1 \exp(\beta U) \sum_\mu \sum_\nu \dot{x}_\mu \dot{x}_\nu f_{\mu\nu} \Big/ \Big(\sum_\mu \dot{x}_\mu^2 \Big)^{1/2} \qquad \text{(A4.3.12)}$$

The corresponding Euler equations are

$$0 = \frac{d}{dt} \left\{ \frac{\partial}{\partial \dot{x}_k} \left[\exp(\beta U) \sum_\mu \sum_\nu \dot{x}_\mu \dot{x}_\nu f_{\mu\nu} \Big/ \Big(\sum_\lambda \dot{x}_\lambda^2 \Big)^{1/2} \right] \right\} - \frac{\partial}{\partial x_k} \left[\exp(\beta U) \sum_\mu \sum_\nu \dot{x}_\mu \dot{x}_\nu f_{\mu\nu} \Big/ \Big(\sum_\lambda \dot{x}_\lambda^2 \Big)^{1/2} \right] \qquad \text{(A4.3.13)}$$

It should be noted that $f_{\mu\nu}$ is in general anisotropic and a function of the coordinates. Following Berkowitz *et al.*, a simple case will therefore be considered at this stage.

Uniform Isotropic Friction

This quantity is defined by

$$f_{\mu\nu} = f\delta_{\mu\nu} = \text{const} \qquad \text{(A4.3.14)}$$

In this example, Berkowitz *et al.*[367] obtain after a short calculation

$$\frac{d^2 x_\lambda}{ds^2} + \frac{dx_\lambda}{ds} \frac{d}{ds} (\beta U) - \frac{\partial}{\partial x_\lambda} (\beta U) = 0 \qquad \text{(A4.3.15)}$$

or, in vector form,

$$\kappa \mathbf{n} + \mathbf{t}(\nabla \beta U \cdot \mathbf{t}) - \nabla \beta U = 0 \qquad \text{(A4.3.16)}$$

where κ is the curvature of the path while $\mathbf{n}$ and $\mathbf{t}$ are the principal unit vectors normal and tangential to the path, respectively. Equation (A4.3.16) describes the optimum reaction coordinate, which is often instead identified with the steepest-descent path given by the equation

$$-\nabla(\beta U)/|\nabla \beta U| = \mathbf{t} \qquad \text{(A4.3.17)}$$

Comparison of equations (A4.3.16) and (A4.3.17) shows that they describe paths that coincide if and only if the path is a straight line (i.e., $\kappa = 0$).

Nonuniform Isotropic Friction

A second example is that of nonuniform isotropic friction defined by

$$f_{\mu\nu} = f(\mathbf{x})\delta_{\mu\nu} \tag{A4.3.18}$$

The Euler equations now assume the form

$$\begin{aligned}\frac{d}{dt}\left\{\frac{\partial}{\partial \dot{x}_\lambda}\left[\exp(\beta U)f(\mathbf{x})\left(\sum_\mu \dot{x}_\mu^2\right)^{1/2}\right]\right\}\\ = \frac{\partial}{\partial x_\lambda}\left[\exp(\beta U)f(\mathbf{x})\left(\sum_\mu \dot{x}_\mu^2\right)^{1/2}\right]\end{aligned} \tag{A4.3.19}$$

If the friction at some arbitrary reference point $\mathbf{x}_0$ is taken to be $f(\mathbf{x}_0)$ and the function $W(\mathbf{x})$ is defined by

$$W(\mathbf{x}) = U(\mathbf{x}) + k_B T \ln f(\mathbf{x})/f(\mathbf{x}_0) \tag{A4.3.20}$$

then the Euler equations become

$$\begin{aligned}\frac{d}{dt}\left\{\frac{\partial}{\partial \dot{x}_\lambda}\left\{\exp[\beta W(\mathbf{x})]\left(\sum_\mu \dot{x}_\mu^2\right)^{1/2}\right\}\right\}\\ = \frac{\partial}{\partial x_\lambda}\left\{\exp[\beta W(\mathbf{x})]\left(\sum_\mu \dot{x}_\mu^2\right)^{1/2}\right\}\end{aligned} \tag{A4.3.21}$$

This is formally identical to the previous example with $W(\mathbf{x})$ in place of $U(\mathbf{x})$. Proceeding as before, one finds

$$\kappa \mathbf{n} = \nabla\beta W - \mathbf{t}(\mathbf{t}\cdot\nabla\beta W) \tag{A4.3.22}$$

Thus for the case of nonuniform isotropic friction, the system moves subject to an effective potential $W(\mathbf{x})$ that depends on both the true potential $U(\mathbf{x})$ and the friction along the path $f(\mathbf{x})$.

In summary, equation (A4.3.10) provides a variational expression for the optimum reaction coordinate of a diffusional process. The determination of the optimum reaction coordinate is seen to be an essential first step for both the detailed description of the mechanism and the efficient calculation of the rate constant of the reaction.[162]

APPENDIX 4.4
ANISOTROPIC SCREENING OF AN ADATOM NEAR A METAL SURFACE

The main topic of this Appendix, following the work of McCaskill and March in conjunction with Miglio and Tosi,[368] is anisotropic screening of an adatom near a metal surface. Two models, having different regimes of validity, will be set up to describe aspects of this shielding by conduction electrons. In order to avoid ambiguity when including electronic effects in the density matrix and adatom potential of equation (4.10.7), it is found to be essential that one should work with a self-consistently screened potential. The full linear-response function $F(\mathbf{r},\mathbf{r}')$ for calculating the charge displaced by a weak external one-body potential has been determined by Moore and March,[317] whose findings are recorded in Appendix 2.1 for the infinite-barrier model of a metal surface. This yields the displaced charge $\Delta\rho$ in the form

$$\Delta\rho(\mathbf{r}) = \int V(\mathbf{r}')F(\mathbf{r},\mathbf{r}')d\mathbf{r}' \qquad \text{(A4.4.1)}$$

with F given by equation (A2.1.10).

If the potential V is sufficiently slowly varying, then one can proceed in the spirit of the Thomas–Fermi approximation[369] and set

$$\Delta\rho(\mathbf{r}) \simeq V(\mathbf{r})\int F(\mathbf{r},\mathbf{r}')d\mathbf{r}' \qquad \text{(A4.4.2)}$$

Rather than appeal to the specific form of F, the following argument parallels that of McCaskill *et al.*[368] in showing that the integral is related to the local density of states at the Fermi surface, $N(z,E_f)$ say, thereby establishing the role of a local screening length $q(z)^{-1}$ in the linear response. Their argument proceeds as follows. If one starts from the Green-function equation representing the effect of a test charge on the Green function in the presence of the surface, say $G_{0s}(\mathbf{r},\mathbf{r}',E)$, then[351]

$$G^{\pm}(\mathbf{r},\mathbf{r}',E) = G^{\pm}_{0s}(\mathbf{r},\mathbf{r}',E) + \int d\mathbf{r}_1 G^{\pm}_{0s}(\mathbf{r},\mathbf{r}_1,E)V(\mathbf{r}_1)G^{\pm}_{0s}(\mathbf{r}_1,\mathbf{r}',E) \qquad \text{(A4.4.3)}$$

where the linear response to the test charge is implicit in that the perturbed Green function does not appear on the right-hand side of this equation.

The electron density is given in terms of G by[351]

$$\rho(\mathbf{r}) = \frac{1}{\pi}\,\mathrm{Im}\int_{-\infty}^{E_f} G^{-}(\mathbf{r},\mathbf{r},E)dE \qquad \text{(A4.4.4)}$$

and hence, after making the above Thomas–Fermi assumption that V can be brought outside the integral, one finds that

$$\Delta\rho = -\frac{1}{\pi} V(\mathbf{r}) \int_{-\infty}^{E_f} dE \int d\mathbf{r}_1 G_{0s}^-(\mathbf{r}, \mathbf{r}_1, E) G_{0s}^-(\mathbf{r}_1, \mathbf{r}, E) \quad \text{(A4.4.5)}$$

The $\mathbf{r}_1$ integration can be evaluated using the eigenfunction expansion of the Green function,[370]

$$\int d\mathbf{r}_1 G_{0s}(\mathbf{r}, \mathbf{r}_1, E) G_{0s}(\mathbf{r}_1, \mathbf{r}, E) = \sum \int d\mathbf{r}_1 \frac{\psi_i^*(\mathbf{r})\psi_i(\mathbf{r}_1)\psi_j^*(\mathbf{r}_1)\psi_j(\mathbf{r})}{(E - E_i)(E - E_j)}$$

$$= -\frac{d}{dE} G_{0s}(\mathbf{r}, \mathbf{r}, E) \quad \text{(A4.4.6)}$$

However, the local density of states is given by

$$N(z, E) = \frac{\partial \rho(\mathbf{r}, \mathbf{r}, E)}{\partial E} = \frac{1}{\pi} G^-(\mathbf{r}, \mathbf{r}, E) \quad \text{(A4.4.7)}$$

which, when substituted in equation (A4.4.5), yields

$$\Delta\rho(\mathbf{r}) = V(\mathbf{r}) N(z, E_f) = \frac{q^2(z)}{4\pi} V(\mathbf{r}) \quad \text{(A4.4.8)}$$

This defines the local screening length $q(z)^{-1}$.

From equation (A2.1.8) for the infinite-barrier model, the local density of states has the explicit form

$$N(z, E_f) = \frac{\sqrt{2E_f}}{2\pi^2} \left[1 - \frac{\sin(2z\sqrt{2E_f})}{2z\sqrt{2E_f}} \right], \qquad z < 0 \quad \text{(A4.4.9)}$$

In order to impose self-consistency, one next employs the Poisson equation

$$\nabla^2 V = 4\pi \Delta\rho_t \quad \text{(A4.4.10)}$$

where the charge density ρ_t includes the bare charge of the adatom. Then one obtains the inhomogeneous equation

$$[\nabla^2 - q^2(z)] V = \delta(\mathbf{r} - \mathbf{r}_0) 4\pi Z e \quad \text{(A4.4.11)}$$

for the screening of a point charge Ze, which if $q^2(z)$ is sufficiently slowly varying has the local screened Coulomb solution

$$V(\mathbf{r}) = Ze\frac{\exp[-q(z_0)|\mathbf{r}-\mathbf{r}_0|]}{|\mathbf{r}-\mathbf{r}_0|} \tag{A4.4.12}$$

the test charge being at position $\mathbf{r}_0$. This spherically symmetric result possesses an oversimplified form, for one must expect the inhomogeneity of the electron gas to lead to an anisotropic screening cloud round the test charge.

Approximations for Adatoms Inside and Outside a Metal Surface

Two approximations are now discussed following McCaskill *et al.*[368]

1. For an adatom inside the metal surface, a first characterization of the importance of anisotropic screening will be contained in the Taylor expansion of $q^2(z)$ around the test-charge position $\mathbf{r}_0$, namely

$$q^2(z) = q^2(z_0) + (z - z_0)\frac{d}{dz}q^2\Big|_{z_0} + \cdots \tag{A4.4.13}$$

and a linear expansion is possible in the gradient $E = (dq^2/dz)_{z_0}$ for the screened potential,

$$\{\nabla^2 - [q_0^2 - E(z - z_0)]\}V = \delta(\mathbf{r} - \mathbf{r}_0) \tag{A4.4.14}$$

where q_0^2 denotes $q^2(z_0)$.

2. For an adatom outside the metal surface, the potential at the surface barrier may be determined classically from image theory and this planar potential used as a cylindrically symmetric boundary condition for the above semiclassical screening, inside the metal. The differential equation (A4.4.11) then separates into Bessel's equation for the radial (parallel to the surface) component, and the z-component Schrödinger equation

$$\frac{d^2Z}{dz^2} - [\kappa + q^2(z)]Z = 0 \tag{A4.4.15}$$

with κ, the separation constant, playing the role of the energy. The solution of equation (A4.4.15) with $q^2(z)$ given by equations (A4.4.8) and (A4.4.9) does not seem tractable analytically. However, in a sense to be outlined below, the step function $q^2(z) = q_0^2H(z)$ provides a self-consistent approximation to the surface screening and one can solve explicitly for the screened potential resulting from an external adatom.

The results of approach (1) above will now be developed. Linearization of equation (A4.4.14) for the screened potential $V = V_0 + V_1 E$ with V_0 given by the homogeneous expression yields

$$V_1(\mathbf{r}) = -\int d\mathbf{r}' \frac{z'}{r'} \exp(-q_0 r') \frac{\exp(-q_0|\mathbf{r}-\mathbf{r}'|)}{|\mathbf{r}-\mathbf{r}'|} \tag{A4.4.16}$$

where z_0 has been set to zero for convenience. By Fourier transform, with θ_k denoting the angle $\mathbf{k}$ makes with the z-axis, McCaskill *et al.*[368] find

$$V_1(\mathbf{k}) = \frac{(4\pi Ze)^2 ik \cos\theta_k}{(k^2+q^2)^{5/2}} \tag{A4.4.17}$$

This expression can be transformed back into $\mathbf{r}$-space to yield

$$V_1(\mathbf{r}) = \frac{2(Ze)^2}{3\pi} zK_0(qr) \tag{A4.4.18}$$

where $K_0(qr)$ is the modified Bessel function defined by

$$K_0(qr) = \int_0^\infty \frac{\cos kr}{(k^2+q^2)^{1/2}} dk \tag{A4.4.19}$$

The large-r behavior is readily found from the asymptotic form of K_0,

$$K_0(qr) \sim \sqrt{\frac{\pi}{2z}} e^{-qr} \left(1 - \frac{1}{8qr} + \frac{9}{2(8qr)^2} - \cdots\right) \tag{A4.4.20}$$

while the small-r behavior is given by

$$K_0(qr) \sim -\ln qr \tag{A4.4.21}$$

The angular potential is of longer range than the spherical term but the exponential decay rate is the same. In the main text a further simplification proves necessary to evaluate the surface resistivity.

The results of method (2) are now considered. In this approach one writes $q^2(z) = q_0^2(z) = q_0^2 H_-(z)$. McCaskill *et al.*[368] note that this has the advantage that the infinite-barrier potential is then itself self-consistent with the displaced charge, the result being readily established by taking the limit of a finite-barrier solution as the barrier height becomes large. One must then solve the z-component equation with $\kappa > 0$:

$$\frac{d^2 Z}{dz^2} - (\kappa^2 + q_0^2)Z = 0 \tag{A4.4.22}$$

with boundary conditions determined by the classical image-theory potential

$$\left.\begin{array}{l} Z_\kappa \xrightarrow[z\to-\infty]{} 0 \\ \left.\dfrac{dZ_\kappa}{dz}\right|_{z=0} = \left.\dfrac{dV_\kappa^{\text{cl}}(z)}{dz}\right|_{z=0} = V_\kappa'^{\text{cl}}(0) \end{array}\right\} \tag{A4.4.23}$$

where the radial transform of the classical potential is

$$V_\kappa^{\text{cl}}(z) = \int V^{\text{cl}}(\mathbf{r})\exp(i\boldsymbol{\kappa}\cdot\boldsymbol{\rho})d\boldsymbol{\rho} \tag{A4.4.24}$$

with $\mathbf{r} = (\rho, \theta, z) = (\boldsymbol{\rho}, z)$. The result is simply

$$V(\mathbf{r}) = \frac{1}{(2\pi)^2}\int d\boldsymbol{\kappa}e^{i\boldsymbol{\kappa}\cdot\boldsymbol{\rho}}\frac{\exp[(q_0^2+\kappa^2)]^{1/2}z}{\sqrt{q_0^2+\kappa^2}}, \qquad z<0 \tag{A4.4.25}$$

demonstrating damping of the image-theory potential into the metal.

The image potential for a single point charge at distance z_0 outside the metal surface is given by

$$V^{\text{cl}}(\mathbf{r}) = \frac{Ze}{|\mathbf{r}-z_0\hat{\mathbf{z}}|} - \frac{Ze}{|\mathbf{r}+z_0\hat{\mathbf{z}}|} \tag{A4.4.26}$$

with derivative

$$\frac{\partial V^{\text{cl}}(\mathbf{r})}{\partial z} = -\frac{Ze}{2}\left[\frac{(z-z_0)}{|\mathbf{r}-z_0\hat{\mathbf{z}}|^3} - \frac{(z+z_0)}{|\mathbf{r}+z_0\hat{\mathbf{z}}|^3}\right] \tag{A4.4.27}$$

and at the infinite barrier

$$V'^{\text{cl}}(\rho) \equiv \left.\frac{\partial V^{\text{cl}}}{\partial z}\right|_{z=0} = \frac{Zez_0}{(\rho^2+z_0^2)^{3/2}} \tag{A4.4.28}$$

so that

$$V'^{\text{cl}}(0) = 2\pi Ze\exp(-z_0\kappa) \tag{A4.4.29}$$

completes the specification of $V(\mathbf{r})$ in equation (A4.4.24):

$$V(\mathbf{r}) = Ze\int_0^\infty d\kappa\,\kappa J_0(\rho\kappa)\frac{\exp[z(q_0^2+\kappa^2)^{1/2}-z_0\kappa]}{\sqrt{q_0^2+\kappa^2}}, \qquad z_0>0 \text{ and } z<0 \tag{A4.4.30}$$

which exhibits exponential decay into the metal of type $\exp(q_0z)$.

In principle, this potential could be used to answer questions about bound states as the adatom approaches the surface, but that matter will not be pursued here. Though the above approach has the merit of being fully analytic, one should not expect it to compete in accuracy with the more sophisticated density functional techniques (see, for example, Lang[371]) employed to good effect in surface calculations of electronic structure, and also the work of Kumamoto *et al.*[372] Rather, equation (A4.4.30) represents the prediction of a rather simple analytic model exhibiting some of the features necessary for a friction-constant calculation for adatoms.

APPENDIX 5.1
SPIN FLUCTUATIONS AND DESORPTION OF HYDROGEN FROM PARAMAGNETIC METALS

This Appendix is based on the work of Suhl *et al.*[373] Their approach appeals to the suggestion of Gomer and Schrieffer,[374] that in chemisorption of hydrogen on strongly paramagnetic metals a type of induced covalency plays a role. Since the energy needed to form a net d-electron spin density in the vicinity of the adatom is quadratic in that density, while its coupling (exchange) to the electron spin of the adatom is negative and linear in the density, formation of a net spin density is energetically favored.

On the basis of this idea, Suhl *et al.*[373] propose that a possible explanation of the high efficiency of some of the d-metals in catalyzing hydrogenation reactions might be found in the interaction of the adsorbate with spin fluctuations, especially when these display paramagnon enhancement.[375] The adatom absorbs thermal spin fluctuations, gaining energy and eventually climbing over the energy barrier characterizing the reaction.

Coupling of the adatom of spin $\mathbf{S}$ to the metal spin density $\mathbf{s}(\mathbf{r})$ at $\mathbf{r}$ is taken by Suhl *et al.* to have the form

$$H' = \int J(\mathbf{R}, \mathbf{r})\mathbf{S} \cdot \mathbf{s}(\mathbf{r})d\mathbf{r} \tag{A5.1.1}$$

where $J(\mathbf{R}, \mathbf{r})$ is an effective exchange energy of the fixed hydrogen orbital with the d-orbitals and $\mathbf{R}$ is the position of the atom. If one ignores the transverse periodic structure, J has the form $J(\mathbf{R}_T - \mathbf{r}_T, Z, z)$ in terms of coordinates transverse and normal to the surface.

Suhl *et al.* discuss the simplest surface reaction, namely desorption of the hydrogen atom, and in particular the desorption rate. The well in which the H atom is bound is estimated by Suhl *et al.* to be about 1 to 3 V deep and 1 to 3 Å wide. Therefore the atom must have between 10 and 200 excited "bound" states. In thermal equilibrium, the rate of departure of the atom from the surface is the chance that it be found in one of the top excited "bound" states (it is presumably somewhat free to move in the plane of the surface in each one of these excited states), multiplied by the transmission coefficient, i.e., the probability of transition from that level to a free state. This process is accompanied by absorption of a "paramagnon"[375] from the metal according to the Hamiltonian (A5.1.1). (Presumably a very similar calculation of the transmission coefficient may be expected to apply to surface reactions involving transfers or rearrangements of chemisorbed hydrogen.)

The calculation of Suhl *et al.*[373] then proceeds as follows:

1. The nth excited bound state of the atom on a given surface area is set up, the periodicity of the potential in the plane of the surface being accounted for approximately by assigning an appropriate (probably rather large) effective mass.
2. For an order-of-magnitude estimate, only the transition rate from the uppermost "bound" state is calculated.
3. $J(\mathbf{R}, \mathbf{r})$ in equation (A5.1.1) is approximated by Ja^3 multiplied by a three-dimensional delta function, a being the range of J.
4. The Golden Rule for the transition probability is expressed in terms of the dynamic spin susceptibility of the d-electrons at the surface. For simplicity, this quantity is then replaced by the susceptibility of the infinite metal.
5. To find the rate of desorption, the above probability has to be convoluted with the probability that the adatom finds itself in the highest bound state and moving along the surface with specified momentum.

The conclusion of Suhl *et al.*[373] is that if the Gomer–Schrieffer induced covalent bond mechanism is responsible for the chemisorption of hydrogen on strongly paramagnetic metals, then the rates of chemical reactions involving the adsorbed hydrogen may be dominated by energy transfers between the bond and spin fluctuations in the metal. For relevant experimental studies the reader may consult Evzerikhin and Lyubarskii.[376]

A somewhat related study by Ilisca[377] suggests that catalytic effects on ferromagnetic substances are aided by interaction with the low-lying excitations of the magnet, namely spin waves.[290]

APPENDIX 5.2
ANOMALIES IN ADSORPTION EQUILIBRIUM NEAR CRITICAL POINTS OF A SUBSTRATE

Collective properties in a substrate, such as long-range magnetic order below the Curie temperature T_c in a ferromagnet, may be reflected in the adsorption equilibrium of an adsorbate, particularly at a phase transition, as demonstrated by Lagos and Suhl.[378]

These workers drew attention to the fact that in heterogeneous catalysis, anomalies in reaction rates are found near phase transitions of the solid substrate, as in the early work of Parravano.[379] An example, among others that may be cited, is that of magnetocatalytic anomalies near the Curie point. On the other hand, anomalies in the adsorption equilibrium are less familiar (e.g., in regard to hydrogen there are limited experimental data; see, however, Itterbeek *et al.*[380]). Yet a quantitative treatment of the effect should reveal interesting information on the coupling mechanism of the adatoms to the order parameter characterizing the phase transition.

The argument below follows closely the treatment of Lagos and Suhl.[378] In terms of the partition function Z_g of a single atom in the gas phase and the partition function Z_A of the adparticle, thermally averaged over all fluctuations of the substrate order parameter, and the gas pressure p, the isobar of Langmuir referred to in Chapter 1 may be written in the form

$$\theta = (1 + Z_g k_B T / p Z_A)^{-1} \tag{A5.2.1}$$

where θ as usual is the fractional coverage. As a result of the coupling of one or more adatom degrees of freedom to the order parameter, anomalous fluctuations in that parameter near a transition result in anomalous behavior of Z_A.

In particular, Lagos and Suhl consider a ferromagnetic transition and suppose that the coupling takes place in the form of a free-energy increment

$$\delta F_s(M) = f_0 + \tfrac{1}{2} A M^2(0) \tag{A5.2.2}$$

where $M(0)$ is the magnetization of the substrate at the adsorption site, f_0 is the M-independent part of the coupling, and A is a model-dependent constant, as discussed by Suhl.

The total free energy of the substrate plus one adatom is

$$F(M) = F_{LG}(M) + \delta F_s(M) \tag{A5.2.3}$$

where $F_{LG}(M)$ is the Landau–Ginzburg expression[290] for the pure substrate, namely

$$F_{LG}(M) = F_0 + \frac{1}{2\Omega} \int d\mathbf{r}[a\,|M(\mathbf{r})|^2 + c\,|\nabla M(\mathbf{r})|^2 + \tfrac{1}{2} b\,|M(\mathbf{r})|^4] \tag{A5.2.4}$$

Ω denoting an atomic volume, so that a and b are energies per electron of the substrate, and a, b, and c are given, for a cubic lattice, by

$$a = k_B(T - T_c), \qquad b = 4I, \qquad c = \tfrac{1}{2}Il^2 \qquad \text{(A5.2.5)}$$

where T is the temperature of the substrate, I the exchange energy per particle between the electrons of the substrate, and l the lattice spacing, equal to $\Omega^{1/3}$.

One can write Z_A as

$$Z_A = \int DM \exp[-\beta F(M)] \Big/ \int DM \exp[-\beta F_{LG}(M)] \qquad \text{(A.5.2.6)}$$

where $\int DM$ denotes functional integration.

Lagos and Suhl show that, with a number of physically reasonably assumptions, one obtains an expression for Z_A of the form

$$Z_A(T) = \exp(-\beta f_0')[1 + \lambda g(x)]^{-1} \qquad \text{(A5.2.7)}$$

where

$$\begin{aligned} \lambda &= A/\pi^2 I, \qquad t = T/T_c - 1 \\ x &= \begin{cases} (6t)^{1/2} & T \gtrsim T_c \\ (12|t|)^{1/2} & T \lesssim T_c \end{cases} \\ g(x) &= 1 - x tg^{-1} x^{-1} \end{aligned} \qquad \text{(A5.2.8)}$$

In Langmuir's model (cf. Chapter 1), f_0 is the adatom energy (negative) when the magnetic coupling is not present, while f_0' in equation (A5.2.7) is given by

$$f_0' = \begin{cases} f_0, & T \gtrsim T_c \\ f_0 + \tfrac{1}{2}A\langle M\rangle^2, & T \lesssim T_c \end{cases} \qquad \text{(A5.2.9)}$$

Equation (A5.2.1) can now evidently be written as

$$\theta(T) = \left\{1 + \frac{Z_g k_B T}{p \exp(-\beta f_0')}[1 + \lambda g(x)]\right\}^{-1} \qquad \text{(A5.2.10)}$$

For the example of hydrogen on Ni, Shanabarger[381] presents measurements that yield $f_0 = -26$ kcal/mol and one can usefully approximate Z_g by $(M_H k_B T/2\pi\hbar^2)$. Lagos and Suhl assume that, solely for simplicity, H_2

is already dissociated before chemisorption, their main purpose being to assess the effect of the phase transition on the adsorption.

More realistic models, such as dissociative adsorption, mobile or immobile adsorption, adsorbate–adsorbate interaction, nonequivalent active sites, etc., will produce of course a different normal ($\lambda = 0$) isobar, but an anomaly of almost identical shape to the one calculated by the assumptions of Lagos and Suhl.

The quantity A in the definition of λ is discussed by Suhl[(382)] for the particular case of a $3d$-band metal. Typical values of $|A|$ range from 0.4 to 1 eV and the exchange energy I is 0.02 eV, so that λ has values ranging from 2 to 5. These estimates, however, are based on a bulk calculation. More realistically, one can take the magnitude and sign of λ (the latter depends on the slope of the density of states at the Fermi energy and the bandwidth) as fitting parameters, to take care of the modifications arising from the presence of the surface. On the other hand, the best fit to available evidence seems to require $\lambda = 0.2$.

In order to check whether this might be due to the use of bulk modes of $M(\mathbf{r})$, unmodified by the different magnetic conditions prevailing at the surface, Lagos and Suhl analyzed one particular model that acounts for the presence of the surface. If the surface is taken to be $z = 0$, one can incorporate the presence of a surface by adding to $M(\mathbf{r})$ the boundary condition given by Mills,[(383)] namely

$$\alpha M(\mathbf{r})\big|_{z=0} = l \frac{\partial}{\partial z} M(r)\big|_{z=0} \tag{A5.2.11}$$

with

$$\alpha = 1 + 4(1 - I_s/I) \tag{A5.2.12}$$

where I_s is the surface exchange energy.

Lagos and Suhl repeated the calculations conducted for the bulk case and found a new form of $g(x)$ in equation (A5.2.8):

$$g(x) = 1 + \lambda \int_0^1 \frac{ds\, s^2}{\alpha^2 + s^2} \ln \frac{1 + x + s^2}{x + s^2} \tag{A5.2.13}$$

As before, $g(x) - 1$ vanishes far from T_c. The shape of the anomaly is similar to the bulk calculation but with two new features:

1. In the bulk case, the tangents to the curve $\theta(T)$ as $T \to T_c$ from below or above are both vertical. Taking the surface into account still gives a cusp-like discontinuity at $T = T_c$ but with tangents of different (and nonvertical) slopes, according as $T \to T_c$ from below or above.

2. As for the magnitude of the effect, this can be expressed in terms of an effective λ defined by

$$\lambda_{\text{eff}} = \lambda \max[g(x)] = \lambda g(0) \quad \text{(A5.2.14)}$$

[for the bulk $g(0) = 1$]. In the presence of the surface, $g(0)$ ranges from about 0.015 to 0.25 as I_s/I varies from 0 to 1.

Thus fitting λ gives information on a possible value of I_s/I. In the treatment of Itterbeek *et al.*[380] λ_{eff} is approximately 0.2, and taking λ equal to about 3 gives I_s/I approximately 0.8.

Lagos and Suhl[378] plot $1 - \theta(t)$ versus t for the normal isobar ($\lambda = 0$) and $\lambda = 0.2$, for $p = 0.25$, 0.5, and 1.0 mm Hg. The reader is referred to their paper for the quantitative results.

APPENDIX 6.1
SYMMETRY RULES BASED ON THE SURFACE ELECTRONIC STRUCTURE OF VARIOUS PLATINUM SURFACES

Bond's picture of the electronic structure of platinum surfaces,[303] referred to in the main text, will be used here as a basis for developing symmetry rules for nitric-oxide decomposition on platinum. It is known that Bond's method leads to rather inaccurate energy levels. However, the more quantitative study of Tersoff and Falicov[384] has resulted in band structures for stepped nickel and copper surfaces. These workers find that the symmetry properties of the occupied d-bands are identical to those found in Bond's calculations. Balasz and Johnson[385] make platinum cluster calculations, and these reveal some small differences due to a lack of translational symmetry. However, the predictions of Banholzer *et al.*[293] are the same if the band structures of Balasz and Johnson are used. The bulk band-structure calculations of Anderson[386] also lead to similar predictions.[293] In fact, it turns out from the study of Banholzer *et al.* that the bond-breaking properties of the surface are largely determined by the symmetry of the d-bands. Therefore, the use of a semiquantitative band structure for purposes of illustration should not be expected to affect the conclusions.

Figure A6.1.1 shows some of the orbitals in the platinum (111) and (100) surfaces. When the (111) surface is cut, the part of the $6s$-band that is exposed is symmetric in all directions. The d-bands are split into t_{2g}- and e_g-like states, which are totally symmetric around the threefold axis perpendicular to the surface, as shown in Figure A6.1.1a. The orbitals are also symmetric around the sets of dihedral planes, which now lie perpen-

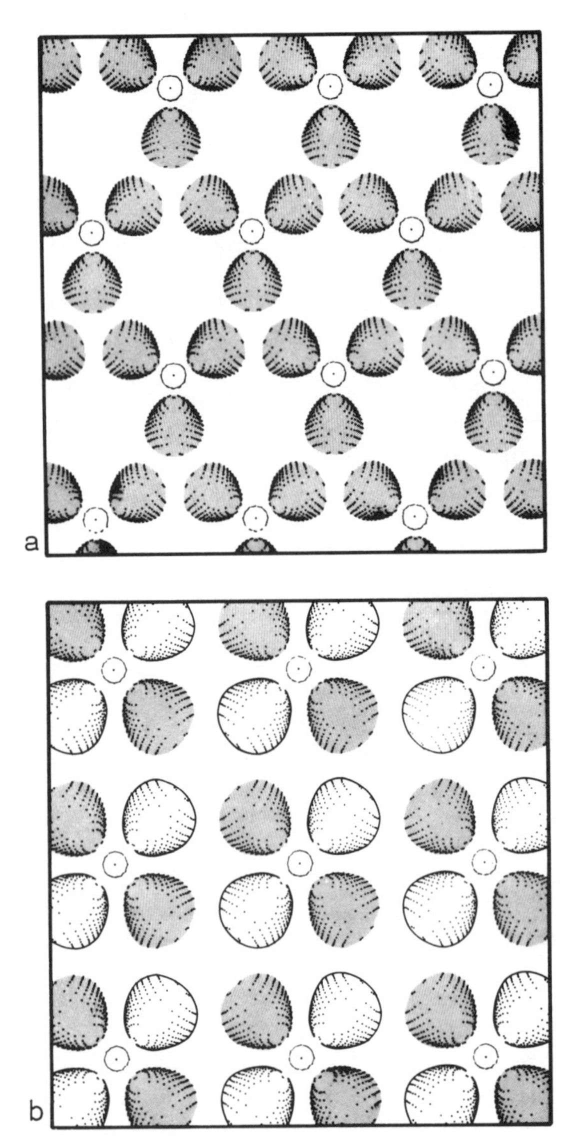

FIGURE A6.1.1. Schematic representation of the protruding orbitals from the t_{2g}-like band in the (a) (111) and (b) (100) surface of platinum, adapted from those of Bond by Banholzer *et al.*[293] Orbital exponents were taken from the work of Clementi *et al.*[293] The orbitals emerge at 35°16″ from the (111) surface and at 45° from the (100) surface. Shaded areas represent positive lobes, clear areas depict negative lobes. Orbitals involved in metal–metal bonding are omitted for reasons of clarity. Orbitals appear larger on the (100) than on the (111) surface, because a lower probability contour was plotted in the (100) case. The e_g orbitals protrude perpendicularly from the (100) surface, while they come out at 35°16″ on the (111) surface, and fill in the spaces (after Banholzer *et al.*[293]).

dicular to the surface. It is noteworthy that none of these orbitals is antisymmetric about a plane perpendicular to the surface. Yet an antisymmetric orbital is needed to fill the $\pi^*_\perp$-orbital on the NO. Hence from arguments such as those outlined above, or those of Moore and Pearson,[387] one would expect a symmetry constraint that inhibits NO decomposition on the (111) surface. These arguments can be reinforced by noting that there is only one symmetry element in Figure 6.9, the mirror plane, denoted plane I. Hence, by group theory,[388] all the symmetry properties of the system can be obtained from the C_s point group. There are two irreducible representations in this point group, an A' representation that includes all symmetric states, and an A'' representation including all antisymmetric states. The d-bands of platinum have states of A' symmetry in the C_s point group while the $\pi^*_\perp$-orbital on the NO has A'' symmetry. Hence a transfer of electrons from the metal to the $\pi^*_\perp$-orbital on the NO has A'' symmetry. Thus a transfer of electrons from the metal to the $\pi^*_\perp$-orbital on the NO would be symmetry-forbidden. As with gas-phase reactions, it is possible in principle for the reaction to proceed via an asymmetric pathway. However, if orbital-symmetry effects are important, then it is clear that one would expect inhibition of the reaction in such a circumstance.

Banholzer *et al.*[293] point out that a similar argument obtains for the (110) surface, and one might expect a reduced rate. But turning to the (100) surface, the situation is markedly different. When the (100) surface is formed, the σ_h planes remain. The portion of the t_{2g} orbital that protrudes from the surface is antisymmetric about planes perpendicular to the surface, as illustrated in Figure A6.1.1b. There are sign changes in the t_{2g} orbitals that mimic those in the $\pi^*_\perp$ on the NO. These orbitals have A'' symmetry in the C_s point group. The e_g and $6s$ bands are symmetric. (They have A' symmetry.) Hence the t_{2g} can donate electrons into the $\pi^*_\parallel$- and σ^*-orbitals. NO bond breaking is symmetry-allowed on the (100) surface. The conclusion is clearly that if symmetry effects dominate, one would anticipate that Pt(100) will be much more active than Pt(111) or Pt(110) for NO bond breaking. The experimental data[293] show that this is indeed the case. Therefore, it seems fairly evident that orbital-symmetry effects are of some importance for the NO decomposition reaction on Pt.

This does not complete the analysis. Banholzer *et al.*[293] note that data indicate there are faces even more active than Pt(100). To explore this they use Pearson's[389,390] formulation in which one can derive symmetry rules based on an examination of overlap integrals. When a reaction proceeds, the rate of reaction is related to the net overlap between the donor and acceptor orbitals. In the case of a true symmetry-forbidden reaction, the net overlap is identically zero. Hence no direct transfer of electrons is possible. However, there are many other reactions where the net overlap is small, but nonzero. Pearson terms these reactions ones with partial-symmetry constraints. In such cases, one would expect the reaction rate to be somewhat inhibited. However, the degree of inhibition is often much smaller than those for a true symmetry-forbidden reaction.

Banholzer *et al.*[293] note that, for the NO decomposition reaction on the (100) surface, one finds a partial-symmetry constraint. The reader is referred to their paper for more details.

Predicted Active Sites: Influence of Step Surfaces

Another aspect of the model proposed by Banholzer *et al.*[293] is that it permits identification of surface structures where the reaction is completely symmetry-allowed. If symmetry effects dominate, then one might anticipate that these structures will be unusually active. Step surfaces are of particular interest. A step lowers the symmetry of the lattice. Motions forbidden elsewhere are allowed at the step. According to the arguments of Banholzer *et al.*[293] set out in their paper, the rate of NO dissociation on the (100) surface is limited by a partial-symmetry constraint to bending motion, due to destabilization of the σ^*- and $\pi^*_\parallel$-orbitals during bending. If there is a step on the surface, then there will be extra lobes at the step that can potentially interact with the NO in a way that will stabilize the orbitals. A search was made by Banholzer *et al.* for step surfaces with favorable geometries (see their Figure 7, for instance). They anticipate that surfaces with (100) terraces and (110) steps will be unusually active.

Measurements have been made[391,392] using a Pt(410) surface. It will be noted that Pt(410) is unusually active. This is in accord with the model discussed above.

It seems that symmetry-conservation (i.e., Woodward–Hoffmann) ideas could be very useful in catalytic systems.

APPENDIX 6.2
LOCAL DENSITY OF STATES, SCATTERING THEORY, AND THE *T*-MATRIX

There is potential interest in the local density of states, as pointed out in Section 6.5. Therefore, an outline will be given here of a framework within which it might be calculated. This Appendix will also serve the purpose of introducing the T-matrix used in Section 6.3 in the theory of the compensation effect in catalysis. We explicitly consider free electrons, with plane-wave functions $\exp(i\mathbf{k}\cdot\mathbf{r})$ and scattered by a one-body potential energy $V(\mathbf{r})$. Formal generalization to Bloch waves suitable for perfect crystals can be readily effected but, of course, the technical difficulties of solving the problem are substantially increased and we shall not embark on a discussion here.

Following March and Murray,[393] the starting point will be the Schrödinger equation

$$\nabla^2\psi + k^2\psi = 2V\psi \tag{A6.2.1}$$

where the energy E has been written as $\frac{1}{2}k^2$. The solution of equation (A6.2.1) with the right-hand side set to zero, and which is everywhere finite, is

$$\psi(\mathbf{r}) = \frac{A \sin k|\mathbf{r}_1 - \mathbf{r}|}{|\mathbf{r}_1 - \mathbf{r}|} \tag{A6.2.2}$$

or, indeed, any function of the form

$$\psi(\mathbf{r}) = \int d\mathbf{r}_1 A(\mathbf{r}_1) \frac{\sin k|\mathbf{r}_1 - \mathbf{r}|}{|\mathbf{r}_1 - \mathbf{r}|} \tag{A6.2.3}$$

A particular integral of equation (A6.2.1) may be obtained using the Green function $(\cos kr)/r$ (compare the solution of Laplace's equation of electrostatics, with Green function $1/r$ obtained by putting $k = 0$ in the factor $\cos kr$). This yields

$$\psi(\mathbf{r}) = -\int d\mathbf{r}_1 2V(\mathbf{r}_1)\psi(\mathbf{r}_1) \frac{\cos k|\mathbf{r}_1 - \mathbf{r}|}{4\pi|\mathbf{r}_1 - \mathbf{r}|} \tag{A6.2.4}$$

Hence equation (A6.2.1) may be written in the form

$$\psi(\mathbf{r}) = -\int d\mathbf{r}_1 V(\mathbf{r}_1)\psi(\mathbf{r}_1) \frac{\cos k|\mathbf{r}_1 - \mathbf{r}|}{2\pi|\mathbf{r}_1 - \mathbf{r}|} + \int d\mathbf{r}_1 A(\mathbf{r}_1) \frac{\sin k|\mathbf{r}_1 - \mathbf{r}|}{|\mathbf{r}_1 - \mathbf{r}|} \tag{A6.2.5}$$

where $A(\mathbf{r}_1)$ may be a function of the potential energy V, to be determined by the boundary conditions. Introducing the generalized electron density $\rho(\mathbf{r}, \mathbf{r}_0, k)$, obtained by an off-diagonal extension of the usual electron density given by the sum of the squares of the normalized ψ functions up to energy $E = k^2/2$, the local off-diagonal density of states $\sigma(\mathbf{r}, \mathbf{r}_0, k)$ is conveniently defined by

$$\sigma(\mathbf{r}, \mathbf{r}_0) = \frac{2\pi^2}{k} \frac{\partial\rho(\mathbf{r}, \mathbf{r}_0, k)}{\partial k} \tag{A6.2.6}$$

from which it is clear that

$$\rho(\mathbf{r}, \mathbf{r}_0, k) = \int_0^k \frac{k}{2\pi^2} \sigma(\mathbf{r}, \mathbf{r}_0, k) dk \tag{A6.2.7}$$

It may readily be shown that $\sigma(\mathbf{r}, \mathbf{r}_0)$ is related to the average wave-function product for energy equal to $\frac{1}{2}k^2$, and hence that it satisfies equation (A6.2.1). For free electrons (singly filled states)

$$\rho_0(\mathbf{r}, \mathbf{r}_0, k) = \frac{k^3}{2\pi^2} \frac{j_1(k|\mathbf{r} - \mathbf{r}_0|)}{k|\mathbf{r} - \mathbf{r}_0|} \tag{A6.2.8}$$

where the first-order spherical Bessel function $j_1(x)$ is given by $(\sin x - \cos x)/x^2$. Therefore, equation (A6.2.6) gives

$$\sigma(\mathbf{r}, \mathbf{r}_0, k) = \frac{\sin k|\mathbf{r} - \mathbf{r}_0|}{|\mathbf{r} - \mathbf{r}_0|} \tag{A6.2.9}$$

On the diagonal, namely $\mathbf{r}_0 = \mathbf{r}$, it is readily verified from equation (A6.2.9) that the free electron density of states $N_0(E) \propto E^{1/2} = k/2^{1/2}$ is regained. If $\psi(\mathbf{r})$ is replaced in equation (6.2.5) by $\sigma(\mathbf{r}, \mathbf{r}_0)$, one obtains the integral equation for the off-diagonal local density of states:

$$\sigma(\mathbf{r}, \mathbf{r}_0, k) = -\int d\mathbf{r}_1 V(\mathbf{r}_1)\sigma(\mathbf{r}_1, \mathbf{r}_0) \frac{\cos k|\mathbf{r}_1 - \mathbf{r}|}{2\pi|\mathbf{r}_1 - \mathbf{r}|} + \int d\mathbf{r}_1 A(\mathbf{r}_1, \mathbf{r}_0) \frac{\sin k|\mathbf{r}_1 - \mathbf{r}|}{|\mathbf{r}_1 - \mathbf{r}|} \tag{A6.2.10}$$

Equation (A6.2.10) may now be solved by iteration in powers of V. Thus one writes[393]

$$\sigma(\mathbf{r}, \mathbf{r}_0) = \sum_{j=0}^{\infty} \sigma_j(\mathbf{r}, \mathbf{r}_0) \tag{A6.2.11}$$

where σ_0 is given by equation (A6.2.9). Insertion of $\sigma_0(\mathbf{r}, \mathbf{r}_0)$ in the right-hand side of equation (A6.2.10) gives

$$\sigma_1(\mathbf{r}, \mathbf{r}_0) = -\int d\mathbf{r}_1 \frac{V(\mathbf{r}_1)}{2\pi} \frac{\sin k|\mathbf{r}_0 - \mathbf{r}_1|}{|\mathbf{r}_0 - \mathbf{r}_1|} \frac{\cos k|\mathbf{r}_1 - \mathbf{r}|}{|\mathbf{r}_1 - \mathbf{r}|} - \int d\mathbf{r}_1 \frac{V(\mathbf{r}_1)}{2\pi} \frac{\cos k|\mathbf{r}_0 - \mathbf{r}_1|}{|\mathbf{r}_0 - \mathbf{r}_1|} \frac{\sin k|\mathbf{r}_1 - \mathbf{r}|}{|\mathbf{r}_1 - \mathbf{r}|} \tag{A6.2.12}$$

the form of the second integral being fixed by requirements of symmetry.

Thus

$$\sigma_1(\mathbf{r}, \mathbf{r}_0) = -\int d\mathbf{r}_1 \frac{V(\mathbf{r}_1)}{2\pi} \frac{\sin k(|\mathbf{r}_0 - \mathbf{r}_1| + |\mathbf{r}_1 - \mathbf{r}|)}{|\mathbf{r}_0 - \mathbf{r}_1|\,|\mathbf{r}_1 - \mathbf{r}|} \qquad \text{(A6.2.13)}$$

The other terms may be obtained similarly and the general term is

$$\sigma_j(\mathbf{r}, \mathbf{r}_0) = \int^j \prod_{l=1}^{j} \left\{ -d\mathbf{r}_l \frac{V(\mathbf{r}_l)}{2\pi} \right\} \frac{\sin k\left(\sum_{l=1}^{j+1} s_l\right)}{\prod_{l=1}^{j+1} s_l} \qquad \text{(A6.2.14)}$$

where $s_l = |\mathbf{r}_l - \mathbf{r}_{l-1}|$ with $\mathbf{r}_{j+1} \equiv \mathbf{r}$. Of course, one can now proceed to the diagonal by putting $\mathbf{r}_0 = \mathbf{r}$, and obtain an infinite series for the local density of states $\sigma(\mathbf{r}, \mathbf{r})$ in terms of the one-body potential $V(r)$.

Formal Summation for Strong Spherical Potential Energy $V(r)$

To date, series (A6.2.11) has not been summed for an arbitrary potential energy $V(\mathbf{r})$. For the central-field case, when $V(\mathbf{r}) \equiv V(|\mathbf{r}|)$, one regains all the results of phase-shift analysis.

In the context of such phase-shift analysis, March[394] has shown how to relate potential theory and phase-shift theory for the scattering cross-section from a finite-range spherical potential of arbitrary strength. One thereby obtains equation (4.10.9), characteristic of inverse-transport theory.[395]

The essential result needed to relate the phase-shift expression (η_l for a partial wave of orbital angular-momentum quantum number l) to the potential theory is

$$\int_0^\infty dr\, r^2 R_{l-1}(r) \frac{\partial V}{\partial r} R_l(r) = \sin(\eta_l - \eta_{l-1}) \qquad \text{(A6.2.15)}$$

where the radial wave function $R_l(r)$ outside the range of V has the form

$$R_l(r) = j_l \cos \eta_l - n_l \sin \eta_l \qquad \text{(A6.2.16)}$$

Equation (A6.2.15) was originally derived by Gerjuoy[396] and has been rediscovered by Gaspari and Gyorffy.[397] In equation (A6.2.16), j_l and n_l denote spherical Bessel and Neumann functions, respectively. Then one obtains, for example, the excess residual resistivity in a metal[398] due to Fermi-surface scattering in terms of the local off-diagonal density of states $\sigma(\mathbf{r}, \mathbf{r}_0, E_f)$, where the Fermi energy $E_f = k_f^2/2$, as proportional to the force–force correlation function F given, apart from unimportant multiplying factors, by[394]

$$F = \int d\mathbf{r}\, d\mathbf{r}_0 \frac{\partial V(\mathbf{r})}{\partial \mathbf{r}} \cdot \frac{\partial V(\mathbf{r}_0)}{\partial \mathbf{r}_0} |\sigma(\mathbf{r}, \mathbf{r}_0, E_f)|^2 \qquad \text{(A6.2.17)}$$

Either this, or the phase shifts at the Fermi energy, $\eta_l(k_f)$, determine the appropriate scattering cross-section for plane waves off a spherical potential V.

Finally a formal expression analogous to equations (A6.2.5) and (A6.2.10) for ψ and σ, respectively, can be written in terms of the Green function,

$$G = G_0 + G_0 V G \qquad \text{(A6.2.18)}$$

where an integration is implied in the notation (A6.2.18). In each case, the desired quantity (ψ, σ, or G) appears in the right-hand side of the corresponding integral equations. A sometimes useful device for (formally) avoiding this is the so-called T-matrix, which one can take as defined by

$$G = G_0 + G_0 T G_0 \qquad \text{(A6.2.19)}$$

where G_0 is evidently the free-particle Green function. Clearly, T has formally summed the iteration of (A6.2.18); for spherical potentials T is readily expressible in terms of phase shifts.[351] It is noteworthy that, in coordinate respresentation, the Green function associated with equation (A6.2.1) can be expressed in terms of the complete set of wave functions ψ_i, with eigenvalues ε_i,

$$G(\mathbf{r}, \mathbf{r}_0, E) = \sum_i{}^{*} \frac{\psi_i(\mathbf{r})\psi_i(\mathbf{r}_0)}{E - \varepsilon_i} \qquad \text{(A6.2.20)}$$

the zero of the denominator being avoided by adding a small imaginary part of appropriate sign.[370]

APPENDIX 6.3
ANALYSIS OF RATE EQUATIONS DESCRIBING CHEMICAL OSCILLATIONS

Following Suhl,[399] equations (6.6.1) and (6.6.2) with $\dot{\theta}_1$ and $\dot{\theta}_2$ equated to zero give a cubic equation for θ_1 alone:

$$f(y) = y^3 - y^2 l + ym - n = 0 \qquad \text{(A6.3.1)}$$

where

$$y = \theta_1 + \frac{c}{k_1}, \qquad l = 1 - \theta_3 + \frac{c}{k_1} - \frac{r_{CO}}{r_0},$$

$$m = \frac{r_{CO}^2}{r_0}\left(1 - \frac{c}{r_{CO}}\right), \qquad n = \frac{r_{CO}^2}{k_1 r_0}\frac{c}{k_1} \tag{A6.3.2}$$

and

$$\theta_2 = \frac{\theta_1 r_{CO}^2}{(\theta_1 + c/k_1)^2 r_0 k_1} \tag{A6.3.3}$$

Insight into the solutions of equation (A6.3.1) can be gained by examining it for all (including unphysical) values of l and m. Suhl constructs the results summarized in Table A6.3.1 for the number of real roots of equation (A6.3.1), a question mark in the table being shorthand for "possibly." As the parameters are varied, real roots of $f(y) = 0$ appear or disappear in pairs. This happens when simultaneously $f(y) = 0$ and $f'(y) = 0$. Elimination of y between these equations gives

$$g(l) = 4nl^3 - l^2m^2 - 18nml + (27n^2 + 4m^3) = 0 \tag{A6.3.4}$$

When $m > 0$, this equation for l has at most two positive real roots and at most one negative real root. Hence, and since $g(0) > 0$ and $g(-\infty) < 0$, it must have exactly one negative real root. The columns $m = 0$ and $m < 0$ of Table A6.3.1 are of no interest as far as oscillations are concerned, because the physical range of y is $1 + c/k > y > c/k$ and a necessary (not sufficient) condition for the bistability type of oscillation to occur obviously needs three branches of the function y versus l to lie in $(c/k, 1 + c/k)$. For $m \leq 0$

TABLE A6.3.1. Number of Real Roots of Equation (A6.3.1) in Various Ranges of Parameters l and m[a]

	$m > 0$	$m = 0$	$m < 0$
$l > 0$	One positive Two more positive?	One positive	One positive Two negative?
$l = 0$	One positive	One positive	One positive Two negative?
$l < 0$	One positive Two negative?	One positive Two negative?	One positive Two negative?

[a] A question mark denotes "possibly."

this is not the case. (In fact, only in the case $l > 0$, $m > 0$ of Table A6.3.1 might oscillations occur.) Further analysis will therefore be restricted to the case $m > 0$. To complete the picture for the roots of $f(y)$ as a function of l (i.e., for θ_1 as a function of θ_3) one needs the roots of equation (A6.3.1) for large positive and negative l. For large negative l, there is one large negative root,

$$y = -|l| + m/|l| + O(1/l^2) \tag{A6.3.5}$$

and two more small roots,

$$y = \pm(n/|l|)^{1/2} + O(1/|l|) \tag{A6.3.6}$$

For large positive l there is only one real root:

$$y = l - m/l + O(1/l^2) \tag{A6.3.7}$$

Thus, both for $l = 0$ (see Table A6.3.1) and for l very large and positive, $f(y) = 0$ has only one real positive root. If it is to have two more positive roots, these must come about either by formation of an "S" in the y versus l curve or by formation of a closed loop. Only the former case is of interest for bistability oscillations. A figure S will first appear when all three roots of $f(y) = 0$ coincide. The condition for this is

$$f(y) = f'(y) = f''(y) = 0 \tag{A6.3.8}$$

which gives

$$l = (3m)^{1/2}, \qquad n = (3^{1/2}/9)m^{3/2}, \qquad y = l/3 \tag{A6.3.9}$$

Since one knows that for $m = 0$ there is only one positive root, by continuity the condition for three positive roots must be

$$n < (3^{1/2}/9)m^{3/2} \tag{A6.3.10}$$

or

$$c/k_1 < (r_{CO}^2/r_0 k_1)^{1/2}(1 - c/r_{CO})^{3/2} \tag{A6.3.11}$$

The value of θ_3 at the first appearance of three roots is given by the first of the conditions (A6.3.9):

$$\theta_3^{\text{crit}} = 1 + \frac{c}{k_1} - \frac{r_{CO}}{r_0} - \left[\frac{3r_{CO}^2}{r_0 k_1}\left(1 - \frac{c}{r_{CO}}\right)\right]^{1/2} \tag{A6.3.12}$$

Since one must have $0 < \theta_3^{\text{crit}} < 1$, a further necessary condition for oscillations is

$$\frac{r_{CO}}{r_0} < 1 + \frac{c}{k_1} - \left[\frac{3r_{CO}^2}{k_1 r_0}\left(1 - \frac{c}{r_{CO}}\right)\right]^{1/2} \qquad \text{(A6.3.13)}$$

which is $\theta_3^{crit} > 0$. The condition $\theta_3^{crit} < 1$ holds automatically because of the inequality (A6.3.11). For $c \ll r_{CO}$ this yields

$$\left(1 - \frac{r_{CO}}{r_0}\right)^2 > \frac{64}{27}\left(\frac{r_{CO}}{r_0}\right)^2 \frac{r_0}{k_1} \qquad \text{(A6.3.14)}$$

The course of the roots in the y–l plane is shown in Figure A6.3.1 following Suhl.[399] It can be seen from Figure A6.3.1B that oscillations might occur if the variables remain in their physical ranges everywhere on the dotted line.

Approximate values for $l_>$ and $l_<$ may be found by setting $n = m^{3/2}\nu$ and $l = m^{1/2}w$ in equation (A6.3.4), when this transforms to a one-parameter equation for w:

$$4\nu w^3 - w^2 - 18\nu w + (27\nu^2 + 4) = 0 \qquad \text{(A6.3.15)}$$

For oscillations one needs $\nu < 0.19245$, and this makes it possible to assume ν small. To about 10% accuracy the roots are given by

$$l_< = m^{1/2}(2 - \nu) = m^{1/2}(2 - nm^{-3/2}) \qquad \text{(A6.3.16)}$$

$$l_> = \frac{m^2}{8n} + \left(\frac{m^4}{64n^2} + \frac{9}{2}m\right)^{1/2}, \qquad l_- = \frac{m^2}{8n} - \left(\frac{m^4}{64n^2} + \frac{9}{2}m\right)^{1/2} \qquad \text{(A6.3.17)}$$

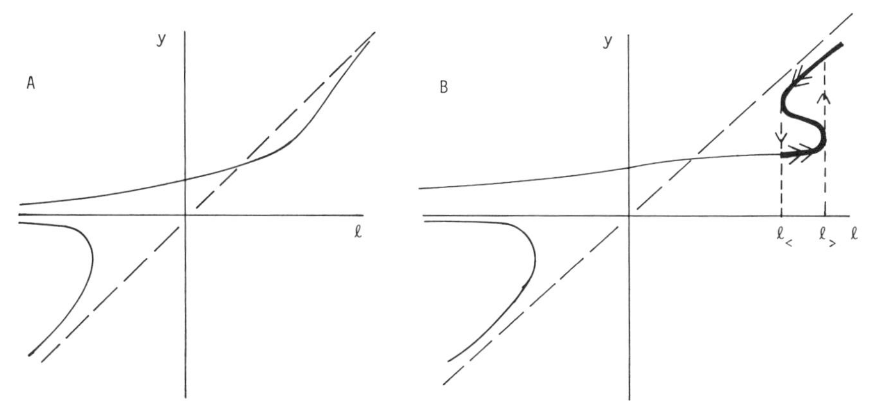

FIGURE A6.3.1. The course of the roots in the y–l plane of equation (A6.3.1): A) $n > (\sqrt{3}/9)m^{3/2}$, B) $n < (\sqrt{3}/9)m^{3/2}$.

The condition $0 < \theta_3 < 1$ applied to $l_<$ and $l_>$, for moderate c/k_1, leads to

$$\frac{4c}{k_1} > \frac{r_{CO}^2}{k_1 r_0}\left(1 - \frac{c}{r_{CO}}\right)^2\left(1 + \frac{c}{k} - \frac{r_{CO}}{r_0}\right) \times\left[\left(1 + \frac{c}{k_1} - \frac{r_{CO}}{r_0}\right)^2 - \frac{9}{2}\frac{r_{CO}^2}{k_1 r_0}\left(1 - \frac{c}{r_{CO}}\right)\right]^{-1} \quad \text{(A6.3.18)}$$

This is guaranteed, while c/k_1 is still small, if r_{CO}/r_0 is small and if

$$1 \gg c/k_1 > \tfrac{1}{4}(r_{CO}/r_0)^2 r_0/k_1 \quad \text{(A6.3.19)}$$

For example, Suhl points out that if c/k_1 is about 10^{-2} and r_0/k_1 about 10^{-1}, then this latter condition requires $r_{CO}/r_0 < 0.63$. If c/k_1 is about 10^{-2} and r_0/k_1 about 1, condition (A6.3.19) requires that r_{CO}/r_0 be less than 0.2. It can evidently be consistent with equation (A6.3.11) for significant ranges of r_{CO}/r_0.

Finally, to make oscillations around the loop in Figure A6.3.1b possible, it is necessary that points at which $\dot{\theta}_3 = 0$ lie outside the range $(l_<, l_>)$. Otherwise the phase point either sticks at such a position (if it is stable) or else never moves beyond it (if it is unstable). At such a point

$$\frac{a}{k_2}\frac{1-\theta_3}{\theta_3} = \frac{\theta_2}{\theta_1} \quad \text{(A6.3.20)}$$

On the upper branch, one has

$$\theta_{1,2} = \tfrac{1}{2}l \pm (l^2 - r_{CO}^2/r_0 k)^{1/2} \quad \text{(A6.3.21)}$$

for small c/k_1. To allow free traversal of the upper branch, equation (A6.3.20) must not have a solution for θ_3 within the range $(\theta_{3<}, \theta_{3>})$ where

$$\theta_{3<,>} = 1 + c/k_1 - r_{CO}/r_0 - l_{>,<} \quad \text{(A6.3.22)}$$

A graphical analysis by Suhl shows that no such solution exists if

$$k_2/a < (1/\theta_{3>} - 1) \quad \text{(A6.3.23)}$$

Since, for small c/k_1, $\theta_{3>} = 1 - r_{CO}/r_0 + 2(r_{CO}^2/r_0 k)^{1/2}$, which can be fairly close to unity, it follows that k_2/a must be somewhat less than unity. On the lower branch, an adequate solution of equation (A6.3.1) is

$$y = m/2l - (m^2/4l - n/l)^{1/2} \quad \text{(A6.3.24)}$$

For very small c/k_1, this gives $\theta_1 = c^2/k_1 r_{CO}$ and θ_1 remains quite small even for moderate c/k_1. Also,

$$\theta_2 = \theta_1 r_{CO}^2 / r_0 k_1 (\theta_1 + c/k_1)^2 \approx r_{CO}/r_0$$

for very small c/k_1. Hence, and since $(k_2/a)(1-\theta_3)/\theta_3$ is of order unity on the lower branch, no stationary point for θ_3 will occur on this particular branch.

Suhl also calculates the times spent on the upper and lower branches and the waveform of the CO_2 output: the interested reader should consult his paper for the details.

APPENDIX 6.4
PROPOSED REACTION SEQUENCE FOR FISCHER–TROPSCH SYNTHESIS

Bell[400] has proposed the following reaction sequence for the Fischer–Tropsch synthesis:

1. $CO + S \rightleftharpoons CO_S$.
2. $CO_2 + S \rightleftharpoons C_S + O_S$.
3. $H_2 + 2S \rightleftharpoons 2H_S$.
4a. $O_S + H_2$ (or $2H_S$) $\rightarrow H_2O + 2S$.
4b. $O_S + CO$ (or CO_S) $\rightarrow CO_2 + 2S$.
5. $C_S + H_S \rightleftharpoons CH_S + S$.
6. $CH_S + H_S \rightleftharpoons CH_{2S} + S$.
7. $CH_{2S} + H_S \rightleftharpoons CH_{3S} + S$.
8. $CH_{3S} + H_S \rightarrow CH_4 + S$.
9. $CH_{3S} + CH_{2S} \rightarrow CH_3CH_2 + S$.
10. $CH_3CH_{2S} + CH_{2S} \rightarrow CH_3CH_2CH_{2S} + S$.
11. $CH_SCH_{2S} + S \rightarrow CH_2CH_2 + H_S$.
12. $CH_3CH_{2S} + H_S \rightarrow CH_3CH_3 + 2S$.

This idealized mechanism would lead[13] to the formation of a Schulz–Flory distribution of straight-chain hydrocarbons. In practice, oxygenation, isomerization, and aromatization side reactions frequently appear to intervene. Unfortunately, at the time of writing, the precise mechanistic roles of these site reactions are not well understood, but it would appear that their control, via suitable catalyst formation, can lead to the selective synthesis of almost any hydrocarbon starting from CO and H_2.

APPENDIX 6.5
DISSOCIATIVE CHEMISORPTION — A PRIMITIVE SURFACE CHEMICAL PROCESS

We summarize below the work of Auerbach *et al.*[401] relating to activated dissociative adsorption of N_2 on W(110). However, it should be noted that in general dissociative chemisorption, the process whereby intramolecular bonds are broken and new bonds formed to surface atoms, is one of the simplest of surface chemical processes. It is a key step in a wide variety of catalytic chemistry and, when accompanied by a substantial activation barrier, is commonly the rate-determining step in such reactions. An understanding of the nature and role of such activation barriers is therefore important to any detailed picture of surface chemical dynamics.

The interaction of H_2 with metal surfaces (for statics, see Chapter 2), and in particular with Cu, has served as a prototype for such studies, beginning with the concept of a one-dimensional barrier (1DB) to adsorption.[402]

Studies of the dependence of the dissociative adsorption probability at zero coverage S_0 on incident kinetic energy E_i and angle of incidence θ_i for H_2/Cu have lent support to this picture,[212] as have observations of noncosine angular distributions and non-Boltzmann kinetic-energy distribution of molecules that recombine and desorb.[215,403,404] A key feature of the 1DB model is that the dissociative adsorption probability scales with the normal component of the incident kinetic energy: $E_n = E_i \cos^2 \theta_i$.

As regards heavier molecules, rather little data are presently available on the dynamic aspects of activated dissociative chemisorption, despite the considerable chemical relevance of such processes. However, Cosser *et al.*[405] have reported a noncosine angular distribution for N_2 desorbing from W(110). This was also interpreted in terms of a 1DB model yielding an activation barrier height of 17 kJ mol^{-1}. In molecular-beam experiments, which Auerbach *et al.*[406,407] performed at $\theta_i = 45°$, they found that S_0 does indeed increase strongly with E_i. However the steepest increase occurred for E_i equal to about 80 kJ mol^{-1} (E_n about 40 kJ mol^{-1}).

Auerbach *et al.*[401] have subsequently reported measurements of the dependence of S_0 on angle of incidence. They concern themselves only with the initial sticking or dissociative adsorption probability, i.e., the value obtained in the limit of zero surface coverage. Their finding is that $S_0(E_i, \theta_i)$ is insensitive to θ_i for $0 < \theta_i < 45°$, with only a weak dependence at larger angles.

Figure A6.5.1 displays their S_0 values at various angles of incidence plotted against E_i. These data, obtained with a sample temperature of 800 K, demonstrate the activated nature of the adsorption process. The sharp increase in adsorption probability is seen to occur when E_i is about 80

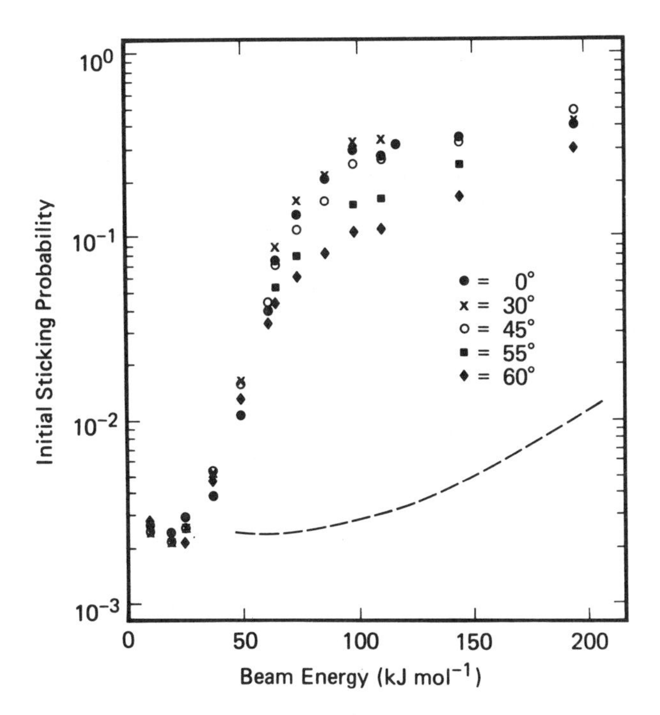

FIGURE A6.5.1. Initial sticking probability for N_2 on W(110) as a function of beam energy at various angles of incidence for a sample temperature of 800 K. The dashed line indicates sticking probabilities predicted for $\theta = 60°$ obtained by assuming normal energy scaling for the $\theta = 0°$ data. (Reproduced with permission from Auerbach *et al.*[401])

kJ mol^{-1}. Auerbach *et al.*[401] note that the increase of S_0 with E_i is far from a step function, contrary to the behavior expected for a single barrier. In this respect the data are similar to those reported[212] for H_2 on Cu, although the increase observed by Auerbach *et al.* is considerably larger than the factor of about 5 observed for that system.

Figure A6.5.1 makes it clear that, for $\theta_i \leq 45°$, S_0 is largely independent of θ_i. Thus the scaling behavior observed for H_2 on Cu is no longer found. Normal energy scaling must result from any 1DB model or indeed from any model in which the molecule-surface potential depends on z, the normal coordinate, and not on x and y. A much stronger dependence of the potential on z than on x and y (weak corrugation) is quite characteristic of low-energy atomic and molecular interactions with low-index planes of metal surfaces.[213] Weak corrugation is found[408] in the Ar–Pt(111) system even at 200 kJ mol^{-1}, the highest energy studied. Rotational excitation of NO in collisions with Ag(111) has been shown to scale with E_n[409,410] up to at least 180 kJ mol^{-1}. The breakdown of normal energy scaling reported by Auerbach *et al.*[401] is surprising in light of their observation that almost perfect normal energy scaling exists for CH_4 dissociative chemisorption on this same crystal surface.[411]

Possible explanations of the observed insensitivity of S_0 to θ_i fall into two classes.[401]

1. Those which consider the surface to be strongly corrugated or rough on an atomic scale.
2. Those which postulate a long-lived surface intermediate or precursor state, which mediates the randomization of the initial energy modes.

APPENDIX 6.6
NONLINEAR RATE EQUATIONS AND THE COMPENSATION EFFECT

The arguments of Sommer and Kreuzer[235] in Section 5.5.1 lead to rate equations on which a kinetic theory of physisorption can be based. These equations take the form

$$\frac{dn_i}{dt} = \sum_j R_{ij} n_j - \sum_j R_{ji} n_i \qquad \text{(A6.6.1)}$$

for the occupation functions $n_i(t)$, where **i** denotes a set of quantum numbers characterizing a bound state or continuum state of a gas particle in the surface potential V_s discussed in the main text. The transition probabilities R_{ij} take into account bound-state–bound-state, bound-state–continuum, and continuum–bound-state transitions.

Calculations have been carried out, assuming that: (1) for a mobile adsorbate, the surface potential can be adequately represented by a Morse potential; and (2) the quantities R_{ij} can be calculated in second-order perturbation theory, which in the case of phonon-mediated processes is equivalent to the use of Fermi's Golden Rule; of isothermal desorption times for a variety of systems, including Xe–W.[226] Also, for gas–solid systems with many surface bound states, kinetic equations of the Fokker–Planck and Kramers type, discussed at some length in Chapter 4, have been derived and studied extensively.[412]

Sommer and Kreuzer obtained the rate equations (A6.6.1), by analyzing the displacement vector $\mathbf{u}(t)$ in equation (5.5.8)

$$\mathbf{u}(t) = (\hbar/2\rho)^{1/2} \sum_J \omega_J^{-1/2}[\mathbf{u}^{(J)} b_J \exp(-i\omega_J t) + \mathbf{u}_J^{(J)*} b_J^\dagger \exp(i\omega_J t)] \quad \text{(A6.6.2)}$$

where $b_J^\dagger$ and b_J are respectively phonon creation and annihilation operators of the phonon mode-J, energy ω_J, while ρ is the mass density of the solid. The index J refers either to the proper surface modes of a semi-infinite elastic solid[413] or, in an approximate way, to bulk modes.

By expanding gas-particle field operators

$$\psi(\mathbf{r}, t) = \sum_i \psi_i(\mathbf{r})\alpha_i(t) \quad \text{(A6.6.3)}$$

in terms of the Hartree–Fock single-particle states, one introduces creation and annihilation operators for particles in these states. The Hamiltonian (5.5.7) then assumes the form

$$H = \sum_i E_i \alpha_i^\dagger \alpha_i + \sum_J \hbar\omega_J b_J^\dagger b_J + L^{-3} \sum_{i,j,J} \omega_J^{-1/2} \mathbf{Y}_{ij} \alpha_i^\dagger \times [b_J^\dagger \exp(i\omega_J t)\mathbf{u}^{(J)} + b_J \exp(-i\omega_J t)\mathbf{u}^{(J)}]\alpha_j + H_{res} \quad \text{(A6.6.4)}$$

where

$$\mathbf{Y}_{ij} = L^3(\hbar/2\rho)^{1/2} \int d\mathbf{r}\psi_i^*(\mathbf{r})\nabla V_s(z)\psi_j(\mathbf{r}) \quad \text{(A6.6.5)}$$

and H_{res} contains all other interactions of H that are not diagonal in the Hartree–Fock basis, in particular, all terms quartic in $\alpha_i^\dagger$ and α_j, and all terms with higher powers of the phonon operators. The above is widely accepted practice for phonon-mediated physisorption kinetics at very low coverage, nevertheless some justification is needed when the adsorbate reaches monolayer coverage. First, the surface loading may change the phonon modes; this has been observed in gas–solid systems with chemisorption of heavy adsorbates, as in the work of Ibach and Bruchmann.[414] Second, at monolayer coverage, desorbing particles might draw the necessary energy not directly from the solid but via collective excitations in the adsorbate itself.[235]

Following Sommer and Kreuzer,[235] the rate equations (A6.6.1) are derived by solving the Heisenberg equations of motion for the particle and phonon operators in second-order perturbation theory and calculating

$$n_i(t) = \mathrm{Tr}[\alpha_i^\dagger(t)\alpha_i(t)\hat{\rho}] \quad \text{(A6.6.6)}$$

where ρ is the statistical operator of a grand canonical ensemble based on the static part of the Hamiltonian (A6.6.4), i.e., the Hartree–Fock and phonon part. Taking the long-time limit (Fermi's Golden Rule) one finds by following, for example, Kreuzer and Summerside[415] that

$$\frac{dn_i}{dt} = \sum_j R_{ij} n_j (1 \pm n_i) - \sum_j R_{ji} n_i (1 \pm n_j) \tag{A6.6.7}$$

where for $E_i < E_j$

$$R_{ij} = \frac{2\pi}{\hbar} \sum_J \omega_J^{-1} |\mathbf{Y}_{ij} \mathbf{u}_J|^2 [n(\omega_J) + 1] \delta(E_i - E_j + \hbar\omega_J) \tag{A6.6.8}$$

with

$$n(\omega_J) = 1/[\exp(\hbar\omega_J / k_B T) - 1] \tag{A6.6.9}$$

the thermal occupation function for phonons. For $E_j < E_i$ one has similarly

$$R_{ij} = \frac{2\pi}{\hbar} \sum_J \omega_J^{-1} | Y_{ij} \mathbf{u}_J |^2 n(\omega_J) \delta(E_j - E_i + \hbar\omega_J) \tag{A6.6.10}$$

The plus sign in equation (A6.6.7) holds for bosons and the minus sign for fermions.

Kreuzer and Summerside[415] have examined a model in which fermions can be trapped into a single bound state in a surface potential. In this case equation (A6.6.7) for the occupation function of this one bound state takes the form

$$\frac{dn_0}{dt} = \sum_j R_{0j} n_j (1 - n_0) - \sum_j R_{j0} (1 - n_j) n_0 \tag{A6.6.11}$$

an equation on which Langmuir-type theories are based, such as $d\theta/dt = r_a - r_d$ in the main text.

In an isothermal-desorption experiment, a gas–solid system is prepared in equilibrium at pressure p and temperature T. At time $t = 0$, the gas pressure is reduced substantially and the time evolution of the adsorbate is measured and described by a simple rate equation

$$\frac{d\theta}{dt} = -\frac{\theta}{t_d(\theta)} \tag{A6.6.12}$$

where $t_d(\theta)$ is the desorption time, which is in general coverage-dependent. Sommer and Kreuzer[235] calculated this quantity as follows:

(1) For a gas–solid system in equilibrium at pressure p and tempera-

ture T, namely with occupation functions

$$n_i = \{\exp[\beta(E_i - \mu)] + 1\}^{-1}; \qquad \beta = (k_B T)^{-1} \qquad \text{(A6.6.13)}$$

for a fermionic gas particle (say ^{3}He) where μ is the chemical potential of an ideal (classical) gas in front of the solid, the Hartree–Fock equations are solved to yield single-particle wave functions $\psi_i(\mathbf{r}) = \phi_i(z)\exp(i\mathbf{q}\cdot\boldsymbol{\rho})$ and energies $E_i = \varepsilon_i + \hbar^2 q^2/2m$ at a coverage $\theta = \Sigma_i n_i/n_i(\max)$, where $n_i(\max)$ is the maximum occupation of the ith state corresponding to monolayer density. To ensure saturation, in their treatment Sommer and Kreuzer[235] introduced a cutoff q_c for the summation over the two-dimensional lateral momentum q in the surface plane.

(2) The transition probabilities R_{ij} are calculated with the aid of equation (A6.6.8).

(3) The removal of the gas phase in an isothermal-desorption experiment is taken into account by omitting continuum–bound-state transitions from the rate equations (A6.6.7), which are integrated for a small increment Δt with the right-hand side determined by the initial conditions.

(4) With the new occupation functions $n_i(\Delta t)$ (all continuum states are empty) corresponding to a reduced coverage $\theta(\Delta t)$, the Hartree–Fock equations are employed [step (1) above] and $\psi_i(\mathbf{r})$ and E_i are recalculated self-consistently, after which one can return to step (2). In this way, the time evolution $\theta(t)$ can be generated and hence one can extract the time scale $t_d(\theta)$ of desorption. The implicit assumption in the above procedure of Sommer and Kreuzer is that the internal readjustment of the adsorbate during the desorption process is much faster than the desorption process itself. It is noteworthy that in addition to the explicit nonlinearity of the rate equations (A6.6.7), there is a much stronger implicit nonlinearity through the dependence of the initial and final states in R_{ij} on the quantity n_i in equation (A6.6.8).

The above calculations can be simplified considerably if one recognizes that in some cases (such as He on graphite) the systems remain in quasi-equilibrium during the desorption process in a temperature regime where the desorption time is much larger than the time characterizing bound-state–bound-state transitions, and that this justifies the use of perturbation theory with the rate equations to calculate the desorption time in the form[412]

$$t_d^{-1} = (\pi/m_s N_s) \sum_i \sum_p \sum_c \omega_p^{-1} \left| \int d\mathbf{r}\, \psi_i^*(\mathbf{r}) \frac{\partial V_s(z)}{\partial z} \psi_c(\mathbf{r}) \right|^2$$

$$\times\, \delta(E_i - E_c + \hbar\omega_p) n_i n_p(\text{Ph}) \Big/ \sum_j n_j \qquad \text{(A6.6.14)}$$

Here **c** is the momentum of a gas particle in the continuum. Sommer and Kreuzer use for a graphical presentation of their result the Frenkel–Arrhenius parametrization

$$t_d = t_d(T, \theta) = \nu^{-1}(T, \theta)\exp[Q(T, \theta)/k_B T] \qquad \text{(A6.6.15)}$$

where p is the gas pressure necessary to maintain a transient coverage at temperature T, and $Q(T, \theta)$ is the isosteric heat of adsorption and ν the prefactor.

In their calculations of ^{3}He on graphite, Sommer and Kreuzer[235] find that the prefactor ν changes in the same direction as the heat of adsorption. In particular,

$$\ln \nu = bQ + c \qquad \text{(A6.6.16)}$$

which is the compensation effect described in Section 6.3 for thermally activated processes and, as detailed there, has been observed in many chemisorbed systems and families of similar catalysts. Sommer and Kreuzer argue that the compensation effect they find has a microscopic explanation in the mean-field binding of an adsorbed particle, particularly around monolayer coverage, to the solid *per se* and the other adsorbed particles, and in the weakening of the effective adsorbate–phonon coupling.

Notes Added in Proof

CHAPTER 1

T. B. Grimley (*Surf. Sci.* **164**, in press) has discussed the theory of the sticking probability, while the sticking of molecular hydrogen on a cold Cu(100) surface has been studied by S. Andersson, L. Wilzen, and J. Harris [*Phys. Rev. Lett.* **55**, 2591 (1985)]. These authors conclude in this specific case that phonon processes dominate the sticking and that sticking does not occur in several steps as is often assumed, but as a result of the initial collision with the lattice.

The book by D. M. Ruthven, *Principles of Adsorption and Absorption Processes*, Wiley–Interscience, New York (1984), discusses the statistical derivation of the Langmuir isotherm. The Brunauer *et al.* isotherm is also derived via the grand partition function, while elsewhere in this book the Dubinin–Polanyi theory is treated. A further valuable source is J. Oscik's *Adsorption*, Wiley–Interscience, New York (1983).

In connection with Section 1.4.2, the peak width at half maximum as well as the peak temperature have been employed by L. D. Schmidt [*Catal. Rev. Sci. Eng.* **9**, 115 (1974)] and by D. Edwards [*Surf. Sci.* **54**, 1 (1976)] to determine the desorption energy. G. Ehrlich [*J. Appl. Phys.* **32**, 4 (1961)] proposed a method of measuring the activation energy as a function of coverage, while currently used methods are those of Taylor and Weinberg[191] and a modification of this method by J. L. Falconer and R. J. Madix [*J. Catal.* **48**, 262 (1977); *Surf. Sci.* **48**, 393 (1975)].

With regard to reaction mechanisms, L.-G. Petersson, H. M. Dannetum, and I. Lundstrom [*Surf. Sci.* **163**, 250 (1985)] discuss water production on Pd in hydrogen–oxygen atmospheres. They are able to describe their findings by a simple Langmuir–Hinshelwood model, assuming that hydrogen has a large lateral mobility and that both hydrogen and oxygen adsorption and dissociation is blocked by adsorbed oxygen. They conclude that both O and H have to be adsorbed and dissociated before the reaction takes place; i.e., one has the Langmuir–Hinshelwood mechanism described in Section 1.4.1.

As a second example, C. T. Campbell *et al.* [*J. Chem. Phys.* **73**, 5862 (1980)] present a molecular-beam study of the catalytic oxidation of CO on a Pt(111) surface. Under all conditions studied, this reaction also follows a Langmuir–Hinshelwood mechanism in which the CO molecule is first strongly adsorbed before reaction takes place. This contrasts with the Rideal–Eley mechanism, which was favored in some of the earlier work; by this mechanism a CO molecule reacts with a chemisorbed oxygen atom either directly upon impact from the gas phase or after momentary trapping in a weakly held, precursor state. These mechanisms are therefore distinguished by the strength of the CO–metal interaction, and hence by the mean residence time of the CO molecule on the surface.

P. Hofman, S. R. Bare, and D. A. King [*Phys. Scr.* **T4**, 118 (1983)] discuss the characterization of a low-temperature surface phase of CO on Pt(110), while P. S. Bagus and W. Muller [*Chem. Phys. Lett.* **115**, 540 (1985)] treat the C–O vibration of chemisorbed DO. Tilted CO on metal surfaces is examined by C. W. Bauschlicher [*Chem. Phys. Lett.* **115**, 535 (1985)] and compared with isoelectronic N_2.

CHAPTER 2

An unusually low stretching frequency for CO adsorbed on Fe(100) has been observed by C. Benndorf, B. Kruger, and F. Thieme [*Surf. Sci.* **163**, L675 (1985)]. This is significant for understanding the mechanism of the Fischer–Tropsch synthesis of hydrocarbons from CO + H_2 on Fe surfaces. In this synthesis, one of the fundamental steps is the dissociation of CO. The results of these workers suggest that the "lying down" binding configuration of CO is the reason for the unusual weakening of the CO bond, consistent with the theory given in Chapter 2. This "lying down" CO on Fe(100) seems likely to be a precursor state for the CO dissociation on iron surfaces; see also A. B. Anderson and D. P. Onwood, *Surf. Sci.* **154**, L261 (1985). Other work on CO is that by D. W. Moon, D. J. Dwyer, and S. L. Bernasek [*Surf. Sci.* **163**, 215 (1985)] on clean and sulfur-modified Fe(100) surfaces. An overview of the electronic and vibration properties of adsorbed CO is presented by S. Ishi, Y. Ohno, and B. Viswanathan [*Surf. Sci.* **161**, 349 (1985)]. Electronic excitations of adsorbed CO are studied by F. P. Netzer, J. U. Mack, E. Bertel, and J. A. D. Matthew [*Surf. Sci.* **160**, L509 (1985)].

O_2 adsorption on Mo(110) has been investigated by Souksassian *et al.* [*J. Phys. C* **18**, 4785 (1985)] while NO chemisorption on Re(001) has been further examined by S. Tatarenko, M. Alnot, and R. Ducros [*Surf. Sci.* **163**,

249 (1985)]. At 100 K, NO is molecularly adsorbed on this metal surface. If adsorption takes place at room temperature, NO is mainly dissociated. Chemisorption of NO and N_2 on Mo is treated by H. Miki *et al.* [*Surf. Sci.* **161**, 446 (1985)], while oxygen chemisorption and corrosion on Cr(100) and Cr(110) single-crystal faces is investigated by J. S. Foord and R. M. Lambert [*Surf. Sci.* **161**, 513 (1985)].

Indirect electronic interactions [see T. L. Einstein, *CRC Crit. Rev. Solid State Mater. Sci.* **7**, 261 (1978)] between H atoms adsorbed on metals are studied by P. Nordlander and S. Holmstrom [*Surf. Sci.* **159**, 443 (1985)] using an extension of effective-medium theory. The adsorption and absorption of hydrogen by metals is reviewed by R. Burch in *Chemical Physics of Solids and their Surfaces*, Volume 8, p. 1, Specialist Periodical Reports, R. Soc. Chem., Burlington House, UK.

CHAPTER 3

An XPS study of water adsorption on Cu(110) is reported by A. Spitzer and H. Lüth [*Surf. Sci.* **160**, 353 (1985)], while evidence for oriented water dimers when water is adsorbed on Ni(110) is presented by C. Nöbl, C. Benndorf, and T. E. Madey [*Surf. Sci.* **157**, 29 (1985)]. Their results suggest that the configuration of H_2O molecules on metal surfaces can be derived from that of H_2O in ice. For ammonia adsorption on an Ag(311) surface, S. T. Cayer and J. T. Yates [*Surf. Sci.* **155**, 584 (1985)] conclude that the C_{3v} axes of the more strongly bound ammonia molecules on the metal surface are perpendicular to the surface and that the NH_3 molecule is bound through the N atom.

D. B. Kang and A. B. Anderson [*Surf. Sci.* **155**, 639 (1985)] have studied theoretically the adsorption and structural rearrangement of acetylene and ethylene on Pt(111). They conclude that olefinic and acetylenic C–H bonds are readily activated by Pt(111). They note how the Pt orbitals participating in hydrocarbon adsorption bonding and CH activation have strong *s–d* hybridization. Neutron-spectroscopic studies of adsorption and decomposition of C_2H_2 and C_2H_4 on Raney Ni have been carried out by R. D. Kelley *et al.* [*Surf. Sci.* **155**, 480 (1985)], while related UPS experiments on Co are reported by P. Tiscione and G. Roirda [*Surf. Sci.* **154**, L255 (1985)]. In the search for reaction pathways that may result in ethylene oxidation, E. M. Stuve and R. J. Madix have studied C_2H_4 on Pd(100) [*Surf. Sci.* **160**, 293 (1985)].

The molecular structure of benzene on transition metals is surveyed by M. Neumann *et al.* [*Surf. Sci.* **155**, 629 (1985)] with particular reference to the Rh(111) surface, while the structure and orientation of benzene and pyridine on Ir(111) has been studied by J. U. Mack, E. Bertel, and F. P.

Netzer [*Surf. Sci.* **159**, 265 (1985)], evidence being presented of molecular distortion of benzene. Pyridine adsorbed on Ag(111) has also been investigated by A. Otto *et al.* [*Surf. Sci.* **163**, 140 (1985)].

Because of its significance for questions in catalysis, D. H. Ehlers, A. Spitzer, and H. Lüth [*Surf. Sci.* **160**, 57 (1985)] have made a study of methanol on Pt(111) using infrared reflection and UV photoemission; see also a paper by B. A. Sexton, A. E. Hughes, and N. R. Avery [*Surf. Sci.* **155**, 366 (1985)].

P. Berlowitz *et al.* [*Surf. Sci.* **159**, 540 (1985)] stress the continuing interest in the properties of the surface analogs of organometallic compounds. Among the various systems, hydrocarbons adsorbed on Pt surfaces have been the most extensively studied [see G. A. Somorjai, *Chemistry in Two Dimensions: Surfaces*, Cornell University Press (1982)], especially ethylene and acetylene. Berlowitz *et al.* point out the interest in comparing the behavior of methylene with that of ethylene and acetylene. To this end, they report a study of the reactions of diazomethane (CH_2N_2).

The structure and bonding of chemisorbed ethylene (and ethylidene) on Pt(111) has been investigated by J. A. Horsley *et al.* [*J. Chem. Phys.* **83**, 3146 (1985)] by near-edge X-ray-absorption fine-structure spectroscopy and multiple-scattering calculations [see also *J. Chem. Phys.* **83**, 6099 (1985)].

CHAPTER 4

Several studies have been made on surface diffusion. In the work of G. Wahnström [*Surf. Sci.* **159**, 311 (1985)], the motion of a single adsorbed atom on a solid surface is studied by a mode-coupling approximation, which removes some restrictive assumptions underlying the use of the usual Fokker–Planck equation. His conclusion is that when the adatom is identical with the substrate atoms, the Fokker–Planck equation gives quite accurate results provided one employs a position-dependent friction coefficient. The situation is found to be different for a lighter adatom; when its characteristic frequency is comparable to or larger than the corresponding value for the substrate atoms, then the Fokker–Planck equation must be transcended. The position dependence of the friction constant is found to be important. Any treatment based on a constant friction coefficient would spoil the agreement either for the diffusion constant or for the strength of the resonance peak in the frequency spectrum of the velocity correlation function. This position dependence, giving considerably greater friction when the adparticle is at its potential minimum compared with the value at the top of the potential barrier, is well adapted for the assumption behind the absolute rate theory for the escape rate. The agreement found for the diffusion constant is in accord with that.

Wahnström restricts his discussion to the phonon excitations in the substrate. For metals, plasmons and electron–hole pair excitations play a role in surface kinetics. The conduction electrons, of course, bring about dielectric screening of pair-interaction potentials, as discussed in Chapter 2. In addition, they make a position-dependent but frequency-independent contribution to the friction constant (see the discussion of Section 4.10.1). Wahnström exposes the relation of his work to that of B. Hellsing and M. Persson [*Phys. Scr.* **29**, 360 (1984)] which considered only the electronic excitations.

Other contributions to surface diffusion have been made by D. Ghaleb [*Surf. Sci.* **137**, L103 (1984)] through a molecular-dynamics study of the diffusion of adatom dimers in the (111) surface of face-centered cubic crystals; R. C. Baetzold [*Surf. Sci.* **150**, 193 (1985)] also emphasizes the usefulness of computer simulation of surface diffusion of adsorbates, a matter of intrinsic importance in heterogeneous catalysis. Directional anisotropy of surface self-diffusion on Pt(110) has been discussed by N. Freyer and H. P. Bonzel [*Surf. Sci.* **160**, L501 (1985)]; see also R. Gomer in *Vacuum* (*GB*) **33**, 537 (1983).

A number of quantum-mechanical studies of diffusion of hydrogen on metal surfaces, a subject of wide interest [see, for example, *The Chemical Physics of Solid Surfaces and Heterogeneous Catalysis* (D. A. King and D. P. Woodruff, eds.), Vol. 4, Elsevier, Amsterdam (1982)], have been carried out [M. J. Puska and R. M. Nieminen, *Surf. Sci.* **157**, 413 (1985); S. M. Valone, A. F. Voter, and J. D. Doll, *Surf. Sci.* **155**, 687 (1985); see also R. DiFoggio and R. Gomer, *Phys. Rev. B* **25**, 3490 (1982)]. V. P. Zhdanov [*Surf. Sci.* **161**, L614 (1985)] treats the effect of conduction electrons on the diffusion of H in bulk or on metal surfaces. He uses the friction constant $\eta = m_e E_f / M_p \hbar$, where E_f is the Fermi energy; η is proportional to the ratio of the electron to proton mass [see E. G. D'Agliano *et al.*, *Phys. Rev. B* **11**, 2122 (1975), and Section 4.10.1 of the main text].

K. G. Lloyd and J. C. Hemminger [*J. Chem. Phys.* **82**, 3858 (1985)] present an analysis of adsorbate vibrations and conclude that dispersion of a vibrational mode as a function of $k_{\parallel}$ can be analyzed and should shed light on lateral interactions between adsorbates. It is noteworthy here that Nos. 1–3 of Vol. 158 of *Surf. Sci.* (1985) are devoted to spectroscopic studies of adsorbates on solid surfaces. The theory of Raman optical activity of molecules adsorbed on metal surfaces is treated by S. Efrima [*J. Chem. Phys.* **83**, 1356 (1985)].

In relation to Section 4.9, P. J. Feibelman [*Surf. Sci.* **160**, 139 (1985)] stresses the importance of energy transfer between a molecule and a surface and focuses on the mechanism in which energy is transferred between a molecule and a surface to the degree that charge transfer is delayed as a molecular-affinity level traverses the Fermi level.

CHAPTER 5

Two important general reviews are (1) J. A. Barker and D. J. Auerbach, *Surf. Sci. Rep.* **4** (1984), Nos. 1/2, on Gas–Surface Interactions and Dynamics: Thermal Energy Atomic and Molecular Beam Studies; and (2) H. J. Kreuzer and Z. W. Gortel, *Physisorption Kinetics*, Springer Series on Surface Science, Vol. 1, Springer-Verlag, Berlin (1985).

Absorption and desorption of NO from Rh(111) and Rh(331) surfaces have been examined by L. A. DeLouise and N. Winograd [*Surf. Sci.* **159**, 199 (1985)] who write: "the catalytic oxidation and reduction of NO forms the basis of many technologically important processes ranging from the manufacturing of explosive to the elimination of toxic automotive exhaust" [see also J. C. Summers and K. Baron, *J. Catal.* **57**, 380 (1979); and the study of adsorption and desorption of NO on W(100) using XPS and AES by E. Pelach, R. E. Viturvo, and M. Folman, *Surf. Sci.* **161**, 553 (1985)].

A temperature-programmed desorption study of acetylene on a clean, H-covered and O-covered Pt(111) surface has been performed by C. E. Megiris *et al.* [*Surf. Sci.* **159**, 184 (1985)]. One of their aims was to use the interaction of acetylene with Pt surfaces to explore the nature of the coupling of a triple bond with the surface. They conclude that the reactions of acetylene on a Pt(111) surface with and without adsorbed H and O can be described by three independent pathways: decomposition, hydrogenation, and oxidation.

Decomposition takes place on the bare Pt surface and is via a vinylidene intermediate. Hydrogenation takes place both on the bare Pt and H-covered Pt. Acetylene adsorbed on a bare Pt atom is hydrogenated less efficiently to form ethylidene and then ethylene. Acetylene adsorbed on an H-covered Pt atom is hydrogenated readily to methane, ethane, and ethylene. The oxidation reaction leads to the production of CO, CO_2, and H_2O and it takes place at the perimeter of islands of adsorbed O atoms, which are rather immobile.

B. A. Sexton [*Surf. Sci.* **163**, 99 (1985)] has reported vibrational (EELS) and thermal desorption spectra (TDS) for the heterocyclic 5-membered ring molecules pyrrole (C_4H_4NH), furan (C_4H_4O), and thiophene (C_4H_4S) adsorbed on Cu(100) at 85–300 K. Adsorption is molecular and reversible with no evidence for dissociation. At low exposures, each molecule is weakly π-bonded to the Cu(100) surface with the molecular ring parallel to the surface plane. More complex models are however needed to explain adsorption at higher exposures.

In relation to physisorption, S. Das Sarma and S. M. Paik have calculated higher-order corrections to the interaction between an adsorbed atom and a metal surface in *Chem. Phys. Lett.* **115**, 525 (1985). An electronic mechanism for desorption has been discussed further by B. Hellsing [*J. Chem. Phys.* **83**, 1371 (1985)].

CHAPTER 6

The neutron techniques referred to in Chapter 4 are discussed in relation to catalyst characterization by C. J. Wright [*Catalysis*, Vol. 7, p. 46 (1985): Specialist Periodical Reports, R. Soc. Chem., Burlington House, London]. In this same volume (p. 149) R. Burch emphasizes that catalytic reactions at a metal surface involve a subtle balance of adsorption forces. Too weak an adsorption will result in the catalyst having low activity; too strong an adsorption will lead to the surface becoming poisoned by adsorbed reactants or products.

S. W. Jorgensen and R. J. Madix [*Surf. Sci.* **163**, 19 (1985)] have studied the mechanism of sulfur poisoning on the Pd(100) surface using TPD, HREELS, and LEED for sulfur coverages ranging from zero to saturation. They conclude that adsorbed sulfur atoms at low coverage poison the adsorption of CO by local CO–S interactions, which prevent the formation of the high-density CO "compression" structures. At increased sulfur coverages, the poisoning proceeds by direct site blocking only. CO–CO interactions appear to be involved in propagating the local CO–S interactions. Sulfur is observed to have a Pd–S stretch at 295 cm^{-1}.

An infrared reflection–absorption study of CO chemisorbed on clean and sulfided Ni(111) has been made by M. Trenary, K. J. Uram, and J. T. Yates [*Surf. Sci.* **157**, 512 (1985)]. The motivation for this, they emphasize, is that the catalytic properties of transition metals are profoundly affected by the presence of certain impurity atoms. For instance, the Fischer–Tropsch synthesis (discussed in Chapter 1) of hydrocarbons from CO and H_2 over Ni and Fe catalysts is promoted by alkali metals but inhibited by sulfur. Because of the technological interest of chalcogens on transition metals in relation to corrosion and catalysis, F. Maca *et al.* [*Surf. Sci.* **160**, 467 (1985)] have also studied adsorption of sulfur on Pd(111). The review by B. E. Koel and G. A. Somorjai [*Surface Structural Chemistry in Catalysis*, in: *Science and Technology*, Volume 38 (J. R. Anderson and M. Boudart, eds.), Springer, New York (1983)] is also a valuable source in this general area; see also B. I. Lundqvist, *Vacuum* (*GB*) **33**, 639 (1983).

Further discussion of the compensation effect in catalysis, treated in Section 6.3 and Appendix 6.6, can be found in the work of B. J. McCoy [*J. Chem. Phys.* **80**, 3627 (1984)], who proposed a model based on vibrational relaxation processes [see also V. P. Zhdanov, *Surf. Sci.* **159**, L416 (1985) and the reply by B. J. McCoy, *ibid.*, L419; and V. B. Kazansky, in: *Proc. 6th Int. Cong. on Catalysis*, Volume 1, p. 50 (1976), and A. Galway, *Adv. Catal.* **26**, 247 (1977)].

In relation to Section 6.6, reaction rate oscillations during H_2 oxidation over polycrystalline Pt have been studied by G. Saidi and T. T. Tsotsis [*Surf. Sci.* **161**, L591 (1985)], and the theory of kinetic oscillations in the catalytic CO oxidation on Pt(100) by R. Imbihl *et al.* [*J. Chem. Phys.* **83**, 1578 (1985)].

References

1. D. O. HAYWARD AND B. M. W. TRAPNELL, *Chemisorption*, 2nd ed. (Butterworths, London, 1964).
2. H. MOESTA, *Chemisorption and Ionization in Metal–Metal Systems* (Springer-Verlag, Berlin, 1968).
3. G. WEDLER, *Chemisorption: An Experimental Approach* (Butterworths, London, 1976).
4. L. PAULING, *Proc. R. Soc. London, Ser. A* **126**, 343 (1949); *The Nature of the Chemical Bond* (Cornell University Press, Ithaca, 1960).
5. D. D. ELEY, *Discuss. Faraday Soc.* **8**, 34 (1950); *Electrochem. Ber. Bern sanga Physik. Chem.* **60**, 797 (1956).
6. D. P. STEVENSON, *J. Chem. Phys.* **23**, 203 (1955).
7. J. H. DE BOER, *The Dynamic Characteristics of Adsorption* (Clarendon Press, Oxford, 1968).
8. A. R. MILLER, *Adsorption of Gases on Solids* (University Press, Cambridge, 1949), p. 57.
9. I. J. LANGMUIR, *J. Am. Chem. Soc.* **40**, 1361 (1918).
10. S. BRUNAUER, P. H. EMMETT, AND E. TELLER, *J. Am. Chem. Soc.* **60**,.309 (1938).
11. T. L. HILL, *J. Chem. Phys.* **14**, 263 (1946); R. H. Fowler, *Proc. Camb. Phil. Soc.* **31**, 260 (1935).
12. P. A. REDHEAD, *Vacuum* **12**, 203 (1962).
13. N. D. SPENCER AND G. A. SOMORJAI, *Rep. Prog. Phys.* **46**, 1 (1983).
14. P. SABATIER AND J. B. SENDERENS, *C.R. Acad. Sci. Paris* **134**, 514 (1902).
15. F. FISCHER AND H. TROPSCH, *Brennst.-Chem.* **7**, 97 (1926).
16. N. V. RICHARDSON AND A. M. BRADSHAW, *Surf. Sci.* **88**, 255 (1979).
17. M. ARAKI AND V. PONEC, *J. Catal.* **44**, 439 (1976).
18. H. CONRAD, G. ERTL, AND E. E. LATTA, *Surf. Sci.* **41**, 435 (1974).
19. F. BOZSO, G. ERTL, M. GRUNZE, AND M. WEISS, *J. Catal.* **49**, 18 (1977).
20. K. HORN, A. M. BRADSHAW, AND K. JACOBI, *Surf. Sci.* **72**, 719 (1978).
21. H. KOLBEL AND D. HANUS, *Chem.-Ing.-Tech.* **46**, 1042 (1974).
22. G. A. MILLS AND F. W. STEFFGEN, *Catal. Rev.* **8**, 159 (1973).
23. M. A. VANNICE, *Catal. Rev. Sci. Eng.* **14**, 153 (1976).
24. M. L. POUTSMA, L. F. ELEK, P. IBARBIA, H. RISCH, AND J. A. RABO, *J. Catal.* **52**, 157 (1978).
25. R. B. ANDERSON, *Adv. Catal.* **5**, 355 (1953).
26. B. A. SEXTON AND G. A. SOMORJAI, *J. Catal.* **46**, 167 (1977).
27. G. A. SOMORJAI, *Catal. Rev. Sci. Eng.* **18**, 173 (1978).
28. M. J. VERAA AND A. T. BELL, *Fuel* **57**, 194 (1975).

29. S. L. Meisel, J. P. McCullough, C. H. Lechshaler, and P. B. Weisz, *Chemtech.* **6**, 86 (1976).
30. D. J. Dwyer and G. A. Somorjai, *J. Catal.* **52**, 291 (1978).
31. P. Biloen, J. N. Helle, and W. M. H. Sachtler, *J. Catal.* **58**, 95 (1979).
32. J. G. McCarty and H. Wise, *J. Catal.* **57**, 406 (1970).
33. J. A. Rabo, A. P. Risch, and M. L. Poutsma, *J. Catal.* **53**, 295 (1978).
34. H. Pichler, *Adv. Catal.* **4**, 271 (1952).
35. J. T. Kummer and P. H. Emmett, *J. Am. Chem. Soc.* **75**, 5177 (1953).
36. W. K. Hall, R. J. Kokes, and P. H. Emmett, *J. Am. Chem. Soc.* **82**, 1027 (1960).
37. A. T. Bell, *Catal. Rev. Sci. Eng.* **23**, 203 (1981).
38. J. Mahanty and N. H. March, *J. Phys. C* **9**, 2905 (1976); *ANU Canberra Rpt.* (1976).
39. H. Margenau and N. R. Kestner, *Theory of Intermolecular Forces*, 2nd ed. (Pergamon Press, Oxford, 1971), Chap. 2.
40. J. Mahanty and B. W. Ninham, *J. Phys. A* **6**, 1140 (1973).
41. J. Mahanty and B. W. Ninham, *J. Chem. Phys.* **59**, 6157 (1973).
42. J. Mahanty and B. W. Ninham, *Dispersion Forces* (Academic Press, New York, 1976).
43. F. Flores, N. H. March, and I. D. Moore, *Surf. Sci.* **69**, 133 (1977).
44. H. Eyring, J. Walter, and G. E. Kimball, *Quantum Chemistry* (Wiley, New York, 1944).
45. J. R. Smith, S. C. Ying, and W. Kohn, *Phys. Rev. Lett.* **30**, 610 (1973).
46. K. H. Lau and W. Kohn, *Surf. Sci.* **75**, 69 (1978).
47. T. B. Grimley, *Proc. Phys. Soc.* **90**, 751 (1967); *ibid.* **92**, 776 (1967).
48. T. L. Einstein and J. R. Schrieffer, *Phys. Rev. B* **7**, 3629 (1973).
49. T. L. Einstein, *Surf. Sci.* **75**, L161 (1978).
50. L. Pauling, *The Nature of the Chemical Bond* (Cornell University Press, Ithaca, 1960).
51. F. Flores, I. Gabbay, and N. H. March, *Chem. Phys.* **63**, 391 (1981).
52. S. L. Altmann, C. A. Coulson, and W. Hume-Rothery, *Proc. R. Soc. London, Ser. A* **240**, 145 (1957).
53. V. L. Moruzzi, J. F. Janak, and A. R. Williams, *Calculated Electronic Properties of Metals* (Pergamon, Oxford, 1978).
54. W. Hume-Rothery, H. M. Irving, and R. J. P. Williams, *Proc. R. Soc. London, Ser. A* **208**, 431 (1951).
55. K. A. Gschneider Jr., *Solid State Phys.* **16**, 275 (1964).
56. G. Brodén, T. N. Rhodin, C. Brucker, R. Benbow, and Z. Hurych, *Surf. Sci.* **59**, 593 (1976).
57. W. Gordy and W. J. O. Thomas, *J. Chem. Phys.* **24**, 439 (1956).
58. T. A. Delchar and G. Erlich, *J. Chem. Phys.* **42**, 2686 (1965).
59. M. Grunze, *Surf. Sci.* **81**, 603 (1979).
60. P. Legare, L. Hilaire, N. Sotto, and G. Maire, *Surf. Sci.* **91**, 175 (1980).
61. M. A. Chesters and G. A. Somorjai, *Surf. Sci.* **52**, 21 (1975).
62. T. N. Rhodin and G. Ertl, *The Nature of the Surface Chemical Bond* (North-Holland, Amsterdam, 1979).
63. H. A. C. M. Hendrickx, A. Hoek, and B. E. Nieuwenhuys, *Surf. Sci.* **135**, 81 (1983).
64. H. Albers, W. J. J. van der Wal, O. L. J. Gijzeman, and G. A. Bootsma, *Surf. Sci.* **77**, 1 (1979).
65. M. Wilf and P. T. Dawson, *Surf. Sci.* **60**, 561 (1976).
66. K. Horn, J. Dinardo, W. Eberhardt, H. J. Freund, and E. W. Plummer, *Surf. Sci.* **118**, 465 (1982).
67. B. E. Nieuwenhuys, *Surf. Sci.* **105**, 505 (1981).
68. R. A. Shigeishi and D. A. King, *Surf. Sci.* **62**, 379 (1977).

69. C. W. Bauschlicher and P. S. Bagus, *J. Chem. Phys.* **80**, 944 (1984), and references there.
70. T. B. Grimley and E. E. Mola, *J. Phys. C* **9**, 3437 (1976).
71. T. B. Grimley and C. Pisani, *J. Phys. C* **7**, 2831 (1974).
72. N. H. Tolk, J. C. Tully, W. Heiland, and C. W. White, eds., *Inelastic Ion-Surface Collisions* (Academic Press, New York, 1977).
73. A. Otto, in *Light Scattering in Solids*, eds. M. Cardona and G. Guntherodt (Springer, New York, 1983), Vol. IV.
74. Ph. Avouris, Ph., N. J. DiNardo and J. E. Demuth, *J. Chem. Phys.* **80**, 491 (1984).
75. E. Umbach and Z. Hussain, *Phys. Rev. Lett.* **52**, 457 (1984).
76. H. Siegbahn, L. Asplund, and P. Kelfve, *Chem. Phys. Lett.* **35**, 330 (1975).
77. M. Grunze, R. K. Driscoll, G. N. Burland, J. C. L. Cornish, and J. Pritchard, *Surf. Sci.* **89**, 381 (1979).
78. M. Barber, J. C. Vickerman, and J. Wostenholme, *Surf. Sci.* **68**, 130 (1977).
79. S. Andersson, *Solid State Commun.* **21**, 75 (1977).
80. J. C. Bertolini and B. Tardi, *Surf. Sci.* **102**, 131 (1981).
81. J. B. Benziger, *Appl. Surf. Sci.* **6**, 105 (1980).
82. T. E. Madey and J. T. Yates, *Chem. Phys. Lett.* **51**, 77 (1977).
83. J. E. Demuth, *Surf. Sci.* **76**, L603 (1978).
84. A. D. Walsh, *J. Chem. Soc.* 2260 (1953).
85. N. H. March, *J. Chem. Phys.* **74**, 2973 (1981).
86. N. V. Sidgwick and H. M. Powell, *Proc. R. Soc. London, Ser. A* **176**, 153 (1940).
87. R. J. Gillespie and R. S. Nyholm, *Q. Rev. Chem. Soc.* **11**, 339, (1957).
88. F. Flores, I. Gabbay, and N. H. March, *Surf. Sci.* **107**, 127 (1981).
89. F. Flores, I. Gabbay, and N. H. March, *Phys. Lett. A* **85**, 433 (1981).
90. W. A. Goddard and L. B. Harding, *Ann. Rev. Phys. Chem.* **29**, 363 (1978).
91. J. H. Van Vleck and P. C. Cross, *J. Chem. Phys.* **1**, 357 (1933).
92. A. B. F. Duncan and J. A. Pople, *Trans. Faraday Soc.* **49**, 217 (1953).
93. I. A. Howard and G. Dresselhaus, *Surf. Sci.* **136**, 229 (1984).
94. C. W. Seabury, T. N. Rhodin, R. J. Purtell, and R. P. Merrill, *Surf. Sci.* **93**, 117 (1980).
95. R. J. Purtell, R. P. Merrill, C. W. Seabury, and T. N. Rhodin, *Phys. Rev. Lett.* **44**, 1279 (1980).
96. B. A. Sexton and G. E. Mitchell, *Surf. Sci.* **99**, 523 (1980).
97. N. A. Surplice and W. Brearly, *Surf. Sci.* **52**, 62 (1975).
98. M. Grunze, F. Bozso, G. Ertl, and M. Weiss, *Appl. Surf. Sci.* **1**, 241 (1978).
99. Y. Yoshida and G. A. Somorjai, *Surf. Sci.* **75**, 46 (1978).
100. M. Abon, G. Bergeret, and B. Tardy, *Surf. Sci.* **68**, 305 (1977).
101. M. Wilf and M. Folman, *J. Chem. Soc., Farday Trans. 1* **72**, 1165 (1976).
102. P. J. Estrup and J. Anderson, *J. Chem. Phys.* **49**, 523 (1968).
103. Y. Fukuda, F. Honda, and J. W. Rabalais, *Surf. Sci.* **99**, 289 (1980).
104. J. E. Demuth, *Surf. Sci.* **76**, L603 (1978).
105. T. E. Felter and W. H. Weinberg, *Surf. Sci.* **103**, 265 (1981).
106. P. J. Hiett, F. Flores, P. J. Grout, N. H. March, A. Martin Rodero, and G. Senatore, *Surf. Sci.* **140**, 400 (1984).
107. R. P. Messmer, in *The Physical Basis for Heterogeneous Catalysis*, eds. E. Drauglis and R. I. Jaffee (Plenum Press, New York, 1975), p. 261.
108. F. A. Cotton and G. Wilkinson, *Advanced Inorganic Chemistry*, 4th ed. (Wiley, New York, 1980), p. 81.
109. H. Ibach, H. Hopster, and B. Sexton, *Appl. Surf. Sci.* **1**, 1 (1977).

110. J. E. Demuth, *Surf. Sci.* **84**, 315 (1979).
111. R. J. Koestner, J. C. Frost, P. C. Stair, M. A. Van Hove, and G. A. Somorjai, *Surf. Sci.* **116**, 85 (1982).
112. H. Froitzheim, in *Electron Spectroscopy for Surface Analysis*, ed. H. Ibach (Springer-Verlag, Berlin, 1977), p. 205.
113. C. R. Brundle, in *Electronic Structure and Reactivity of Metal Surfaces*, eds. E. G. Derouane and A. A. Lucas, NATO Advanced Study Series (Plenum Press, New York, 1976), p. 389.
114. J. E. Demuth, *IBM J. Res. Dev.* **22**, 265 (1978).
115. A. D. Baker, C. Baker, C. R. Brundle, and D. W. Turner, *Int. J. Mass Spectrom. Ion Phys.* **1**, 285 (1968).
116. D. G. Streets and A. W. Potts, *J. Chem. Soc., Faraday Soc. II* **70**, 1505 (1974).
117. D. M. Newns, *Phys. Rev.* **178**, 1123 (1969).
118. P. W. Anderson, *Phys. Rev.* **124**, 41 (1961).
119. S. Cerny and V. Ponec, *Catal. Rev.* **2**, 249 (1969).
120. C. Mariani and K. Horn, *Surf. Sci.* **126**, 279 (1983).
121. F. P. Netzer and T. E. Madey, *Surf. Sci.* **127**, L102 (1983); *ibid.* **117**, 549 (1982).
122. D. Dahlgren and J. C. Hemminger, *Surf. Sci.* **134**, 836 (1983).
123. A. Gavezzotti and M. Simonetta, *Surf. Sci.* **134**, 601 (1983).
124. D. E. Williams, *J. Chem. Phys.* **45**, 3770 (1966).
125. Ph. Avouris and J. E. Demuth, *J. Chem. Phys.* **75**, 4783 (1981).
126. Ph. Avouris, N. J. DiNardo, and J. E. Demuth, *J. Chem. Phys.* **80**, 491 (1984).
127. P. G. Hall and C. J. Wright, *Chemical Physics of Solids and their Surfaces* (Royal Society of Chemistry, London, 1978), p. 89.
128. J. P. McTague, M. Nielsen, and L. Passell, *Crit. Rev. Solid State Mater. Sci.* **8**, 135 (1978).
129. D. L. King, *J. Catal.* **61**, 77 (1980).
130. M. Chesters, *Rep. Prog. Phys.*, to appear.
131. W. Marshall and S. W. Lovesey, *Theory of Thermal Neutron Scattering* (Clarendon Press, Oxford, 1971).
132. H. Stiller, *NATO Adv. Study Inst. Ser. B* **56**, 415 (1980).
133. C. G. Windsor and C. J. Wright, *New Sci.* **88**, 714 (1980).
134. C. J. Wright and C. M. Sayers, *Rep. Prog. Phys.* **46**, 773 (1983).
135. C. J. Wright, *J. Chem. Soc., Faraday Trans. 2* **73**, 1497 (1977).
136. H. Kono, A. R. Ziv, and S. H. Lin, *Surf. Sci.* **134**, 614 (1983).
137. R. F. Wallis, *Prog. Surf. Sci.* **4**, 233 (1973).
138. T. B. Grimley, *Proc. Phys. Soc.* **79**, 1203 (1962).
139. J. Mahanty, D. D. Richardson, and N. H. March, *J. Phys. C* **9**, 3421 (1976).
140. R. Stockmeyer, H. Stortnik, I. Nathaniec, and J. Mayer, *Ber. Bunsenges. Phys. Chem.* **84**, 79 (1980).
141. F. O. Goodman, *Surf. Sci.* **116**, 573 (1982).
142. J. E. Black, *Surf. Sci.* **116**, 240 (1982).
143. V. P. Zhdanov and K. I. Zamarev, *Catal. Rev. Sci. Eng.* **24**, 373 (1982).
144. C. J. Wright, C. F. Sampson, D. Fraser, R. B. Moyes, P. B. Wells, and C. Riekel, *J. Chem. Soc., Faraday Trans. 1* **76**, 1585 (1980).
145. C. M. Sayers, *J. Phys. C* **14**, 4969 (1981).
146. J. E. Black, *Surf. Sci.* **105**, 59 (1981).
147. C. M. Sayers, *J. Phys. C* **16**, 2381 (1983); *ibid.* **14**, 4969 (1981).
148. R. Stockmeyer, H. M. Conrad, A. Renouprez, and P. Foullioux, *Surf. Sci.* **49**, 549 (1975).

149. C. J. Wright et al., private communication; see Ref. 139.
150. R. Dutton and B. N. Brockhouse, *Can. J. Phys.* **50**, 2915 (1972).
151. C. C. Matthai, P. J. Grout, and N. H. March, *Int. J. Quantum Chem., Symp.* **12**, 443 (1978); *J. Phys. Chem. Solids* **42**, 317 (1981).
152. A. A. Maradudin, E. W. Montroll, G. H. Weiss, and F. P. Inatova, *Theory of Lattice Dynamics in the Harmonic Approximation* (Academic Press, New York, 1971), p. 381.
153. M. Wagner, *Phys. Rev.* **161**, 2520 (1963); *Phys. Rev. A* **133**, 750 (1964).
154. M. Sachdev and J. Mahanty, *J. Phys. C* **3**, 1225 (1970).
155. E. G. Thomas, *J. Phys. C* **9**, 2857 (1976).
156. J. W. Gadzuk, *J. Chem. Phys.* **79**, 6341 (1983).
157. J. L. Lebowitz and E. Rubin, *Phys. Rev.* **131**, 2881 (1963).
158. A. T. Bullock, G. G. Cameron, and P. M. Smith, *J. Chem. Soc., Faraday Trans.* **70**, 1202 (1974).
159. B. Gavish and M. M. Werber, *Biochemistry* **18**, 1269 (1979).
160. D. Beece, L. Eisenstein, and H. Frauenfelder, *Biochemistry* (to appear).
161. L. E. Reichl, *J. Chem. Phys.* **77**, 4199 (1982).
162. H. A. Kramers, *Physica* **7**, 284 (1940).
163. S. Chandrasekhar, *Rev. Mod. Phys.* **15**, 1 (1943).
164. E. G. d'Agliano, W. L. Schaich, D. Kumar, and H. Suhl, *Nobel Symp.* **24**, 200 (1973).
165. W. L. Schaich, *Solid State Commun.* **15**, 357 (1974); *J. Chem. Phys.* **60**, 1087 (1974).
166. W. Brenig, *Z. Phys. B* **48**, 127 (1982).
167. R. Brako and D. M. Newns, *Solid State Commun.* **33**, 713 (1980).
168. G. P. Brivio and T. B. Grimley, *Surf. Sci.* **89**, 226 (1979).
169. J. S. Rousseau, J. C. Stoddart, and N. H. March, *J. Phys. C* **5**, L175 (1972).
170. N. H. March, in *Linear and Nonlinear Transport in Solids*, eds. J. T. Devreese and V. E. van Doren (Plenum Press, New York, 1975), p. 131; W. Huberman and G. V. Chester, *Adv. Phys.* **24**, 489 (1975).
171. J. S. McCaskill and N. H. March, *J. Phys. Chem. Solids* **45**, 215 (1984).
172. J. S. McCaskill and N. H. March, *Surf. Sci.* **131**, 34 (1983); including an Appendix with L. Miglio and M. P. Tosi.
173. P. Kumar and R. S. Sorbello, *Thin Solid Films* **25**, 25 (1975).
174. D. B. Jack and H. J. Kreuzer, *Phys. Rev. B* **26**, 6516 (1982).
175. N. G. van Kampen, *Stochastic Processes in Physics and Chemistry* (North-Holland, Amsterdam, 1981).
176. C. R. Helms and R. J. Madix, *Surf. Sci.* **52**, 677 (1975).
177. P. A. Redhead, *Trans. Faraday Soc.* **57**, 641 (1961).
178. J. B. Pendry, *Low Energy Diffraction*, Techniques of Physics Series (Academic Press, London, 1974).
179. L. A. Harris, *J. Appl. Phys.* **39**, 1419 (1968).
180. L. D. Schmidt, *Catal. Rev. Sci. Eng.* **9**, 115 (1975).
181. S. Glasstone, K. J. Laidler, and H. Eyring, *The Theory of Rate Processes* (McGraw-Hill, New York, 1941), p. 347.
182. K. J. Laidler, in *Catalysis*, Vol. 1, ed. P. H. Emmett (Reinhold, New York, 1954), Chap. 5.
183. J. D. Doll, *J. Chem. Phys.* **68**, 3158 (1978).
184. E. K. Grimmelmann, J. C. Tully, and E. Helfand, *J. Chem. Phys.* **74**, 5300 (1981).
185. B. J. Garrison D. J. Diestler, and S. A. Adelman, *J. Chem. Phys.* **67**, 4317 (1977).
186. G. S. De, U. Landman, and M. Rasolt, *Phys. Rev. B* **21**, 3256 (1980).
187. S. Efrima, K. F. Freed, C. Jedrzejek, and H. Metiu, *Chem. Phys. Lett.* **74**, 43 (1980).

188. Y. Zeiri, A. Redondo, and W. A. Goddard, *Surf. Sci.* **131**, 221 (1983).
189. H. Pfnür, P. Feulner, H. A. Engelhardt, and D. Menzel, *Chem. Phys. Lett.* **59**, 481 (1978).
190. H. Ibach, W. Erley, and H. Wagner, *Surf. Sci.* **92**, 29 (1980).
191. J. L. Taylor and W. H. Weinberg, *Surf. Sci.* **78**, 259 (1978).
192. J. E. Adams and J. D. Doll, *J. Chem. Phys.* **74**, 1467 (1981).
193. S. A. Adelman and J. D. Doll, *J. Chem. Phys.* **61**, 4242 (1974); *ibid.* **63**, 4908 (1975); **64**, 2375 (1976).
194. S. A. Adelman and J. D. Doll, *Acc. Chem. Res.* **10**, 378 (1977).
195. J. C. Tully, *Ann. Rev. Phys. Chem.* **31**, 319 (1980).
196. S. A. Adelman, *J. Chem. Phys.* **71**, 4471 (1979).
197. B. J. Garrison and S. A. Adelman, *J. Chem. Phys.* **67**, 2379 (1977).
198. S. Andersson, Ref. 79 and *Chem. Phys. Lett.* **55**, 185 (1978).
199. L. Schmidt and R. Gomer, *J. Chem. Phys.* **42**, 3573 (1965).
200. L. M. Kahn and S. C. Ying, *Solid State Commun.* **16**, 799 (1975).
201. G. Ehrlich, *Adv. Catal.* **14**, 255 (1963).
202. A. Redondo, Y. Zeiri, J. J. Low, and W. A. Goddard, *J. Chem. Phys.* **79**, 6410 (1983).
203. A. Redondo, Y. Zeiri, and W. A. Goddard, *Phys. Rev. Lett.*, **49**, 1847 (1982).
204. J. Reyes, I. Romero, and F. O. Goodman, *J. Chem. Phys.* **79**, 5906 (1983).
205. F. O. Goodman and N. Garcia, *Surf. Sci.* **120**, 251 (1982).
206. F. O. Goodman, *J. Chem. Phys.* **78**, 1582 (1983).
207. F. O. Goodman, *Phys. Rev. B* **27**, 6478 (1983).
208. P. Taborek, *Phys. Rev. Lett.* **48**, 1737 (1982).
209. G. Comsa, R. David, and K. D. Rendulic, *Phys. Rev. Lett.* **38**, 775 (1977).
210. G. Comsa, R. David, and B. J. Schumacher, *Surf. Sci.* **95**, L210 (1980).
211. J. N. Smith and R. L. Palmer, *J. Chem. Phys.* **56**, 13 (1972).
212. M. Balooch, M. J. Cardillo, D. R. Miller, and R. E. Stickney, *Surf. Sci.* **46**, 358 (1974).
213. F. O. Goodman and H. Y. Wachman, *Dynamics of Gas–Surface Scattering* (Academic Press, New York, 1976), Chap. 2.
214. Kenney et al., Quarterly Report No. 102, (MIT Research Laboratory of Electronics, 1971), p. 39.
215. W. Van Willigen, *Phys. Lett. A* **28**, 80 (1968).
216. G. Comsa and R. David, *Chem. Phys. Lett.* **49**, 512 (1977).
217. W. L. Schaich, *Phys. Lett. A* **64**, 133 (1977).
218. W. L. Schaich, *J. Phys. C* **11**, 2519 (1978).
219. N. Cabrera, V. Celli, F. O. Goodman, and J. R. Manson, *Surf. Sci.* **19**, 67 (1970).
220. F. O. Goodman, *Surf. Sci.* **30**, 1 (1972).
221. Z. W. Gortel, H. J. Kreuzer, and R. Teshima, *Phys. Rev. B* **22**, 512 (1980).
222. J. M. Horne and D. R. Miller, *Phys. Rev. Lett.* **41**, 511 (1978).
223. T. L. Bradley, A. E. Dabiri, and R. E. Stickney, *Surf. Sci.* **29**, 590 (1972).
224. Z. W. Gortel, H. J. Kreuzer, M. Schaff, and G. Wedler, *Surf. Sci.* **34**, 577 (1983).
225. Z. W. Gortel et al., to appear.
226. Z. W. Gortel, H. J. Kreuzer, and R. Teshima, *Phys. Rev. B* **22**, 5655 (1980).
227. Z. W. Gortel and H. J. Kreuzer, *Int. J. Quantum Chem., Symp.* **14**, 617 (1980).
228. Z. W. Gortel, H. J. Kreuzer, and D. Spaner, *J. Chem. Phys.* **72**, 234 (1980); see also *Surf. Sci.* **116**, 33 (1982).
229. S. A. Cohen and J. G. King, *Phys. Rev. Lett.* **31**, 703 (1973).
230. H. J. Kreuzer and G. M. Obermair, unpublished work; see Ref. 227.
231. B. Bendow and S. C. Ying, *Phys. Rev. B* **7**, 622, 637 (1973).
232. A. Clark, *The Theory of Adsorption and Catalysis* (Academic Press, New York, 1970).
233. F. C. Tompkins, *Chemisorption of Gases on Metals* (Academic Press, New York, 1978).

234. G. Wedler, *Chemisorption: An Experimental Approach* (Butterworths, London, 1976).
235. E. Sommer and H. J. Kreuzer, *Phys. Rev. B* **26**, 4094 (1982).
236. E. Bauer, F. Bonczek, H. Poppa, and G. Todd, *Surf. Sci.* **53**, 87 (1975).
237. H. Pfnür, P. Fehlner, H. A. Engelhardt, and D. Menzel, *Chem. Phys. Lett.* **59**, 481 (1978).
238. E. Bertel and F. P. Netzer, *Surf. Sci.* **97**, 409 (1980).
239. V. P. Zhdanov, *Surf. Sci.* **111**, L662 (1981).
240. J. C. Slater, *Phys. Rev.* **81**, 385 (1951).
241. J. E. Lennard-Jones and C. Strachan, *Proc. R. Soc. London, Ser. A* **150**, 442 (1935).
242. J. E. Lennard-Jones and A. Devonshire, *Proc. R. Soc. London, Ser. A* **156**, 6, 29 (1936).
243. R. Opila and R. Gomer, *Surf. Sci.* **112**, 1 (1980).
244. A. Redondo, Y. Zeiri, and W. A. Goddard, *Surf. Sci.* **136**, 41 (1984).
245. J. A. Barker, *J. Chem. Phys.* **61**, 3081 (1974).
246. R. Gomer, *Proc. DIET-I*, eds. N. H. Tolk, M. M. Traum, J. C. Tully, and T. E. Madey (Springer-Verlag, Berlin, 1983), p. 40.
247. W. Brenig, as in Ref. 246, p. 90.
248. D. Menzel and R. Gomer, *J. Chem. Phys.* **41**, 3311 (1964).
249. P. A. Redhead, *Can. J. Phys.* **42**, 886 (1964).
250. R. Gomer, see Ref. 246.
251. K. F. Freed, *Surf. Sci.* **122**, 317 (1982).
252. M. L. Knotek and P. J. Feibelman, *Phys. Rev. Lett.* **40**, 964 (1978); *Phys. Rev. B* **18**, 6531 (1978).
253. M. L. Knotek, *Rep. Prog. Phys.* **47**, 1499 (1984).
254. T. E. Madey, R. Stockbauer, J. F. Van Der Veen, and D. E. Eastman, *Phys. Rev. Lett.* **45**, 187 (1980).
255. T. E. Madey, Inelastic Particle–Surface Collisions, *Springer Series Chem. Phys.* **17**, 80 (1981).
256. H. Niehus, *Appld. Surf. Sci.* **13**, 292 and references given there.
257. J. I. Gersten, R. Janow, and N. Tzoar, *Phys. Rev. Lett.* **36**, 610 (1976).
258. W. L. Clinton, *Phys. Rev. Lett.* **39**, 965 (1977).
259. W. L. Clinton, to appear.
260. D. E. Ramaker, C. T. White and J. S. Murday, *Phys. Lett.* **89A**, 211 (1982).
261. J. Berzelius, *Jahres-Bericht Uber Die Fortschritte Der Physischen Wissenschaften* (Tubingen, 1836), p. 243.
262. H. Ibach, W. Erley, and H. Wagner, *Surf. Sci.* **92**, 29 (1980).
263. T. Engl and G. Ertl, *Adv. Catal.* **28**, 1 (1979).
264. J. C. Buchholz and M. G. Lagally, *Phys. Rev. Lett.* **35**, 442 (1975).
265. W. D. Gillespie, R. K. Herz, E. E. Petersen, and G. A. Somorjai, *J. Catal.* **70**, 147 (1981).
266. N. D. Spencer, R. C. Schoonmaker, and G. A. Somorjai, *J. Catal.* **74**, 129 (1982).
267. I. Toyoshima and G. A. Somorjai, *Catal. Rev. Sci. Eng.* **19**, 105 (1979).
268. D. A. King, *Surf. Sci.* **47**, 384 (1975).
269. S. T. Ceyer and G. A. Somorjai, *Ann. Rev. Phys. Chem.* **28** (1977).
270. E. Umbach and D. Menzel, *Surf. Sci.* **135**, 199 (1983).
271. J. L. Skinner and P. G. Wolynes, *J. Chem. Phys.* **69**, 2143 (1978).
272. W. Brenig, H. Müller and R. Sedlmeier, *Phys. Lett.* **A54**, 109 (1975).
273. W. T. Tysoe, G. L. Nyberg, and R. M. Lambert, *Surf. Sci.* **135**, 128 (1983).
274. J. E. Demuth, *Chem. Phys. Lett.* **45**, 12 (1977).
275. J. E. Demuth, *Surf. Sci.* **84**, 315 (1979).
276. J. A. Gates and L. L. Kesmodel, *J. Chem. Phys.* **76**, 4281 (1982).

277. T. Upton and W. A. Goddard, *J. Am. Chem. Soc.* **100**, 321 (1978).
278. H. Kobayashi, H. Kato, K. Tamara, and K. Fukui, *J. Catal.* **49**, 294 (1977).
279. J. K. Kochi, *Organometallic Mechanisms and Catalysis* (Academic Press, New York, 1978), Chap. 14.
280. J. E. Germain, *Catalytic Conversion of Hydrocarbons* (Academic Press, London, 1969) Chap. 4.
281. C. W. Spangler, *J. Org. Chem.* **31**, 346 (1966).
282. S. A. R. Knox, R. F. D. Stansfield, F. G. A. Stone, M. J. Winter, and P. Woodward, *J. Chem. Soc., Dalton Trans.* 173 (1982).
283. R. Goudsmit, B. F. G. Johnson, J. Lewis, P. Raithby, and M. J. Rosales, *J. Chem. Soc., Dalton Trans.*, 2257 (1983).
284. W. Meyer and H. Nedel, *Z. Tech. Phys.* **18**, 588 (1937).
285. E. Peacock-López and H. Suhl, *Phys. Rev. B* **26**, 3774 (1982).
286. F. H. Constable, *Proc. R. Soc. London, Ser. A* **108**, 355 (1925).
287. E. Cramer, *Adv. Catal.* **7**, 75 (1956); *J. Chem. Phys.* **47**, 439 (1950).
288. A. K. Galwey, *Adv. Catal.* **26**, 247 (1977).
289. J. H. Sinfelt, *Catal. Rev.* **3**, 175 (1969).
290. N. H. March and M. Parrinello, *Collective Effects in Solids and Liquids* (Adam Hilger, Bristol, 1982).
291. Y. O. Park, R. I. Masel, and K. Stoll, unpublished results cited in Ref. 293.
292. R. B. Woodward and R. Hoffmann, *The Conservation of Orbital Symmetry* (Verlag Chemie, Weinheim, 1969).
293. W. F. Banholzer, Y. O. Park, K. M. Mak, and R. I. Masel, *Surf. Sci.* **128**, 176 (1983); *ibid.* **133**, 623 (1983).
294. R. G. Pearson, *Symmetry Rules of Chemical Reactions* (Wiley, New York, 1976); *Acc. Chem. Res.* 152 (1971).
295. K. Fukui, *Acc. Chem. Res.* **4**, 57 (1971).
296. A. Balazs and K. H. Johnson, *Surf. Sci.* **114**, 197 (1982).
297. F. Mango, *Adv. Catal.* **20**, 291 (1969).
298. R. P. Messmer and A. J. Bennett, *Phys. Rev. B* **6**, 633 (1972).
299. F. A. Cotton, *Chemical Applications of Group Theory* (Wiley, New York, 1971).
300. W. L. Jorgensen and L. Salem, *The Organic Chemist's Book of Orbitals* (Academic Press, New York, 1973).
301. R. G. Pearson, *Symmetry Rules for Chemical Reactions* (Wiley, New York, 1976).
302. R. G. Pearson, *Acc. Chem. Res.* **4**, 152 (1971).
303. G. C. Bond, *Discuss. Faraday Soc.* **41**, 200 (1966).
304. A. C. Balasz and K. H. Johnson, *Surf. Sci.* **114**, 197 (1982).
305. T. D. Halachev and E. Ruckenstein, *Surf. Sci.* **108**, 292 (1982).
306. O. K. Anderson, *Phys. Rev. B* **2**, 883 (1970).
307. P. J. Feibelman and D. R. Hamann, *Phys. Rev. Lett.* **52**, 61 (1984).
308. M. Kiskinova and D. W. Goodman, *Surf. Sci.* **108**, 64 (1981).
309. E. L. Garfunkel, J. E. Crowell, and G. A. Somorjai, *J. Phys. Chem.* **86**, 310 (1982).
310. J. Dauchot and J. Van Cakenberghe, *Nature, Phys. Sci.* **246**, 61 (1973).
311. J. E. Turner, B. C. Sales, and M. B. Maple, unpublished work cited in Ref. 312.
312. H. Huhl, *Surf. Sci.* **107**, 88 (1981).
313. R. Dagonnier and J. Nuyts, *J. Chem. Phys.* **65**, 206 (1976).
314. H. Suhl and R. E. Lagos, in *Proc. AIP Conf.* **61** (Am. Inst. Phys., 1979).
315. J. H. Craig, *Surf. Sci.* **134**, 745 (1983).
316. O. Gurel and D. Gurel, *Topics in Current Chemistry*, Vol. **118** (Springer-Verlag, Berlin, 1983).
317. I. D. Moore and N. H. March, *Ann. Phys. (N.Y.)* **97**, 136 (1976).

318. H. S. Carslaw and J. C. Jaeger, *Conduction of Heat in Solids* (Clarendon Press, Oxford, 1959).
319. J. S. Brown, R. C. Brown, and N. H. March, *Phys. Lett. A* **46**, 463 (1974).
320. J. Bardeen, *Phys. Rev.* **49**, 653 (1936).
321. J. C. Stoddart, N. H. March, and M. J. Stott, *Phys. Rev.* **186**, 683 (1969).
322. F. Flores, N. H. March, Y. Ohmura, and A. M. Stoneham, *J. Phys. Chem. Solids* **40**, 531 (1979).
323. R. Gomer, *Solid State Physics*, Vol. 30, eds. H. Ehrenreich, F. Seitz, and D. Turnbull (Academic Press, New York, 1975), p. 93.
324. T. L. Einstein, *Phys. Rev. B* **11**, 577 (1975).
325. B. I. Lundqvist, in *Vibrations at Surfaces*, eds. R. Caudano, J. M. Gilles, and A. A. Lucas (Plenum Press, New York, 1982).
326. N. D. Lang and A. R. Williams, *Phys. Rev. B* **18**, 616 (1978).
327. H. H. Hjelmberg, B. I. Lundqvist, and J. K. Norskov, *Phys. Scripta* **20**, 192 (1979).
328. S. G. Louie, *Phys. Rev. Lett.* **42**, 476 (1979).
329. G. P. Kerker, M. T. Yiu, and M. L. Cohen, *Phys. Rev. B* **12**, 4940 (1975).
330. M. Baldo, F. Flores, A. Martin-Rodero, G. Piccitto, and R. Pucci, *Surf. Sci.* **128**, 237 (1983).
331. D. M. Newns, *Phys. Rev.* **178**, 1123 (1969); *Surf. Sci.* **128**, 237 (1983).
332. B. Bell and A. Madhukar, *Phys. Rev. B* **14**, 4281 (1976).
333. J. Avery, *Creation and Annihilation Operators* (McGraw-Hill, New York, 1976).
334. N. H. March, W. H. Young, and S. Sampanthar, *The Many-Body Problem in Quantum Mechanics* (University Press, Cambridge, 1967).
335. W. Brenig and K. Schönhammer, *Z. Phys.* **267**, 201 (1974).
336. J. F. Cornwell, in *Group Theory and Its Applications in 'Techniques of Physics,'* eds. N. H. March and H. N. Daglish (Academic Press, London, 1984).
337. H. Kuhn, D. Mobius, and H. Bucher, *Physical Methods of Chemistry*, Part 3B, eds. A. Weissberger and B. W. Rossitier (Wiley–Interscience, New York, 1972).
338. M. Babiker and N. H. March, Internal Report (International Centre for Theoretical Physics, Trieste, 1976).
339. M. Babiker, *J. Phys. A* **9**, 799 (1976).
340. H. F. Hameka, *J. Chem. Phys.* **47**, 2728 (1967).
341. K. R. Painter, P. J. Grout, N. H. March, and M. P. Tosi, *Nuovo Cimento B* **66**, 202 (1981).
342. L. E. Firment and G. A. Somorjai, *Surf. Sci.* **84**, 275 (1979).
343. L. E. Firment and G. A. Somorjai, *Surf. Sci.* **55**, 413 (1976).
344. G. B. Fisher and J. L. Gland, *Surf. Sci.* **94**, 446 (1980).
345. F. C. Frank and J. H. van der Merwe, *Proc. R. Soc. London, Ser. A* **198**, 205, 216 (1949).
346. R. C. Ecob and B. Ralph, *Proc. Natl. Acad. Sci. U.S.A.* **77**, 1749 (1980).
347. L. Pauling, *J. Am. Chem. Soc.* **57**, 2680 (1935).
348. N. Bjerrum, *K. Dan. Vidensk. Selsk. Mat. Fys. Medd.* **27**, 3 (1951).
349. H. Ibach and S. Lehwald, *Surf. Sci.* **91**, 187 (1980).
350. J. E. Demuth, *IBM J. Res. Dev.* **22**, 265 (1978).
351. W. Jones and N. H. March, *Theoretical Solid State Physics*, Vol. 1 (Wiley–Interscience, London, 1973), p. 343; also Dover reprint series (1985).
352. Y. Takahashi, *J. Phys. Soc. Jpn.* **43**, 1342 (1977).
353. J. R. Banavar, M. H. Cohen, and R. Gomer, *Surf. Sci.* **107**, 113 (1981).
354. R. J. Donnelly and P. H. Roberts, *Proc. R. Soc. London, Ser. A* **312**, 519 (1969).
355. M. Berkowitz, J. D. Morgan, J. A. McCammon, and S. H. Northrup, *J. Chem. Phys.* **79**, 5563 (1983).

356. J. L. Lebowitz and E. Rubin, *Phys. Rev.* **131**, 2381 (1963).
357. H. Grad, in *Encyclopedia of Physics*, Vol. XII, ed. S. Flugge (Springer-Verlag, Berlin, 1958), p. 205.
358. H. Mori, I. Oppenheim, and J. Ross, in *Studies in Statistical Mechanics*, eds. J. de Boer and G. E. Uhlenbeck (North-Holland, Amsterdam, 1962).
359. M. C. Wang and G. E. Uhlenbeck, *Rev. Mod. Phys.* **17**, 323 (1945).
360. L. Van Hove, *Physica* **21**, 517, 901 (1955); *ibid.* **22**, 343 (1956); *ibid.* **23**, 441 (1957).
361. I. Prigogine, *Nonequilibrium Statistical Mechanics* (Interscience, New York, 1962).
362. J. G. Kirkwood, *J. Chem. Phys.* **14**, 180 (1946).
363. J. Ross, *J. Chem. Phys.* **24**, 375 (1956).
364. J. L. Lebowitz, H. L. Frisch, and E. Helfand, *Phys. Fluids* **3**, 1 (1960).
365. W. Kohn and J. M. Luttinger, *Phys. Rev.* **108**, 590 (1957).
366. E. Montroll, *Suppl. Nuovo Cimento* 16 (1960).
367. See M. Berkowitz et al., Ref. 355 and other earlier references given there.
368. J. S. McCaskill and N. H. March, *Surf. Sci.* **131**, 34 (1983), with an Appendix written together with L. Miglio and M. P. Tosi.
369. N. H. March, in *Theory of the Inhomogeneous Electron Gas*, eds. S. Lundqvist and N. H. March (Plenum Press, New York, 1983), p. 1.
370. G. Rickayzen, in *Green Functions in Condensed Matter in 'Techniques of Physics,'* eds. N. H. March and H. N. Daglish (Academic Press, London, 1983).
371. N. D. Lang, in *Theory of the Inhomogeneous Electron Gas*, eds. S. Lundqvist and N. H. March (Plenum Press, New York, 1983), p. 309.
372. D. Kumamoto, J. E. Van Himbergen, and R. Silbey, *Chem. Phys. Lett.* **68**, 189 (1979).
373. H. Suhl, J. H. Smith, and P. Kumar, *Phys. Rev. Lett.* **25**, 1442 (1970).
374. R. Gomer and J. R. Schrieffer, see Ref. 373 and, for instance, Ref. 233.
375. S. Doniach, in Proc. of Int. School of Physics, Enrico Fermi, Course XXXVII, ed. W. Marshall (Academic Press, New York, 1967).
376. E. I. Evzerikhin and G. D. Lyubarskii, in *Scientific Selection of Catalysts*, Vol. XI, eds. A. A. Balandin *et al.* (Jerusalem: Israel Program for Scientific Research, Jerusalem, 1968).
377. E. Ilisca, *Phys. Rev. Lett.* **24**, 797 (1970).
378. R. E. Lagos and H. Suhl, *Surf. Sci.* **75**, L801 (1978); *ibid.* **81**, 657 (1979).
379. G. Parravano, *J. Chem. Phys.* **20**, 342 (1952).
380. A. Van Itterbeek, P. Mariens, and O. Van Paerrel, *Ann. Phys.* **18**, 135 (1943).
381. M. R. Shanabarger, *Solid State Commun.* **14**, 1015 (1974).
382. H. Suhl, *Phys. Rev. B* **11**, 2011 (1975).
383. D. L. Mills, *Phys. Rev. B* **3**, 3887 (1971).
384. J. Tersoff and L. M. Falicov, *Phys. Rev. B* **24**, 754 (1981).
385. A. C. Balasz and K. H. Johnson, *Surf. Sci.* **114**, 197 (1982).
386. O. K. Anderson, *Phys. Rev. B* **2**, 883 (1970).
387. J. W. Moore and R. G. Pearson, *Kinetics and Mechanism* (Wiley, New York, 1981), p. 181.
388. F. A. Cotton, *Chemical Applications of Group Theory* (Wiley, New York, 1971).
389. R. G. Pearson, *Symmetry Rules for Chemical Reactions* (Wiley, New York, 1976).
390. R. G. Pearson, *Acc. Chem. Res.* **4**, 152 (1971).
391. W. F. Banholzer and R. I. Masel, see Ref. 293.
392. Y. O. Park, R. I. Masel, and K. Stolt, see Ref. 293.
393. N. H. March and A. M. Murray, *Proc. R. Soc. London, Ser. A* **261**, 119 (1961).
394. N. H. March, *Phil. Mag.* **32**, 497 (1975); see also *J. Math. Phys.* **19**, 2023 (1978).
395. J. S. Rousseau, J. C. Stoddart, and N. H. March, *J. Phys. C* **5**, L175 (1972).
396. E. Gerjuoy, *J. Math. Phys.* **6**, 993 (1965).

397. G. D. Gaspari and B. L. Gyorffy, *Phys. Rev. Lett.* **28**, 801 (1972).
398. K. Huang, *Proc. Phys. Soc.* **60**, 161 (1948).
399. H. Suhl, *Surf. Sci.* **107**, 88 (1981).
400. A. T. Bell, *Catal. Rev. Sci. Eng.* **23**, 203 (1981).
401. D. J. Auerbach, H. E. Pfnür, C. T. Rettner, J. E. Schlaegel, J. Lee, and R. J. Madix, *J. Chem. Phys.* **81**, 2515 (1984).
402. J. E. Lennard-Jones, *Trans. Faraday Soc.* **28**, 333 (1932).
403. M. Balooch, M. J. Gardillo, and R. E. Stickney, *Surf. Sci.* **44**, 310 (1974); **50**, 263 (1975).
404. G. Comsa, R. David, and B. J. Schumacher, *Surf. Sci.* **85**, 45 (1979).
405. R. C. Cosser, S. R. Bare S. M. Francis, and D. A. King, *Vacuum* **31**, 503 (1981).
406. D. J. Auerbach, J. E. Schlaegel, J. Lee, and R. J. Madix, *J. Vac. Sci. Technol. A* **1**, 1271 (1983).
407. J. Lee, R. J. Madix, J. E. Schlaegel, and D. J. Auerbach, *Surf. Sci.*, **143**, 626 (1984).
408. J. E. Hurst, L. Wharton, K. C. Janda, and D. J. Auerbach, *J. Chem. Phys.* **78**, 1559 (1983).
409. A. W. Kleyn, A. C. Luntz, and D. J. Auerbach, *Surf. Sci.* **117**, 33 (1982).
410. G. D. Kubiak, J. E. Hurst, H. G. Rennagel, G. M. McClelland, and R. N. Zare, *J. Chem. Phys.* **79**, 5163 (1983).
411. C. T. Rettner, H. E. Pfnür, and D. J. Auerbach, see Ref. 401.
412. Z. W. Gortel, H. J. Kreuzer, R. Teshima, and L. A. Turski, *Phys. Rev.* **B24**, 4456 (1981); H. J. Kreuzer and R. Teshima, *ibid.* **24**, 4470 (1981).
413. E. Goldys, Z. W. Gortel, and H. J. Kreuzer, *Surf. Sci.* **116**, 33 (1982); *Solid State Commun.* **40**, 963 (1981).
414. H. Ibach and D. Bruchmann, *Phys. Rev. Lett.* **44**, 36 (1980).
415. H. J. Kreuzer and P. Summerside, *Surf. Sci.* **111**, 102 (1981).

Index